Measurement, analysis, and modeling of extreme precipitation events linked to floods are vital in understanding the world's changing climate. This book examines and documents the impacts of climate change and climate variability on extreme precipitation events, providing methods for assessment of the trends in these events, and their impacts. It also provides a basis to develop procedures and guidelines for climate-adaptive hydrologic engineering. Topics covered include approaches for assessment of hydrometerological floods, recent developments in hydrologic design for flood mitigation, and applications and limitations of improved precipitation forecasts, using information about internal modes of climate variability (teleconnections). State-of-the-art methodologies for precipitation analysis, estimation, and interpolation are included, and exercises for each chapter, supported by modeling software and computational tools available online at www.cambridge.org/teegavarapu, enable the reader to apply and engage with the innovative methods of assessment.

This is an important resource for academic researchers in the fields of hydrology, climate change, meteorology, environmental policy, and risk assessment, and will also be invaluable to professionals and policy-makers working in hazard mitigation, water resources engineering, and adaptation to weather and climate.

This volume is the first in a collection of four books within the International Hydrology Series on flood disaster management theory and practice within the context of anthropogenic climate change. The other books are:

2 – Floods in a Changing Climate: Hydrologic Modeling *by P. P. Mujumdar and D. Nagesh Kumar*
3 – Floods in a Changing Climate: Inundation Modelling *by Giuliano Di Baldassarre*
4 – Floods in a Changing Climate: Risk Management *by Slodoban Simonović*

RAMESH S. V. TEEGAVARAPU is an Associate Professor in the Department of Civil, Environmental, and Geomatics Engineering at Florida Atlantic University, and is leader of the Hydrosystems Research Laboratory in that department. His main area of specialization is water resources with focuses on climate change modeling, precipitation extremes, and hydrologic processes. He is a member of the Water Resources Management Committee of the International Association for Hydro-Environment Engineering and Research. Dr. Teegavarapu currently serves on the editorial board of two international journals, and has published over 100 articles in journals and conference proceedings. He has convened, chaired, and moderated over 50 sessions at several international conferences, and served as vice-chair for task committees related to radar rainfall and uncertainty analysis approaches under the Surface Water Hydrology Technical Committee of the American Society of Civil Engineers.

INTERNATIONAL HYDROLOGY SERIES

The **International Hydrological Programme** (IHP) was established by the United Nations Educational, Scientific and Cultural Organization (UNESCO) in 1975 as the successor to the International Hydrological Decade. The long-term goal of the IHP is to advance our understanding of processes occurring in the water cycle and to integrate this knowledge into water resources management. The IHP is the only UN science and educational programme in the field of water resources, and one of its outputs has been a steady stream of technical and information documents aimed at water specialists and decision-makers.

The **International Hydrology Series** has been developed by the IHP in collaboration with Cambridge University Press as a major collection of research monographs, synthesis volumes, and graduate texts on the subject of water. Authoritative and international in scope, the various books within the series all contribute to the aims of the IHP in improving scientific and technical knowledge of freshwater processes, in providing research know-how and in stimulating the responsible management of water resources.

EDITORIAL ADVISORY BOARD
Secretary to the Advisory Board
Mr Shabhaz Khan *Division of Water Science, UNESCO, 1 rue Miollis, Paris 75015, France*

Members of the Advisory Board
Professor B. P. F. Braga Jr *Centro Technológica de Hidráulica, São Paulo, Brazil*
Professor G. Dagan *Faculty of Engineering, Tel Aviv University, Israel*
Dr J. Khouri *Water Resources Division, Arab Centre for Studies of Arid Zones and Dry Lands, Damascus, Syria*
Dr G. Leavesley *US Geological Survey, Water Resources Division, Denver Federal Center, Colorado, USA*
Dr E. Morris *Scott Polar Research Institute, Cambridge, UK*
Professor L. Oyebande *Department of Geography and Planning, University of Lagos, Nigeria*
Professor S. Sorooshian *Department of Civil and Environmental Engineering, University of California, Irvine, California, USA*
Professor K. Takeuchi *Department of Civil and Environmental Engineering, Yamanashi University, Japan*
Professor D. E. Walling *Department of Geography, University of Exeter, UK*
Professor I. White *Fenner School of Environmental Society, Australian National University, Canberra, Australia*

TITLES IN PRINT IN THIS SERIES
M. Bonell, M. M. Hufschmidt and J. S. Gladwell *Hydrology and Water Management in the Humid Tropics: Hydrological Research Issues and Strategies for Water Management*
Z. W. Kundzewicz *New Uncertainty Concepts in Hydrology and Water Resources*
R. A. Feddes *Space and Time Scale Variability and Interdependencies in Hydrological Processes*
J. Gibert, J. Mathieu and F. Fournier *Groundwater/Surface Water Ecotones: Biological and Hydrological Interactions and Management Options*
G. Dagan and S. Neuman *Subsurface Flow and Transport: A Stochastic Approach*
J. C. van Dam *Impacts of Climate Change and Climate Variability on Hydrological Regimes*
D. P. Loucks and J. S. Gladwell *Sustainability Criteria for Water Resource Systems*
J. J. Bogardi and Z. W. Kundzewicz *Risk, Reliability, Uncertainty, and Robustness of Water Resource Systems*
G. Kaser and H. Osmaston *Tropical Glaciers*
I. A. Shiklomanov and J. C. Rodda *World Water Resources at the Beginning of the Twenty-First Century*
A. S. Issar *Climate Changes during the Holocene and their Impact on Hydrological Systems*
M. Bonell and L. A. Bruijnzeel *Forests, Water and People in the Humid Tropics: Past, Present and Future Hydrological Research for Integrated Land and Water Management*
F. Ghassemi and I. White *Inter-Basin Water Transfer: Case Studies from Australia, United States, Canada, China and India*
K. D. W. Nandalal and J. J. Bogardi *Dynamic Programming Based Operation of Reservoirs: Applicability and Limits*
H. S. Wheater, S. Sorooshian and K. D. Sharma *Hydrological Modelling in Arid and Semi-Arid Areas*
J. Delli Priscoli and A. T. Wolf *Managing and Transforming Water Conflicts*
H. S. Wheater, S. A. Mathias and X. Li *Groundwater Modelling in Arid and Semi-Arid Areas*
L. A. Bruijnzeel, F. N. Scatena and L. S. Hamilton *Tropical Montane Cloud Forests*
S. Mithen and E. Black *Water, Life and Civilization: Climate, Environment and Society in the Jordan Valley*
K. A. Daniell *Co-Engineering and Participatory Water Management*
R. Teegavarapu *Floods in a Changing Climate: Extreme Precipitation*
P. P. Mujumdar and D. Nagesh Kumar *Floods in a Changing Climate: Hydrologic Modeling*
G. Di Baldassarre *Floods in a Changing Climate: Inundation Modelling*
S. Simonović *Floods in a Changing Climate: Risk Management*

Floods in a Changing Climate

Extreme Precipitation

Ramesh S. V. Teegavarapu

Florida Atlantic University

CAMBRIDGE
UNIVERSITY PRESS

University Printing House, Cambridge CB2 8BS, United Kingdom

One Liberty Plaza, 20th Floor, New York, NY 10006, USA

477 Williamstown Road, Port Melbourne, VIC 3207, Australia

4843/24, 2nd Floor, Ansari Road, Daryaganj, Delhi - 110002, India

79 Anson Road, #06-04/06, Singapore 079906

Cambridge University Press is part of the University of Cambridge.

It furthers the University's mission by disseminating knowledge in the pursuit of education, learning and research at the highest international levels of excellence.

www.cambridge.org
Information on this title: www.cambridge.org/9781108446747

© Ramesh S. V. Teegavarapu and UNESCO 2012

This publication is in copyright. Subject to statutory exception and to the provisions of relevant collective licensing agreements, no reproduction of any part may take place without the written permission of Cambridge University Press.

First published 2012
First paperback edition 2017

A catalogue record for this publication is available from the British Library

Library of Congress Cataloging in Publication data
Teegavarapu, Ramesh S. V., 1970–
Floods in a changing climate. Extreme precipitation / Ramesh S. V. Teegavarapu.
    pages cm. – (International hydrology series)
Includes bibliographical references and index.
ISBN 978-1-107-01878-5
1. Precipitation (Meteorology) – Measurement.   2. Floods – Mathematical models.
3. Hydrologic models.   4. Climatic extremes.   5. Climatic changes – Environmental
aspects.   I. Title.   II. Title: Extreme precipitation.
QC925.T44   2012
551.48′9011 – dc23   2012015661

ISBN 978-1-107-01878-5 Hardback
ISBN 978-1-108-44674-7 Paperback

Additional resources for this publication at www.cambridge.org/teegavarapu

Cambridge University Press has no responsibility for the persistence or accuracy of URLs for external or third-party internet websites referred to in this publication, and does not guarantee that any content on such websites is, or will remain, accurate or appropriate.

Contents

Foreword	*page* ix
Preface	xi
List of abbreviations	xv

1 Precipitation and climate change — 1
1.1 Introduction — 1
1.2 Climate change and variability — 1
1.3 Precipitation processes and floods — 1
1.4 Impacts of climate change — 5
1.5 Internal modes of climate variability: teleconnections — 6
1.6 Extreme precipitation and floods in a changing climate: main issues — 7
1.7 Conclusions and summary — 8
Exercises — 8

2 Precipitation measurement — 10
2.1 Introduction — 10
2.2 Precipitation measurement in a historical context — 10
2.3 Ground-, radar-, and satellite-based measurements — 10
2.4 Measurement methods, errors, and accuracy — 11
2.5 Configurations of rain gages — 11
2.6 Radar measurement of precipitation — 13
2.7 Weather radar and the theory of reflectivity — 14
2.8 Evaluation of exponents and coefficient values in a Z–R power relationship — 21
2.9 Formulation for optimal coefficients and exponents — 22
2.10 Bias evaluation and corrections — 25
2.11 Evaluation of methods — 28
2.12 Weighting functions — 29
2.13 Performance evaluations with multiple stations — 30
2.14 Optimal parameters for weighting methods — 30
2.15 Bias corrections with limited rain gage data — 31
2.16 Satellite-based rainfall estimation — 31
2.17 Precipitation monitoring networks — 34
2.18 Clustering of rain gages — 34
2.19 Optimal density — 36
2.20 Optimal monitoring networks — 36
2.21 Methods for design — 36
2.22 Recommendations for rain gage placements — 37
2.23 Global precipitation data sets — 38
2.24 Global precipitation data sets: availability and formats — 41
2.25 Evaluation of observed gridded precipitation data sets — 42
2.26 Monitoring networks for extreme events — 45
2.27 Precipitation measurements in the future — 45
2.28 Summary and conclusions — 46
Exercises — 46
Websites for data acquisition and resources — 47

3 Spatial analysis of precipitation — 48
3.1 Spatial analysis of precipitation data — 48
3.2 Missing data estimation — 49
3.3 Spatial interpolation — 50
3.4 Deterministic and stochastic interpolation methods — 51
3.5 Revisions to the inverse distance weighting method — 57
3.6 Integration of the Thiessen polygon approach and inverse distance method — 57
3.7 Correlation coefficient weighting method — 58
3.8 Inverse exponential weighting method — 58
3.9 Regression models — 58
3.10 Trend surface models using local and global polynomial functions — 59
3.11 Example for trend surface models — 60
3.12 Thin-plate splines — 62
3.13 Natural neighbor interpolation — 63
3.14 Normal ratio method — 63
3.15 Nearest neighbor weighting method — 63
3.16 Variants of multiple linear regression methods — 65
3.17 Regression models using auxiliary information — 65

3.18 Geostatistical spatial interpolation   66
3.19 Optimal functional forms   70
3.20 Structure of optimization formulations   73
3.21 Emerging interpolation techniques   78
3.22 Artificial neural networks   81
3.23 Universal function approximation-based
      kriging   81
3.24 Classification methods   83
3.25 Distance metrics as proximity measures   84
3.26 Distance metrics for precipitation data   84
3.27 Boolean distance measures for precipitation
      data   86
3.28 Optimal exponent weighting of proximity
      measures   88
3.29 Optimal K-nearest neighbor classification
      method   88
3.30 Optimal K-means clustering method   89
3.31 Proximity measures: limitations   90
3.32 Use of radar data for infilling precipitation data   90
3.33 Geographically weighted optimization   91
3.34 Single and multiple imputations of missing
      data   92
3.35 Temporal interpolation of missing data   94
3.36 Data set selection for model development and
      validation   95
3.37 Performance measures   96
3.38 Qualitative evaluation   98
3.39 Model selection and multi-model comparison   99
3.40 Surface generation   100
3.41 Geo-spatial grid-based transformations of
      precipitation data   101
3.42 Statistics preserving spatial interpolation   106
3.43 Data for model development   107
3.44 Optimization issues: solvers and solution
      methods   107
3.45 Spatial analysis environments and interpolation   108
3.46 Data filler approaches: application in real time   108
3.47 Local and global interpolation: issues   108
3.48 Under- and overestimation   109
3.49 Main issues and complexities of spatial
      analysis of precipitation data   109
3.50 Spatial interpolation for global gridded
      precipitation data sets   109
3.51 Spatial interpolation of extreme precipitation
      data   110
3.52 Applicability of methods   110
3.53 RAIN: Rainfall Analysis and Interpolation
      Software   110
3.54 Use and application of RAIN software   111
3.55 Conclusions and summary   111
      Exercises   112

4   Extreme precipitation and floods   115
4.1   Introduction   115
4.2   Hydrometeorological aspects of precipitation   115
4.3   Larger-scale precipitation systems   115
4.4   Convective patterns   116
4.5   Precipitation and river regimes   116
4.6   Hydrometeorological aspects of floods: review
      of case studies   116
4.7   Probable maximum precipitation   118
4.8   Precipitation-based drivers and mechanisms
      influencing extreme floods   120
4.9   Flooding mechanisms   120
4.10 Flooding and shallow groundwater levels   120
4.11 Soil moisture contributions to flooding   121
4.12 Spatial and temporal occurrence of extreme
      events: dependence analysis   123
4.13 Joint probability analysis   127
4.14 Partial duration series analysis: peaks over
      thresholds   130
4.15 Baseflow separation methods   131
4.16 Extreme precipitation and flash floods   133
4.17 Precipitation thresholds and floods   133
4.18 Temporal difference in occurrence of peaks   134
4.19 Cyclonic precipitation: episodic events   135
4.20 Desk study approach   135
4.21 Regression analysis   137
4.22 Extreme precipitation events and peak
      flooding: example   138
4.23 Assessment from dependence analysis   140
4.24 Statistical analysis of peak discharge and
      precipitation data   141
4.25 Floods in a changing climate: issues   144
4.26 Conclusions and summary   145
      Exercises   145

5   Climate change modeling and precipitation   148
5.1   Downscaling precipitation   148
5.2   Downscaling methods   148
5.3   Downscaling at spatial level   148
5.4   Downscaling at temporal level   149
5.5   Statistical downscaling techniques   149
5.6   Weather generators   151
5.7   Regional climate model: dynamic
      downscaling   151
5.8   Other approaches   152
5.9   Statistically downscaled climate change
      projections: concept example   152
5.10 Weather generator: concepts   162
5.11 Downscaling precipitation: major issues   167
5.12 Conclusions and summary   167
      Exercises   167

Useful websites 168
Resources for students 168

6 Precipitation variability and teleconnections 169
6.1 Introduction 169
6.2 Southern Oscillation 170
6.3 El Niño Southern Oscillation 170
6.4 Decadal oscillations 175
6.5 Teleconnections and extreme precipitation 177
6.6 ENSO and precipitation 185
6.7 Combined influence of AMO–ENSO phases 187
6.8 Pacific Decadal Oscillation 187
6.9 North Atlantic Oscillation 187
6.10 Forecasts based on teleconnections 189
6.11 Precipitation and teleconnections: global impacts 189
6.12 Conclusions and summary 191
Exercises 191
Useful websites 192

7 Precipitation trends and variability 193
7.1 Historical and future trends 193
7.2 Global precipitation trends 193
7.3 USA precipitation changes 194
7.4 Assessment of extreme precipitation trends: techniques 194
7.5 Fitting probability distributions for extreme rainfall data 195
7.6 Statistical distributions 197
7.7 Parameter estimation 197
7.8 Frequency factors 198
7.9 Parametric and non-parametric tests 199
7.10 Regional frequency analysis 201
7.11 Illustrative examples 201
7.12 Value of fitting a parametric frequency curve 206
7.13 Extreme rainfall frequency analysis in the USA 207
7.14 Uncertainty and variability in rainfall frequency analysis 208
7.15 Assessment of sample variances 211
7.16 Non-parametric methods 211
7.17 Homogeneity 212
7.18 Partial duration series 213
7.19 Incorporating climate variability and climate change into rainfall frequency analysis 213
7.20 Future data sources 213
7.21 Statistical tests and trend analysis: example of extreme precipitation analysis in South Florida 214
7.22 Different tests: moving window approaches 215

7.23 Implications of infilled data 215
7.24 Descriptive indices for precipitation extremes 217
7.25 Rare extremes 221
7.26 Trends based on GCM model simulations 222
7.27 Software for evaluation of extreme precipitation data 222
7.28 Conclusions and summary 222
Exercises 222
Useful website 224

8 Hydrologic modeling and design 225
8.1 Precipitation and climate change: implications on hydrologic modeling and design 225
8.2 Emerging trends in hydrologic design for extreme precipitation 225
8.3 Methodologies for hydrologic design 226
8.4 Hydrologic design 227
8.5 Adaptive hydrologic infrastructure design 228
8.6 Hydrologic design example 231
8.7 Example of water balance model 233
8.8 Water budget model software 235
8.9 Infrastructural modifications and adaptation to climate change 236
8.10 Conclusions and summary 238
Exercises 238

9 Future perspectives 241
9.1 Future hydrologic design and water resources management 241
9.2 Uncertain climate change model simulations 241
9.3 Future of hydrologic data for design 242
9.4 Tools for climate-sensitive management of water resources systems 243
9.5 Example: generation of compromise operating policies for flood protection 243
9.6 Impacts of climate change on reservoir operations: example from Brazil 245
9.7 Climate change and future hydrologic engineering practice 246
9.8 Floods: stationarity and non-stationarity issues 247
9.9 Extreme precipitation: issues for the future 247
9.10 Institutional changes and adaptation challenges 247
9.11 Conclusions and summary 248
Exercises 248

Glossary 249
References 253
Index 266
See color plates between pp. 110 and 111.

Foreword

Professor Siegfried Demuth

Chief Hydrological Systems and Global Change Section
Division of Water Sciences
Natural Sciences Sector
UNESCO, Paris

Flooding is the greatest water-related natural disaster known to the human race – its human, material, and ecological costs can be devastating for sustainable development. Floods affect an estimated 520 million people across the world yearly, resulting in up to 25,000 annual deaths. Along with other water-related disasters, they cost the world economy some $50 to $60 billion a year. An estimated 96 percent of deaths related to natural disasters in the past decade occurred in developing countries with limited capacity to forecast and manage these disasters. The number of people vulnerable to a devastating flood is expected to rise due to large-scale urbanization, population growth in natural floodplains, every increasing rate of deforestation, climate change, and rising sea levels. New disaster risk reduction approaches are needed now to build the necessary capacity to address these challenges faced by some of the poorest in the world.

On the other hand, floods are natural phenomena, which contribute to the biodiversity and sustainability of ecosystems and to many human activities. Both developed and developing countries have benefited from economic development in areas prone to flooding. Close to one billion people – one-sixth of the global population, the majority of them among the world's poorest inhabitants – now live on the floodplains. Developing countries with mainly agricultural economies depend largely on their fertile floodplains for food security and poverty alleviation. The nutrient rich deltas of many river systems favor low-tech agricultural practices and provide livelihoods for millions. The wetlands in floodplains contribute to biodiversity and also create employment opportunities.

UNESCO and WMO, aware of both the significant achievements made in flood management in the recent years and of the existing opportunities to develop practical solutions within this context, launched the International Flood Initiative (IFI) in close cooperation with other partners such as the United Nations University (UNU), the International Association of Hydrological Sciences (IAHS) and the International Strategy for Disaster Reduction (ISDR). The initiative has developed an enhanced knowledge system on all flood-related activities, such as monitoring, network design, improving statistical analysis of floods, real-time forecasting and flood modeling, and risk management.

In 2010 UNESCO's Division of Water Sciences has launched in the framework of the International Hydrological Programme (IHP) a new Book Series on floods as a contribution to the International Flood Initiative (IFI). The objective of the book series is to provide sound knowledge to the theory of flood disaster management and practice under the current climate change conditions. Best practices around the world and state-of-the-art knowledge are presented along with a set of contemporary computational tools and techniques for managing flood disasters including remote sensing, spatial precipitation analysis, distributed hydrologic modeling and fuzzy risk analysis. The IFI Book Series comprises of four books:

1. ***Floods in a Changing Climate: Extreme Precipitation***
2. ***Floods in a Changing Climate: Hydrologic Modeling***
3. ***Floods in a Changing Climate: Inundation Modelling***
4. ***Floods in a Changing Climate: Risk Management***

All four books focus on various aspects of floods in a changing climate. The first book examines and enhances our understanding of the impacts of climate change on extreme precipitation events and quantifies the uncertainties and develops procedures and guidelines for risk-based decision-making in the presence of the impacts of climate change. The second book focuses on practical tools for hydrologic modeling of extreme water-related events under climate change conditions. This book concentrates on a selection of proper hydrologic modeling approaches for the estimation of floods. The third book presents hydraulic tools in public domain and focuses on the use of GIS technology for floodplain mapping under climate change conditions and finally the fourth book puts emphasis on different methodologies of flood risk management presenting fuzzy set approaches to water related disaster risk management. All books contain practical worked examples, case studies, annexes and supporting electronic materials which will be accessible on an accompanying website. The book series will help water managers, decision makers to deal with floods in a holistic manner. It will train students and will help educators to provide a fresh look at various aspects of integrated flood management.

UNESCO wishes to thank Ramesh S. V. Teegavarapu, P. P. Mujumdar, D. Nagesh Kumar, Giuliano Di Baldassarre and Slobodan Simonovic for their enthusiasm and strong commitment to provide the books on a very tight time frame. We are especially very grateful to Professor Simonovic for his role as scientific coordinator of the IFI Books series; he kept close contact with the authors during the writing process over the last two years and provided guidance to the authors. I also would like to thank my colleague Biljana Radojevic for initiating the book project. Finally I would like to extend my gratitude to Cambridge University Press who has accepted the IFI Book Series as part of their wider International Hydrology Series, and namely Laura Clark and Susan Francis for facilitating and overlooking the production of the series.

Preface

This book was prepared under the responsibility and coordination of Dr. Siegfried Demuth, International Hydrological Programme, Chief of the Hydrological Systems and Global Change Section and scientist responsible for the International Flood Initiative, and Dr. Biljana Radojevic, Division of Water Sciences, UNESCO. Dr. S. P. Simonovic and Dr. S. Demuth conceived and developed the idea of the four-volume series dealing with extreme precipitation, hydrologic modeling, inundation modeling, and risk management. Their overall vision for the book series, and feedback and support during different stages of development is greatly appreciated.

The book examines and documents the impacts of climate change and variability on extreme precipitation events in an effort to enhance our understanding of the impacts of climate change on extreme precipitation events; quantify the uncertainties associated with extreme precipitation events; and develop procedures and guidelines for risk-based decision making in the presence of the impacts of climate change. Practical tools for spatial precipitation analysis and simulation (such as weather generators, spatial interpolation of missing data, use of multiple sensors in estimating precipitation, and others) are included. This book distinguishes itself from already available books that deal with climate variability and extremes by placing a strong emphasis on extreme precipitation events and their links to floods under climate change conditions. The book documents innovative emerging methodologies and state-of-the-art approaches for precipitation estimation and interpolation. New approaches for understanding hydrometeorological floods, recent developments in hydrologic design for floods under changing climate scenarios, and utility of improved seasonal forecasts of precipitation using information about teleconnections are also discussed. The book will serve as a valuable reference document and also as a specialized monograph to hydrologists, water resources modelers, students, professionals, researchers, and members of governmental and non-governmental agencies interested in changing climate and its impact on one of the main components of the hydrologic cycle. The exercises, application software, and content-specific computational tools included in the book will promote learning of concepts within academic environments while serving as a textbook for graduate students. The book

also contains brief explanations of the content-specific words or phrases used, climate information web-links, and data sources that are of immense use to modelers, practitioners, and researchers. The explanations used in the glossary conform to standard definitions provided and used in publications by UNESCO and WMO. A detailed list of acronyms is also included. Several websites included in the book provide links to organizations involved in data collection and climate change studies and documents provided by them. Public domain software listings and information on the public domain computational codes are also included.

The contents are divided into four major themes and they include (1) precipitation processes and measurement; (2) extreme precipitation, floods, and climate change; (3) precipitation variability, teleconnections and trends, and hydrologic modeling; and (4) hydrologic design in a changing climate. These contents are described next.

The initial part of the book introduces the first main theme with emphasis on precipitation as a major driver for floods, and assessment of this major input and its influence on future floods under non-stationary climate conditions. Figures that precede the chapter descriptions provide the gist of the contents of the chapter visually.

Chapter 1

Chapter 1 provides an introduction to extreme precipitation and climate change, precipitation variability, and teleconnections. This chapter discusses major issues that are relevant to floods in a changing climate and provides details of global extreme precipitation events and factors responsible for such events.

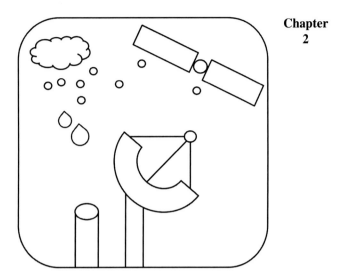

Chapter 2

Precipitation processes and measurement methods are discussed exhaustively in **Chapter 2**. Rain gage, and radar- and satellite-based measurements are described in this chapter. The main focus of the chapter is on radar-based precipitation measurements. Design of an optimal rain gage network, and global precipitation data available at different spatial and temporal resolutions in the form of gridded precipitation data sets from different organizations are also discussed.

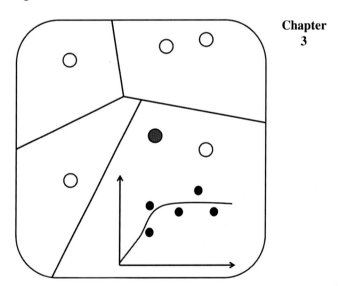

Chapter 3

Spatial analysis of precipitation and interpolation methods are discussed in great detail in **Chapter 3**. Stochastic and deterministic, data-driven, optimization and data mining-based spatial interpolation methods are described and discussed exhaustively. Several numerical examples are included. Emerging new methods that use concepts from numerical taxonomy for spatial

interpolation are also introduced. This chapter also discusses methods for precipitation estimation using radar, surface generation, and estimation of missing data. The RAIN (Rainfall Analysis and INterpolation) tool developed by the author is also introduced.

Chapter 4

Chapter 4 covers a major part of the second theme of book with emphasis on topics related to extreme precipitation and floods in the context of climate change. Hydrometeorological aspects of floods along with exhaustive details of probable maximum precipitation (PMP) and probable maximum floods (PMF) concepts are discussed. Extreme flooding events caused by cyclonic precipitation are common in many parts of the world and are gaining much attention due to the increased frequency of these events in the past decade. Discussion of these events is provided. A joint probability approach dealing with extreme precipitation and peak discharges forms the main focus of this chapter.

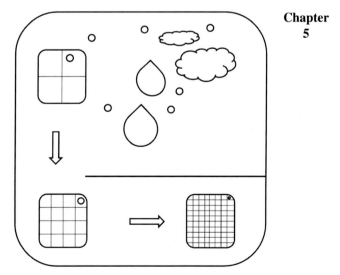

Chapter 5

Chapter 5 focuses on downscaling methods used for precipitation and their limitations. Descriptions of different downscaling methods and an example of a bias corrected spatial disaggregation (BCSD) procedure are provided.

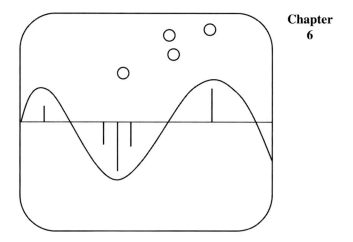

Chapter 6

Precipitation variability and the oscillatory behavior of teleconnections and their use for seasonal forecasts for operation of water resources systems are discussed in **Chapter 6**. Two major teleconnections (oscillations), the Atlantic Multi-decadal Oscillation (AMO) and ENSO (El Niño Southern Oscillation), are described exhaustively. Spatial and temporal variability of precipitation extremes affected by different phases of these oscillations are evaluated.

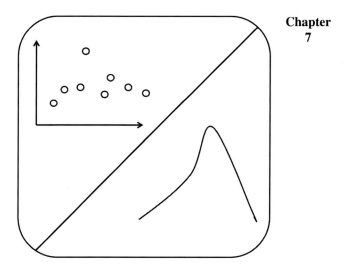

Chapter 7

Chapter 7 details the global trends in precipitation and statistical methods appropriate for trend analysis. Future hydrologic modeling and design in a changing climate are major concerns for hydrologists, water resources modelers, and managers. This chapter also introduces the statistical analysis tool developed by the author to aid analysis of precipitation time series data. **Chapters 6** and 7 cover the third major theme of the book.

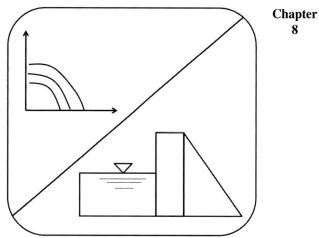

Chapter 8

Chapter 8 provides an exhaustive look at changing trends in hydrologic modeling and design considering climate change. Hydrologic infrastructure design under changing climate is discussed. The topics in this chapter focus especially on sustainable and climate change sensitive management approaches.

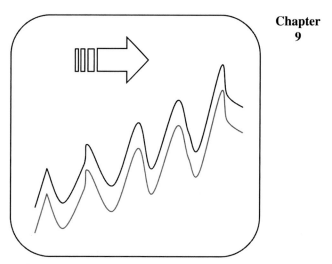

Chapter 9

Chapter 9 discusses future perspectives on hydrologic engineering and practice under future climate change conditions, along with challenges faced by water resource managers. This chapter also provides discussion on non-stationarity issues in hydrologic modeling and future hydrologic design. Chapters 8 and 9 address the fourth theme of the book.

The book also includes three tools developed by the author. These tools will aid in analysis and solution of several problems provided in the exercises at the end of each chapter. The tools are:

1. RAIN: Rainfall Analysis and INterpolation
2. SEEP: Statistical Evaluation of Extreme Precipitation
3. WBM: Water Balance Model

RAIN is a precipitation analysis and data processing tool mainly aimed at spatial interpolation for estimation of missing precipitation data. The tool, along with several data sets, is provided on the website accompanying this book. The tool was developed in a Windows environment and incorporates interpolation schemes for precipitation data analysis. Additional scripts within the tools are also included for climate data (precipitation data) processing. SEEP is a useful tool for statistical analysis of extreme precipitation time series data. It includes several modules to analyze and fit statistical distributions to extreme values of precipitation. A few data sets are included on the website for analysis. The WBM tool is a four-parameter water balance model (Thomas model) that uses precipitation and potential evapotranspiration data for estimation of different components of water budget in a watershed. The website contains the tools arranged in different folders, along with the required data. User manuals are provided for each tool.

Many individuals have helped me in the writing of this book. I would like to thank several of my graduate students who have worked with me on research studies related to radar-rainfall, extreme precipitation, and climate variability and change. These studies have resulted in various publications that are discussed and described in this book. I thank my graduate students Mr. Aneesh Goly, Mr. Husayn El Sharif, Mr. Andre Ferreira, Mr. Delroy Peters, Mr. Chia-hung Lin, Mr. Kandarp Pattani, and Mr. Ricardo Brown for their enormous help during the book preparation. Mr. Goly and Mr. El Sharif helped me tremendously with the data analysis and illustrations. Dr. Noemi Gonzalez helped me with some sections of Chapter 4. Mr. Tadesse Meskele, who worked as a summer intern on a radar-rainfall study, helped me with two sections in Chapter 2. I have greatly benefited from research collaborations with Dr. Chandra Pathak, Dr. Miguel Marino, Dr. Lindell Orsmbee, Dr. Baxter Vieux, Dr. Pradeep Behera, Dr. Chandramouli Viswanathan, Dr. Jayantha Obeysekera, Dr. Leonard Berry, Dr. Young-Oh Kim, and Ms. Angela Prymas. I would like to extend my sincere thanks to Dr. Simonovic and Dr. Siegfried Demuth for their meticulous planning, feedback from time to time, and help throughout the preparation of the book. Suggestions from other authors of the books in this series, Dr. Mujmudar and Dr. Nagesh Kumar, and Dr. Di Baldassarre, are greatly appreciated. Special thanks go to Ms. Barbara Lwanga for all her help at UNESCO headquarters in Paris. Finally, I would like to extend my sincere thanks and appreciation to Ms. Laura Clark and other production staff at Cambridge University Press.

Abbreviations

AC	Anomaly Correlation	DOD	Department of Defense (USA)
AE	Absolute Error	DSA	Desk Study Approach
AEI	Annual Exceedence Interval	DSD	Drop Size Distribution
AEP	Annual Exceedence Probability	EDA	Exploratory Data Analysis
AIC	Akaike's Information Criterion	ENSO	El Niño Southern Oscillation
AMC	Antecedent Moisture Condition	EPA	Environmental Protection Agency (USA)
AMO	Atlantic Multi-decadal Oscillation	ERSST	Extended Reconstructed Sea Surface Temperature
AMS	Annual Maximum Series	ETCCDI	Expert Team on Climate Change Detection and Indices
ANN	Artificial Neural Network		
AO	Arctic Oscillation	FFSGAM	Fixed Function Set Genetic Algorithm Method
API	Antecedent Precipitation Index	FLP	Fuzzy Linear Programming
AR4	Fourth Assessment Report (IPCC)	FTP	File Transfer Protocol
ARI	Annual Recurrence Interval	GA	Genetic Algorithm
ARM	Association Rule Mining	GAMS	General Algebraic Modeling System
ASCII	American Standard Code for Information Interchange	GCM	General Circulation Model
		GCOS	Global Climate Observing System
ASE	Averaged Standard Error	GEV	Generalized Extreme Value
BCSD	Bias-Correction and Spatial Disaggregation	GHCN	Global Historical Climatology Network
BFI	Baseflow Index	GIS	Geographical Information System
BIC	Bayesian Information Criterion	GME	Gage Mean Estimator
BOA	Barnes Objective Analysis	GPCC	Global Precipitation Climatology Centre
BVN	Bivariate Normal Distribution	GPCP	Global Precipitation Climatology Project
CCA	Canonical Correlation Analysis	GPM	Global Precipitation Measurement
CCI	Climate Change Information	GRG	Generalized Reduced Gradient
CCWM	Correlation Coefficient Weighting Method	GTN-H	Global Terrestrial Network-Hydrology
CDD	Consecutive Dry Days (RR <1 mm)	GWO	Geographically Weighted Optimization
CDF	Cumulative Density Function	HAS	Holographic data storage system Access System
CME	Climatological Mean Estimator	HDSS	Holographic Data Storage System
CMIP	Coupled Model Inter-comparison Project	HEC-HMS	Hydrologic Engineering Center – Hydrologic Modeling System
CN	Curve Number		
CRU	Climatic Research Unit, University of East Anglia, UK	HFM	High Flow Magnitude
		HFT	High Flow Timings
		HRAP	Hydrologic Research Analysis Project
CWD	Consecutive Wet Days	HSS	Heidke Skill Score
DCP	Downscaled Climate Projections	HUC	Hydrologic Unit Code
DDC	Data Distribution Center (IPCC)	HWT	High Water Table
DEFRA	Department of Environment, Food and Rural Affairs, UK	ICOADS	International Comprehensive Ocean–Atmosphere Data Set
DICOPT	DIscrete Continuous OPTimizer		
DMC	Deep Moist Convection	IDF	Intensity Duration Frequency

IDWM	Inverse Distance Weighting Method	NPI	North Pacific Index
IETD	Inter-Event Time Definition	NRCS	Natural Resources Conservation Service (USA)
IEWM	Inverse Exponential Weighting Method	NSEC	Nash–Sutcliffe Efficiency Coefficient
IPCC	Intergovernmental Panel on Climate Change	NWS	National Weather Service (USA)
IPO	Inter-decadal Pacific Oscillation	NWSCPC	National Weather Service Climate Prediction Center
ISRO	Indian Space Research Organization		
JAXA	Japanese Aerospace Exploration Agency	OK	Ordinary Kriging
KDE	Kernel Density Estimation	ONI	Oceanic Niño Index
K-NN	K-Nearest Neighbor	PDO	Pacific Decadal Oscillation
KS	Kolmogorov–Smirnov	PDS	Partial Duration Series
LARS-WG	Long Ashton Research Station Weather Generator	PE	Potential Evapotranspiration
LEPS	Linear Error in Probability Space	PMM	Probability Matching Method
LFM	Low Flow Magnitude	PMP	Probable Maximum Precipitation
LFT	Low Flow Timings	PNA	Pacific North American
LID	Low Impact Development	POT	Peaks Over Thresholds
LP	Linear Programming	PRCPTOT	Total precipitation in wet days (>1 mm)
LS-SVR	Least Squares-Support Vector Regression	PRISM	Precipitation-elevation Regression on Independent Slopes Model
LWOM	Linear Weight Optimization Method		
MADI	Mean Absolute Deviation Index	PW	Pre-Whitening
MAE	Mean Absolute Error	PWM	Probability Weighted Moments
MAR	Missing At Random	R10mm	Count of precipitation days with RR greater than 10 mm
MAUP	Modifiable Areal Unit Problem		
MB	Multiplicative Bias	R20mm	Count of precipitation days with RR greater than 20 mm
MCAR	Missing Completely At Random		
MCC	Mesoscale Convective Complexes	R95pTOT	Total precipitation due to wet days (>95th percentile)
ME	Mean Error		
MFB	Mean Field Bias	R99pTOT	Total precipitation due to extremely wet days (>99th percentile)
MIDWM	Modified Inverse Distance Weighting Method		
MINLP	Mixed Integer Non-Linear Programming	RAIN	Rainfall Analysis and INterpolaton
MJO	Madden–Julian Oscillation	RCM	Regional Climate Model
MK	Mann–Kendall Statistic	RMSE	Root Mean Squared Error
ML	Maximum Likelihood	RMSF	Root Mean Squared Factor
MLE	Maximum Likelihood Estimation	Rnnmm	Count of days where RR greater than a threshold value
MLR	Multiple Linear Regression		
MNAR	Missing Not At Random	RR	Recorded daily Rainfall
MOC	Meridional Overturning Circulation	RSS	Residual Sum of Squares
MOM	Method Of Moments	RVAR	Ratio of the VARiance
MPE	Multi-sensor Precipitation Estimates	RX1day	Maximum 1-day precipitation
MRE	Mean Relative Error	RX5day	Maximum 5-day precipitation
MSDI	Mean Square Deviation Index	SBC	Single Best Classifier
MSE	Mean Squared Error	SBE	Single Best Estimator
MSE2	Mean Standard Error	SCS	Soil Conservation Service
MSLP	Mean Sea Level Pressure	SD	Spatial Downscaling
MSRE	Mean Squared Reduced Error	SDII	Simple Daily Intensity Index
NAO	North Atlantic Oscillation	SEEP	Statistical Evaluation of Extreme Precipitation
NCDC	National Climate Data Center	SFWMD	South Florida Water Management District
NCEP	National Centers for Environmental Prediction	SMA	Soil Moisture Accounting
NetCDF	Network Common Data Form	SO	Southern Oscillation
NEXRAD	NEXt generation RADar	SOI	Southern Oscillation Index
NLS	Non-negative Least Squares	SRES	Special Report on Emissions Scenarios
NOAA	National Oceanic and Atmospheric Administration	SS	Sen's Slope Estimator

SST	Sea Surface Temperature
SVM	Support Vector Machine
SVMLRC	Support Vector Machine–Logistic Regression-based Copula
SVR	Support Vector Regression
TFPW	Trend Free Pre-Whitening
TMPA	TRMM Multi-satellite Precipitation Analysis (see TRMM)
TNI	Trans-Niño Index
TR-55	Technical Report-55 (US Department of Agriculture)
TRMM	Tropical Rainfall Measuring Mission
UCAR	University Corporation for Atmospheric Research
UNEP	United National Educational Programme
UNESCO	United Nations Educational Scientific and Cultural Organization
UNFCCC	United Nations Framework Convention on Climate Change
UOK	Universal function approximation Ordinary Kriging
USAID	United States Agency for International Development
USEPA	United States Environmental Protection Agency
USGS	United States Geological Survey
USHCN	United States Historical Climatology Network
VASClimO	Variability Analysis of Surface Climate Observations
WBM	Water Balance Model
WCRP	World Climate Research Programme
WCT	NOAA Weather Climate Toolkit
WDC	World Data Center
WEKA	Waikato Environment for Knowledge Analysis
WG	Weather Generator
WMO	World Meteorological Organization
Z–R	Reflectivity (Z)–Rainfall rate (R)

1 Precipitation and climate change

1.1 INTRODUCTION

Precipitation events of a specific magnitude that lead to catastrophic floods are under scrutiny by many research studies aimed at understanding the influence of climate change and variability on these events. According to the Intergovernmental Panel on Climate Change (IPCC, 2007d, 2008), increases in the frequency of heavy precipitation events (frequency or proportion of total rainfall from heavy falls) over most areas are indicated as *likely* based on the trends in the later part of twentieth century, especially after 1960. The human contribution to the trends in these events is described as *most likely*. Also, the IPCC (2007a) suggests that increases in the frequency of these events are *very likely* based on the projections from different climate change models for the twenty-first century, considering the Special Report on Emissions Scenarios (SRES).

1.2 CLIMATE CHANGE AND VARIABILITY

Variations in future extreme precipitation events are linked to climate change and variability. Therefore, it is imperative to evaluate the frequency, magnitude, and duration of extreme precipitation events under these situations. According to the World Meteorological Organization (WMO), climate change refers to a statistically significant variation either in the mean state of the climate or in its variability, persisting for an extended period (typically decades or longer). Climate change may be due to natural internal processes or external forcings, or to persistent anthropogenic changes in the composition of the atmosphere or in land use. The United Nations Framework Convention on Climate Change (UNFCCC), in its Article 1, defines climate change as: A change of climate which is attributed directly or indirectly to human activity that alters the composition of the global atmosphere and which is in addition to natural climate variability observed over comparable time periods. A clear distinction between climate change attributable to anthropogenic activities altering the atmospheric composition, and climate variability attributable to natural causes was made by UNFCCC.

Again, WMO defines climate variability as variations in the mean state and other statistics (variations, the occurrence of extremes, frequency) of the climate on all temporal and spatial scales beyond that of individual episodic events. Variability may be due to natural internal processes within the climate system (internal modes of variability), or to variations in natural or anthropogenic external forcing (external variability). Climate change and climate variability have implications on our understanding of extreme precipitation events and their influences on global and regional hydrologic processes.

1.3 PRECIPITATION PROCESSES AND FLOODS

1.3.1 Precipitation extremes

Precipitation extremes or extreme events in general are referred to as X-events by Jentsch *et al.* (2008). X-events also constitute those events that may not be completely natural in origin. An evaluation of historical world extreme precipitation totals from available instrumental records is possible. Table 1.1 provides details of the world's greatest rainfall totals for different durations. The data set was compiled by WMO (1994) in their report.

Figure 1.1 shows the log–log plot of time (in minutes) and precipitation depth (in millimeters) based on the data provided in Table 1.1. A linear relationship (WMO, 2010) given by Equation 1.1, using a least squares fit, relates time in hours and precipitation depth in millimeters:

$$P = 422(t)^{0.475} \quad (1.1)$$

The simple linear relationship suggests that a scaling law that is somewhat universal between rainfall totals and accumulated periods is possible (Hense and Friederichs, 2006). A similar functional form (Equation 1.1) is obtained when different units are used for precipitation depth and time. However, the law is not representative of the rainfall extremes occurring across the world. Also,

Table 1.1 *World's greatest precipitation totals for different durations*

Duration		Precipitation depth (mm)	Location
1	Minutes	38	Barot, Guadeloupe
8		126	Fussen, Bavaria
15		198	Plumb Point, Jamaica
20		206	Curtea-de-Arges, Romania
42		305	Holt, USA
60		401	Shangdi, Nei Monggol, China
2.17	Hours	483	Rockport, USA
2.75		559	D'Hanis, USA
4.5		782	Smethport, USA
6		840	Muduocaidang, China
9		1,087	Belouve, La Réunion
10		1,400	Muduocaidang, China
18.5		1,689	Belouve, La Réunion
24		1,825	Foc Foc, La Réunion
2	Days	2,467	Aurere, La Réunion
3		3,130	Aurere, La Réunion
4		3,721	Cherrapunji, India
5		4,301	Commerson, La Réunion
6		4,653	Commerson, La Réunion
7		5,003	Commerson, La Réunion
8		5,286	Commerson, La Réunion
9		5,692	Commerson, La Réunion
10		6,028	Commerson, La Réunion
11		6,299	Commerson, La Réunion
12		6,401	Commerson, La Réunion
13		6,422	Commerson, La Réunion
14		6,432	Commerson, La Réunion
15		6,433	Commerson, La Réunion
31		9,300	Cherrapunji, India
2	Months	12,767	Cherrapunji, India
3		16,369	Cherrapunji, India
4		18,738	Cherrapunji, India
5		20,412	Cherrapunji, India
6		22,454	Cherrapunji, India
11		22,990	Cherrapunji, India
12		26,461	Cherrapunji, India
2	Years	40,768	Cherrapunji, India

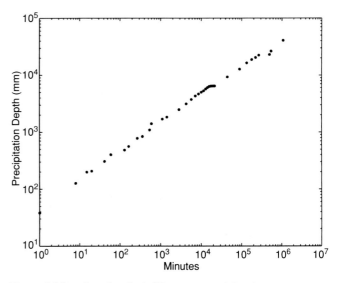

Figure 1.1 Log–log plot of rainfall extremes and durations.

over mountainous regions, especially in the Cascade and Olympic mountains in the USA.

Evaluation of short and long duration precipitation events is critical to our understanding of flooding and its impacts on natural and built environments. Also, it is important to note that some extreme precipitation events from the instrument records are not the typical convective or winter frontal events that are generally considered the main drivers for flooding in different parts of the world. Cronin (2009), using the Clausius–Clapeyron equation for saturation vapor pressure of water, suggests that climatic warming will be accompanied by an enhanced hydrologic cycle and changing precipitation patterns in future. Evidence of changes to the global hydrologic cycle were suggested by Huntington (2006) as ongoing intensification of the water cycle and increased rainfall in the past few decades (Wentz *et al.*, 2007). Recent increases in tropical cyclone activity have been linked to human-induced ocean warming (Emanuel, 2005; Cronin, 2009). Also, paleo-hurricane and paleo-precipitation studies indicate evidence of intense hurricane and precipitation storm event activities. Precipitation extremes are also influenced by internal modes of climate variability manifested as multi-year and decadal or longer period atmospheric oscillations. Anthropogenic influences via emissions of aerosols on precipitation locally and globally have been studied by Lohmann (2008). Lohmann indicates that aerosols have been shown to have both increasing and decreasing effects on convective precipitation, and the present-day general circulation models (GCMs) cannot account for all the microphysical processes that are relevant to aerosol–cloud interactions.

1.3.2 Floods and flooding risks

An evaluation of the world's largest floods and their causes during the Quaternary period (i.e., the period that extends from about

the historical precipitation extremes seem to have been confined to regions that are dominated by deep convection (Hense and Friederichs, 2006). The greatest rainfall amounts for durations in excess of one month occurring at Cherrapunji, India, are mostly influenced by orographic precipitation events. The influence of climate change and variability of extreme rainfall due to orographic precipitation events is an active area of research. Recent studies (e.g., Salathe *et al.*, 2008) have established links between changes in storm intensity and distribution of precipitation events

Table 1.2 *Largest meteorological floods from river basins across the world*

River basin	Country	Basin area (10^3 km^2)	Peak discharge (m^3/s)	Flood type
Amazon	Brazil	5,854	370,000	Rainfall
Nile	Egypt	3,826	13,200	Rainfall
Congo	Zaire	3,699	76,900	Rainfall
Mississippi	USA	3,203	70,000	Rainfall
Amur	Russia	2,903	38,900	Rainfall
Parana	Argentina	2,661	43,070	Rainfall
Yenisey	Russia	2,582	57,400	Snowmelt
Ob-Irtysh	Russia	2,570	44,800	Snowmelt
Lena	Russia	2,418	189,000	Snowmelt/Ice-jam
Niger	Niger	2,240	27,140	Rainfall
Zambezi	Mozambique	1,989	17,000	Rainfall
Yangtze	China	1,794	110,000	Rainfall
Mackenzie	Canada	1,713	30,300	Snowmelt
Chari	Chad	1,572	5,160	Rainfall
Volga	Russia	1,463	51,900	Snowmelt
St. Lawrence	Canada	1,267	14,870	Snowmelt
Indus	Pakistan	1,143	33,280	Rain/Snowmelt
Syr Darya	Kazakhstan	1,070	2,730	Rain/Snowmelt
Orinoco	Venezuela	1,039	98,120	Rainfall
Murray	Australia	1,032	3,940	Rainfall
Ganges	Bangladesh	976	74,060	Rain/Snowmelt
Shatt al Arab	Iraq	967	7,366	Rain/Snowmelt
Orange	South Africa	944	16,230	Rainfall
Huanghe	China	894	36,000	Rainfall
Yukon	USA	852	30,300	Snowmelt
Senegal	Senegal	847	9,340	Rainfall
Colorado	USA	808	7,080	Rainfall
Rio Grande	USA	805	17,850	Rain/Snowmelt
Danube	Romania	788	15,900	Snowmelt
Mekong	Vietnam	774	66,700	Rainfall
Tocantins	Brazil	769	38,780	Rainfall
Columbia	USA	724	35,100	Snowmelt
Darling	Australia	650	2,840	Rainfall
Brahmaputra	Bangladesh	650	81,000	Rain/Snowmelt
Sao Francisco	Brazil	615	15,890	Rainfall
Amu Darya	Kazakhstan	612	6,900	Rain/Snowmelt
Dnieper	Ukraine	509	23,100	Snowmelt

Source: O'Connor and Costa (2004)

1.8 million years ago to the present) was provided by O'Connor and Costa (2004). They indicated that only 4 of the 27 largest documented floods were primarily the result of meteorological conditions and atmospheric water sources, especially due to rainfall events. The magnitude of the floods evaluated during this period was extremely large and was equal to 100,000 m^3/s. The paucity of meteorological floods in the list of largest Quaternary floods does not indicate lesser significance of meteorological floods. Indeed, meteorological floods are by far the most common of the types of floods in human experience, affecting parts of the globe every year (O'Connor and Costa, 2004). Table 1.2 summarizes the largest floods occurring in river basins larger than about 500,000 square kilometers. The data set reported in Table 1.2 was compiled by O'Connor and Costa (2004) from several sources. The following statements summarize the floods, magnitudes, and causes based on the report by O'Connor and Costa (2004).

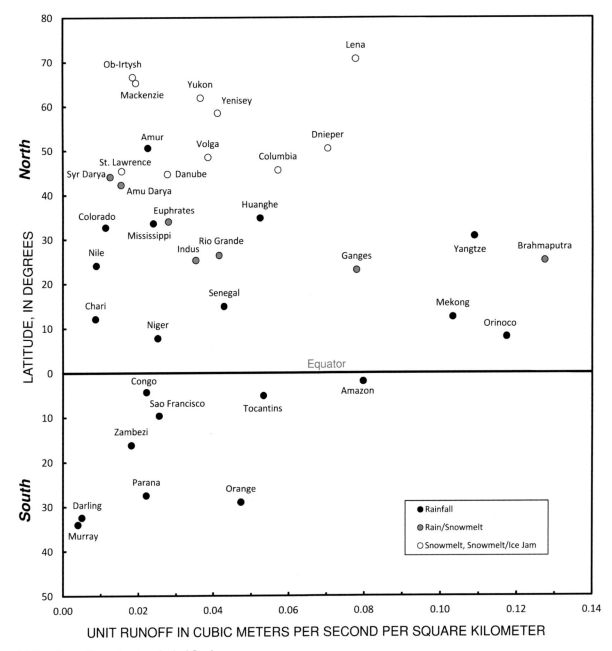

Figure 1.2 Locations of largest meteorological floods.

- Most of the largest documented floods in the past 100,000 years resulted from failures of natural dams or large ice-dam failures at the margins of continental ice sheets.
- The largest floods in large basins within the tropics are primarily derived from rainfall within areas affected by tropical cyclones or strong monsoonal airflow. The largest recorded meteorological floods have been in tropical regions, especially where monsoon moisture falling on large and high-relief drainage basins produces immense volumes of runoff.
- Nearly all of the largest floods caused by rainfall have occurred in basins south of 40 degrees north latitude.

North of that, snowmelt- and ice-jam-related floods have predominated.
- The largest meteorological floods from river basins larger than about 500,000 square kilometers are mainly caused by extreme rainfall events.

Figure 1.2 shows the locations of peak flooding events with a classification based on rainfall, rain/snowmelt, snowmelt, and snowmelt/ice-jam categories. The figure is based on the information available in the report by O'Connor and Costa (2004).

Flood frequency and magnitude are projected to increase in the regions experiencing increases in precipitation intensity, while drought frequency is projected to increase in many regions, in particular those where reduction of precipitation is projected (Kundzewicz *et al.*, 2008). Future flooding risks associated with extreme hydrometeorological conditions can be evaluated by the source–pathway–receptor models (DETR, 2000; Thome *et al.*, 2007). The fourth book in the International Flood Initiative (IFI) four-volume series, by Simonovic, deals with flood risk management and all these issues in much detail.

Changes in all aspects of precipitation (amount, intensity, duration, location, and clustering) will impact flood regimes (Reynard, 2007). Two other factors that affect the flooding risks in a watershed are physical landscape and antecedent hydrologic conditions. The antecedent conditions generally reflect the soil moisture state influenced by available moisture in the soil column or recent storm events. Extreme precipitation events qualify as sources (weather events or sequence of events) that cause flooding; the pathways are the mechanisms that convey floodwaters, and the receptors are the people or built environments that may be impacted by flooding. The third book in this IFI four-volume series, by Di Baldassarre, deals with issues of floodplain analysis and inundation modeling.

Thome *et al.* (2007) define drivers as those processes that change the state of the flooding system. Catchment responses are different to different extreme precipitation events. Reynard (2007) highlights two major types of catchment responses: (1) quick and sensitive to short-duration rainfall events for smaller and steep-sided catchments; and (2) slow and long-term response to precipitation events for large and rural catchments dominated by large groundwater storage. A conceptually simple model proposed by Wigley and Jones (1985) and discussed by Dingman (2008) provides an easy way to determine the impact of changes in precipitation and evapotranspiration on runoff due to climate changes. The following equations describe their water balance model (WBM).

$$R = P - ET \qquad (1.2)$$

$$\alpha = R/P \qquad (1.3)$$

$$ET = (1 - \alpha) P \qquad (1.4)$$

$$P_f = \gamma P \qquad (1.5)$$

$$ET_f = \beta ET \qquad (1.6)$$

$$R_f = \delta R \qquad (1.7)$$

The variables P, R, ET are current precipitation, runoff, and evapotranspiration; β is the runoff ratio; P_f, ET_f are the future precipitation and evapotranspiration; and α, γ are the factors by which the precipitation and evapotranspiration change due to climate change. The variable R_f is the future runoff due to climate change and δ is the runoff factor. Using Equations 1.2–1.7, the relative change in runoff can be calculated by:

$$\delta = \frac{\gamma - \{(1 - \alpha)\beta\}}{\alpha} \qquad (1.8)$$

An example application of this model can help with understanding the change in runoff by using the following variable values.

Increase in future precipitation: 10% ($\gamma = 1.1$)
Runoff ratio (α) = 0.4
Decrease in future evapotranspiration: 10% ($\beta = 0.9$)
The relative change in runoff (δ) is given by

$$\delta = \frac{1.1 - \{(1 - 0.4)\,0.9\}}{0.4} = 1.4 \qquad (1.9)$$

When increase in precipitation is 20% then the δ value is 1.65. Wigley and Jones (1985) indicate that changes in runoff are more sensitive to changes in precipitation than to changes in evapotranspiration, and in general large increases in runoff are expected due to predicted warming unless compensated by increases in evapotranspiration. Expected future increases in precipitation, especially increases in the frequency of high intensity events (IPCC, 2007a), may lead to similar conclusions in some parts of the world. Hydrologic modeling under changing climate and analysis of impacts are dealt with extensively in the second book of this IFI four-volume series, by Mujumdar and Kumar.

1.4 IMPACTS OF CLIMATE CHANGE

1.4.1 Climate change trends and projections

The IPCC has concluded that increases in the amount of precipitation are very likely in the high latitudes, while decreases are likely in most subtropical land regions (IPCC, 2007b). Flood magnitudes and frequencies will very likely increase in most regions – mainly as a result of increased precipitation intensity and variability – and increasing temperatures are expected to intensify the climate's hydrologic cycle and melt snowpacks more rapidly (IPCC, 2007c; USEPA, 2008). Changes in temperature and radiative forcing alter the hydrologic cycle, especially the characteristics of precipitation (amount, frequency, intensity, duration, type) and extremes (Trenberth *et al.*, 2007). Climate change models suggest an increase in global average annual precipitation during the twenty-first century, although changes in precipitation may vary from one region to another (IPCC, 2007b; USEPA, 2008). A recent Australian study (Jacob *et al.*, 2009) indicated that (1) GCMs have some skill in simulating the spatial pattern for the mean of R10mm (i.e., the number days for which precipitation amount $\geq$ 10 mm) and RX5day (i.e., highest precipitation amount in 5-day period); and (2) the models have very limited ability in simulating the mean for the most extreme rainfall index (R95pT: total precipitation due to wet days > 95th percentile). The lack of agreement in model trends based on projections of GCMs could be due to lack of skill in simulating the large-scale mechanisms affecting Australian rainfall, such as ENSO.

1.4.2 Anthropogenic influences and climate change: extreme precipitation

Many researchers argue and agree that stationarity is dead (e.g., Milly *et al.*, 2008) due to possible anthropogenic changes of Earth's climate thereby altering the magnitudes and occurrence of extreme events. Use of traditional flood frequency analysis and estimation methods is based on the assumption that the underlying processes are stationary. However, use of these methods is not appropriate to predict future flood frequencies (DEFRA, 2002). Also, the current procedures for designing water-related infrastructure must be revised. Otherwise, systems will be wrongly conceived, under- or over-designed, resulting in either inadequate performance or excessive costs (Kundzewicz *et al.*, 2008). Changes in precipitation, evapotranspiration, and rates of discharge of rivers are documented by IPCC (2007b; 2007c). It is concluded by many research studies that warming augments atmospheric humidity and water transport. This process increases precipitation where prevailing atmospheric water-vapor fluxes converge (Held and Soden, 2006). Rising sea level induces gradually heightened risk of contamination of coastal freshwater supplies. Glacial melt water temporarily enhances water availability, but glacier and snowpack losses diminish natural seasonal and interannual storage.

A recent study by Min *et al.* (2011) showed that human-induced increases in greenhouse gases have contributed to the observed intensification of heavy precipitation events found over approximately two-thirds of the data-covered parts of Northern Hemisphere land areas. Their study results were based on a comparison of observed and multi-model simulated changes in extreme precipitation over the latter half of the twentieth century. The analysis was carried out using an optimal fingerprinting technique. Min *et al.* (2011) indicate that changes in extreme precipitation projected by models, and thus the impacts of future changes in extreme precipitation, may be underestimated because models seem to underestimate the observed increase in heavy precipitation with warming. To arrive at this conclusion, Min *et al.* (2011) used the annual maxima of daily (RX1D) and 5-day consecutive (RX5D) precipitation amounts as indices to evaluate observed and simulated changes in extreme precipitation for the second half of the twentieth century. These indices and several others are discussed in Chapter 7 of this book.

1.5 INTERNAL MODES OF CLIMATE VARIABILITY: TELECONNECTIONS

1.5.1 Oscillatory behaviors

Several oscillatory behaviors of climate are discussed by Rosenzweig and Hillel (2008), indicating the variability about mean climate and their influences on precipitation and temperature across the globe. Quantifying the human fingerprint on climate change and predicting future changes are two of the greatest challenges facing all scientists who are involved in understanding the variability of hydrologic cycles in different regions around the globe (Cronin, 2009). Detection and attribution, which deal with identification of trends in essential climatic variables and address the causes respectively, generally lead the current climate change and impact studies. Research in the climate change arena should aim at understanding the fundamental problem of separating climate change from natural variability over different temporal scales.

Natural variability of the climate manifested through internal modes or phases has implications on global weather patterns and precipitation events. It is imperative to establish whether a certain climatic pattern is stochastic or whether there is an identifiable regularity, or periodicity, to atmospheric or oceanic variability (Cronin, 2009). Once these patterns are identified, the next fundamental question to answer is how do these patterns or teleconnections (or distant connections) affect extreme episodic climatic events that influence hydrologic and critical water infrastructure design? The stationarity assumption used all along in the past for hydrologic design is no longer valid and new statistical methodologies are required to assess hydrologic variables under changing climate conditions. Evaluation of extreme precipitation events at different temporal and spatial scales with clear emphasis on interconnection between these events and teleconnections needs to be carried out in future studies. Many regions in the world are heavily influenced by different climatic oscillations along with episodic cyclonic storms or hurricane event landfalls. An example of such a region is the southern part of the USA, especially the South Florida region. These regions need special attention due to these internal modes of climate variability and extremes.

Prominent global climatic oscillations, such as ENSO, Atlantic Multi-decadal Oscillation (AMO), and Pacific Decadal Oscillation (PDO), influence hydrologic processes, available water resources, and critical infrastructure in a region. Links to increases in Atlantic hurricane activity during the warm phases of AMO are already documented (USGS, 2009). It is essential to evaluate extreme precipitation events (with relevance to duration, frequency, and temporal and spatial scales) and their links to internal modes of climate variability, to develop statistical methods that combine parametric and non-parametric techniques to understand and confirm these links at different temporal scales, and also to assess the joint influence of these phases on extreme precipitation events. Cronin (2009) rightly points out that interpretation of all observed climatic variable trends including precipitation strongly depends on the still poorly understood decadal and longer time period internal modes of climate variability. The establishment of long-term instrument and paleo-records may improve our understanding about these modes.

1.5.2 Precipitation extremes and associated impacts

The oscillatory behaviors of climate variability over different time scales have enormous impact on regional hydrologic processes. A recent study by Goly and Teegavarapu (2011) documented the spatial and temporal variability of precipitation extremes for different durations in a tropical region influenced by ENSO and AMO. Increased frequency of extreme precipitation (with reference to duration and intensity) will lead to a number of changes in various water resources sectors (USEPA, 2008). Some of the changes indicated are universal and are not limited to a specific region. Increased precipitation leads to an increase in pollution and erosion and sedimentation due to runoff (USEPA, 2008); increases in heavy precipitation, combined with land use changes (i.e., impervious surfaces), could increase urban flood risks and create additional design challenges and costs for stormwater management (Field *et al.*, 2007).

Increased rainfall variability is likely to compromise wetlands through shifts in the timing, duration, and depth of water levels; given the limited capacity of the wetlands for adaptation, they are considered among the most vulnerable ecosystems to climate change (IPCC, 2008). Extreme precipitation events also lead to flooding events that impact water bodies by transporting increased loads of pollutants, overloading stormwater and wastewater systems, and facilitating ecological disturbances and biological invasions (IPCC, 2008). It is important to note that increased precipitation amounts may also lead to augmentation of water supply systems, which can be considered as a benefit in some sense. Kundzewicz *et al.* (2007) indicate that water quality changes may be observed in the future as a result of overloading the capacity of water and wastewater treatment plants during extreme rainfall.

According to the IPCC (2007b), the frequency of heavy precipitation events has increased over most land areas, consistent with warming and observed increases of atmospheric water vapor. Also based on a range of models, it is likely that future tropical cyclones (typhoons and hurricanes) will become more intense, with greater peak wind speeds and more heavy precipitation associated with ongoing increases of tropical sea surface temperatures (SST) (IPCC, 2007b). The WMO (2006) indicated that though there is evidence both for and against the existence of a detectable anthropogenic signal in the tropical cyclone climate record to date, no firm conclusion can be made on this point. It was also suggested by WMO (2006) that no individual tropical storms or hurricanes can be attributed directly to the recent warming of the global oceans. However, it is possible that global warming may have affected the 2004–5 groups of events in the mid-Atlantic as

a whole. An extensive review of these storms was provided by Farris *et al.* (2007).

1.6 EXTREME PRECIPITATION AND FLOODS IN A CHANGING CLIMATE: MAIN ISSUES

Evaluation of extreme precipitation events in relation to intensity, duration, and influences of climate variability and change is critical to address the issues of floods and flooding mechanisms under climate change scenarios. Precipitation measurement, instrumentation for estimation, monitoring network design, availability of data from different measurement devices, and assessment of long-term point and gridded data sets are essential for understanding variability of precipitation in space and time. The limitations of emerging precipitation measurement devices along with systematic and random errors need to be evaluated before data can be used for any analysis. Spatial precipitation analysis for continuous estimation is essential for establishment of homogeneous serially complete precipitation data sets that are used for climate change and trend analyses. Approaches that provide uncertainties in the estimates of missing data, multiple imputation mechanisms, and emerging methods that use data mining and other techniques to preserve site and regional statistics are critical. Assessment of hydrometeorological floods with links to extreme precipitation events, joint characterization of extreme precipitation and flooding events, and evaluation of the influence of climate change combined with anthropogenic activities are needed to address the causes of floods under changing climate.

Climate change projections of precipitation from GCMs, assessment of these projections, model selection, performance measures, and uncertainty assessment are important from the perspective of evaluating the most critical input to the hydrologic cycle. Internal modes of climate variability understood by teleconnections or oscillatory behaviors and their links to variability in extreme precipitation events need to be understood for effective intra-annual seasonal forecasts of hydrologic variables and for use in hydrologic design. Statistical data analysis using methods addressing the non-stationarity issues are essential for local, regional, and global variability of trends in precipitation extremes. Hydrologic design and water resources management that are climate change sensitive and sustainable are essential to mitigate the effects of catastrophic flooding mainly linked to extreme precipitation events. Stationarity is a contentious issue and has a profound impact on future hydrologic design and flood frequency analysis. The essential activities of agencies involved in flood management (O'Connor and Costa, 2004) should include (1) developing new ways to measure and monitor floods and to map flood hazards, investigating the complex relations between floods and persistent climatic patterns such as the ENSO and the PDO; and

Table 1.3 *World's greatest precipitation totals for different durations*

Duration	Precipitation depth (mm)	Observational period	Location
1 minute	31	1948–present	Unionville, MD, USA
1 hour	305	–	Holt, MO, USA
12 hours	1,143	1966–1990	Foc-Foc, La Réunion
24 hours	1,824	1966–1990	Foc-Foc, La Réunion
48 hours	2,466	1952–1980/2004–present	Aurère, La Réunion
72 hours	3,929	1968–present	Cratère Commerson, La Réunion
96 hours	4,869	1968–present	Cratère Commerson, La Réunion
12 months	26,467	1851–present	Cherrapunji, India

(2) understanding the role of flood processes in forming and maintaining landscapes and ecologic systems.

1.7 CONCLUSIONS AND SUMMARY

Precipitation as an important component of the hydrologic cycle bears a significant influence on hydrologic design and water resources management. The uncertainties associated with future climate change coupled with our inability to quantify these uncertainties from climate change models introduces additional complications on how we can adapt to future precipitation extremes and related drivers that influence future flooding mechanisms. Measurements, estimation over different spatial and temporal scales, and understanding of the spatial and temporal variability of extreme precipitation events are therefore critical for future hydrologic design. Use of subtle connections between these events and climatic oscillations or teleconnections and development of statistical methodologies that address the issues of stationarity and sustainable design using information about future climate change need to be addressed. This chapter provides an introduction to extreme precipitation and climate change and sets the stage for all the other issues discussed in different chapters of this book.

EXERCISES

1.1 Identify the main sources of climate variability in your region and evaluate the possible impacts of this variability on extreme precipitation events.

1.2 Download precipitation data at different temporal scales (e.g., day, month, and year) and decipher any trends in the historical data.

1.3 Using the data provided in Table 1.3, compare the extreme precipitation totals for specific durations from your region to the world's greatest precipitation totals (source: http://wmo.asu.edu/#global).

1.4 Classify extreme precipitation events according to the hydrometeorological processes that link to these extreme events.

1.5 Adopt different time slices from the historical data sets of precipitation available in your region and evaluate the statistics of data in different time slices.

1.6 Extreme precipitation data for different durations for the USA (http://www.nws.noaa.gov/oh/hdsc/record_precip/record_precip_us.html) are provided in Table 1.4.

Table 1.4 *Extreme precipitation data for the USA*

Duration		Precipitation depth (mm)	Location
1	Minutes	31	Unionville, MD
5		52	Alamogordo Creek, NM
12		58	Embarrass, WI
15		100	Galveston, TX
30		178	Cambridge, OH
40		235	Guinea, VA
42		305	Holt, MS
60		305	Holt, MS
2	Hours	381	D'Hanis, TX
2.17		483	Rockport, WV
2.75		559	D'Hanis, TX
3		724*	Smethport, PA
4.5		782*	Smethport, PA
12		871	Smethport, PA
18		925	Thrall, TX
24		1,092*	Alvin, TX
4	Days	1,575	Kukaiau Hamakua, HI
8		2,083	Kukaiau Hamakua, HI
31		2,718	Puu Kukui, Maui, HI
1	Year	18,771	Kukui, Kauai, HI

*estimated

Use these data to develop a relationship between duration and precipitation totals. Compare the relationship with the world's greatest precipitation totals.

1.7 Discuss precipitation-based drivers causing peak flooding events in your region.

1.8 Identify and create an inventory of critical infrastructure elements that would be affected by climate change in your region.

1.9 Obtain the storm tracks and information about all the major tropical and cyclonic storms (hurricanes or typhoons) available for your region and evaluate the extreme precipitation depths recorded during these storms.

1.10 A simple water balance model described by Wigley and Jones (1985) evaluates the effects of changes in precipitation and evapotranspiration due to climate change on runoff, if the runoff ratio is 0.2 and the increase in precipitation and evapotranspiration is 10%. For a constant increase in evapotranspiration, assess the future runoff conditions for different values of runoff ratios and future increases in precipitation values.

2 Precipitation measurement

2.1 INTRODUCTION

Precipitation is the key hydrologic variable linking the atmosphere with land surface processes, and plays a dominant role in the terrestrial climate system. Accurate measurements of precipitation on a variety of space and time scales are important not only to weather forecasters and climate scientists, but also to a wide range of decision makers, including hydrologists, agriculturalists, and industrialists (Ebert *et al.*, 2007). Excessive and insufficient precipitation can cause significant damage to life and property through hydrologic extremes such as floods and droughts respectively. Prediction of such phenomena largely depends on how accurate the estimation of precipitation is and its resolution at different spatial and temporal scales. Even though rain-gage-based observations are the conventional method to measure rainfall directly, the very sparse network or zero coverage over oceans and remote areas limit their utility over these regions. The absence of gages in remote regions clearly indicates the need for remotely sensed (radar and satellite) measurements by which rainfall estimation can be monitored as well. This provides the advantage of availability of data in real time and complete area coverage irrespective of terrain or climate. Owing to these reasons and the complex error structure, as well as high uncertainty exhibited by intermittent measurement through sparse rain gages, identifying true rainfall fields has been recognized to be difficult. To overcome these issues, ground-based radar measurements adjusted with rain gage data were found to be more representative of the true rainfall field. As an alternative, satellite-based precipitation estimation techniques have been developed (Huffman *et al.*, 2003; Adler *et al.*, 2003; Hong *et al.*, 2004; Sene, 2009). The widespread availability and easy accessibility of satellite-based precipitation estimates have also enhanced hydrologic modeling and forecasting procedures at the watershed scale. Before the satellite-based precipitation products can be accepted, rainfall estimates from this source have to be assessed to understand their strengths and limitations so that they can be interpreted correctly. Future precipitation measurements will be dominated by radar- and satellite-based observations with dependence on ground truth (i.e., rain gage). In this chapter, an exhaustive description of rain gage and radar-based precipitation measurements is provided. Satellite-based estimates of precipitation are also discussed and compared with those from rain gage and radar systems. Details of global precipitation measurements and the gridded data sets are provided.

2.2 PRECIPITATION MEASUREMENT IN A HISTORICAL CONTEXT

The first written reference to the measurement of precipitation appears to have come from India in 350 BC (in a book titled *Arthasasthra*, by Kautilya). The rain gage was referred to as "Varshanana" and described as a 20-inch diameter bowl, and measurements were taken regularly as a way of estimating the annual crop to be sown. Strangeways (2007) provides a comprehensive review of the historical development of rain gages and the most recent developments in their design. A few countries have led the design of rain gages similar to those in use today. These countries include Palestine, China, and Korea. According to Soon-Kuk (2006), rainfall observations were made in Korea using the world's first rain gage some time in 1441. Strangeways (2007) noted that the reinvention of the rain gage and quantitative measurement of rain in China and Korea are probably the greatest achievements in the fields of meteorology and hydrology during the 1300 years since the isolated Palestinian and Indian measurements.

2.3 GROUND-, RADAR-, AND SATELLITE-BASED MEASUREMENTS

Ground-based measurements refer to the conventional and direct ways of measuring precipitation and these are generally carried out by rain gages. Radar measurements qualify as ground-based measurements even though the operation is primarily intended to obtain data of spatially varying precipitation. Radar- and satellite-based measurements are emerging methods for estimating precipitation. Sene (2009) documents the available satellite-based precipitation estimates derived from different algorithms.

2.4 MEASUREMENT METHODS, ERRORS, AND ACCURACY

The main focus of this chapter is the measurement of precipitation with identification of three methods of measurement. These methods include ground-, radar-, and satellite-based measurements. All measurements are prone to errors and have their advantages and disadvantages. Rain gage measurements are point measurements and are not free from errors (either instrument or human error). Furthermore, rain gages or any measurement devices may have both systematic and random errors. Weather radar-based and satellite-based measurements are known to provide surrogate estimates of precipitation amounts. These two methods provide measurements that are not confined to a specific point in space. They provide information about the temporal and spatial variation of precipitation in a region. These two methods have biases and they need to be corrected considering the ground truth (i.e., the rain gage). A traditional target model (showing the number of arrows hitting a specific region) can be applied in the current context to discuss measurement accuracy and precision of some of these measurements. Drosg (2007) discusses accuracy and precision in a similar context. Error assessment and determination of accuracy are usually carried out by gage consistency analysis. Gage consistency analysis requires quality assessed and quality controlled data from other sources (i.e., rain gages) to obtain a preliminary assessment of errors.

2.4.1 Systematic and random errors

Precipitation measurements, like any type of measurements, are prone to systematic and random errors. The field of metrology (science of measurement) provides an exhaustive treatment of different types of errors and their manifestations in different measurement settings. Rabinovich (2005) provides the following definitions for systematic and random errors:

> **Systematic error**: a component of the inaccuracy of measurement that, in the course of several measurements of the same measurand, remains constant or varies in a predictable way.

> **Random error**: a component of the inaccuracy of a measurement that, in the course of several measurements of the same measurand, under the same conditions, varies in an unpredictable way.

It is evident from the above definitions that systematic errors are mainly due to measuring instruments (instrumental systematic errors) (Rabinovich, 2005). These errors can be corrected but cannot be eliminated completely. In some situations, systematic errors are qualified as personal systematic errors (Rabinovich,

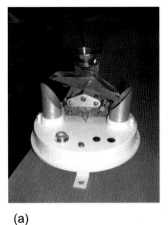 

(a)   (b)

Figure 2.1 (a) Tipping bucket rain gage; (b) the telemetry station housing the rain gage (source: C. Skinner).

2005) and they are mainly due to the individual characteristics of the observer.

2.5 CONFIGURATIONS OF RAIN GAGES

Rain gages come in different configurations and designs (sizes and shapes). Recent improvements in designs are aimed at reducing instrument-specific errors and gage catch. This section provides a description of the rain gages most commonly used throughout the world. Readers are referred to an extensive review of rain gages by Strangeways (2007) and WMO reference documents (e.g., WMO, 2008).

2.5.1 Recording and non-recording gages

Recording gages measure precipitation at discrete intervals of time and thus provide the time distribution of rainfall for a given storm event or within a time frame of consideration. Typically, recording gages are electronically controlled and the information about rainfall measurements is transmitted through telemetric devices. Human intervention is minimal in measurement and collection of the data. Examples of this type of gages include tipping bucket, float, and weighing bucket gages. Descriptions of these gages are provided in the next few sections. Non-recording gages are typically standard gages and measure precipitation without providing a time distribution of rainfall. Generally, these are manually operated and human involvement is critical in the collection and reporting of measured precipitation.

2.5.2 Tipping bucket rain gage

A tipping bucket precipitation gage, as shown in Figure 2.1, is a recording rain gage that works by measuring water volume

(a)   (b)

Figure 2.2 (a) Weighing bucket rain gage; (b) the casing housing the rain gage with recording instrument (source: C. Skinner).

(a)   (b)

Figure 2.3 (a) Float-type rain gage; (b) the casing housing the rain gage with recording instrument (source: SFWMD).

(a)   (b)

Figure 2.4 (a) Standard rain gage; (b) the casing housing the rain gage with recording instrument (source: C. Skinner).

through the use of a lightweight, dual-compartment tipping device. The apparatus, which has two equally sized buckets, is balanced about a horizontal axis such that one bucket is in the upright fill position, and the other is emptying water. Rain that is collected in the first bucket fills the compartment until the weight of the water triggers the container to tip due to instability, causing the second bucket to move into the upright fill position while the first bucket empties below. Each tip of the container is recorded as an electronic signal and corresponds to a specific precipitation amount. The tipping of the bucket creates a mechanism to measure precipitation at discrete intervals of time.

2.5.3 Weighing bucket gage

A weighing bucket rain gage (Figure 2.2), as the name suggests, has a reservoir for collecting precipitation and a recording mechanism to register the amount of precipitation. The reservoir itself rests on a scale and the weight of the collected precipitation is converted into a precipitation amount and is recorded. Some gages come with a recording pen on a clock-driven chart. This type of rain gage allows measurement of accumulated precipitation amounts over a specific period of time, after which the reservoir needs to be emptied.

2.5.4 Float-type gage

A float-type rain gage, shown in Figure 2.3, provides a time distribution of precipitation data through the use of a float mechanism located inside the rainfall collection reservoir. Rainfall enters the collection chamber through a funnel in order to minimize disturbance of the water surface. Additionally, a stilling device is located inside the reservoir in order to lessen oscillations produced by incoming water. The position of the float is recorded by

a pen on a clock-driven chart, thus generating a plot of rainfall over time.

2.5.5 Standard rain gage

The standard rain gage is conceptually the simplest device for measuring rainfall. It consists of a reservoir with a funnel and is obviously a non-recording-type rain gage. A typical standard gage is shown in Figure 2.4. As no automatic recording devices are attached, precipitation measurement readings have to be taken manually at regular time intervals.

2.5.6 Rain gage diversity and distribution

According to Sevruk and Klemm (1989) there are over 150,000 manually read gages throughout the world. There are also several different types of gages that differ in catch area, height, material used, and other aspects. Sevruk and Klemm (WMO, 1989) and Strangeways (2003; 2007) indicate that there are more than 50 types of manually read gages in the world. These estimates may not reflect the current figures, and the variations in the types are not easy to document. However, these numbers and types of gages show the diversity in the configuration of rain gages in the world.

2.5.7 Measurement of snowfall and emerging measurement devices

Measurement of snow using conventional rain gages is prone to several errors. Errors are mainly introduced due to wind, which will affect the catch of the gage. Snow measurement devices can be distinguished based on whether the measurement is done in real time (as the snow falls) or in the short term (after the snow has fallen and settled on the ground). Precipitation amounts based snowmelt volumes are sometimes measured by introduction of anti-freezing chemicals in the rain gages. Snow measurement can be made using recording and non-recording gages provided proper precautions are taken to ensure accurate catch and melt volumes documented by manual or tipping bucket gages with a heating mechanism.

Wind-related errors dominate the errors in snow measurements, and mechanisms or solutions to reduce these errors are recommended by the WMO. Wind shields provided by the use of fences of different configurations (inner and outer regions surrounding a rain gage) are recommended by the WMO. The WMO designated an octagonal, vertical double-fenced shield as the recommended international reference for all rain gage performance comparisons (Strangeways, 2007). Estimation of snowmelt volumes based on undisturbed snow sections from the ground is one of the methods for estimation of snowfall. Other ways of measuring snowfall include snow depth assessment using graduated poles, snowpack weight estimation, and non-destructive (non-disturbing) methods using gamma-rays (Strangeways, 2007).

Optical rain gages and disdrometers belong to another class of precipitation detectors. Disdrometers are often used for understanding cloud physics and the drop size distribution of hydrometeors. Optical precipitation gages can be used for measurement of snow. Strangeways (2007) describes and discusses the advantages of these devices over tipping bucket rain gages. However, the high cost of these devices makes them unappealing for routine or long-term precipitation measurements.

2.5.8 Measurement errors in rain gages

Rain gage measurements of precipitation are prone to systematic and random errors. A clear distinction between these two errors and mechanisms or methods to reduce these errors is possible. However, understanding errors and their manifestation due to site-specific conditions and instrument design should receive

Table 2.1 *Weather radar frequency bands*

Band	Frequency (MHz)	Wavelength (m)
S	1,500–5,200	0.193–0.058
C	3,900–6,200	0.077–0.048
X	5,200–10,900	0.058–0.028

Source: WMO (2008)

emphasis. These errors will lead to incorrect precipitation values and have enormous impact on the output of hydrologic simulation models (Molini *et al.*, 2005).

Strangeways (2007) documents several potential errors induced by external factors in precipitation measurements. Errors in rain gage measurements are introduced due to a variety of reasons and they include (1) inappropriate size of the collector, (2) evaporation loss, (3) out splash, (4) orientation, (5) placement, and (6) wind effects.

Exhaustive discussion about these errors and correction procedures is not provided here. Detailed discussion of errors and possible correction mechanisms can be found in several documents published by WMO (1982, 1984, 1986b, 2008). WMO (2008) recommends the WMO (1982) document that provides details of the models used for adjusting raw precipitation data in Canada, Denmark, Finland, the Russian Federation, Switzerland, and the USA. WMO (1989a) also provides a description of how the errors occur due to a variety of site-specific and instrument-related reasons. Several references on the topic of rain gage errors are available as conference proceedings published by WMO (1986a, 1989b).

2.6 RADAR MEASUREMENT OF PRECIPITATION

The use of radar-based technology to estimate precipitation (rainfall) began in the early 1960s (Pathak *et al.*, 2009; Meischner, 2005). During the early 1990s, the use of this technology proliferated as the National Oceanic and Atmospheric Administration (NOAA) installed many radars across the USA as part of the Weather Surveillance Radar 88-Doppler (WSR 88-D) or NEXt generation RADar (NEXRAD) program initiated by the National Weather Service (NWS). There are several types of weather radars that are in use throughout the world. They are sometimes classified based on their frequency and wavelength characteristics. Table 2.1 provides details of different radar types. The attenuation of the radar's signal is greater at shorter wavelengths, and hence the range is shorter; however, a smaller, cheaper dish and installation is required. Most operational weather radars are either C-band or S-band, although X-band radars are sometimes used to fill in gaps locally in the national network (Sene, 2009).

Figure 2.5 Location of Doppler radar sites in the USA (source: http://radar.weather.gov/).

Exhaustive discussion about radar-based precipitation measurements can be obtained from reference books by Raghavan (2003), Meischner (2005), and Strangeways (2007). Meischner (2005) provides a comprehensive assessment of radar-based precipitation networks, estimation, use of radar data for hydrometeorological applications, and also recent advances in the radar measurement of precipitation. The reader is referred to the works of Collier (1989) and Vieux (2004) for the initial developments of radar-based precipitation networks and their evolution. The textbook by Raghavan (2003) is a valuable reference for radar-based estimation of precipitation.

2.6.1   The US weather radar system (NEXRAD)

The NEXRAD system was initially prototyped in 1988 at the National Severe Storms Laboratory in Norman, Oklahoma, and deployed for use nationally in 1992 under the controlling agencies of the NWS, the US Air Force, and the Federal Aviation Administration. Currently, there are over 160 WSR-88D radars in operation across most of the USA. Figure 2.5 shows the locations of these radar sites. Weather radars can locate and follow precipitation within a range of 200 to 400 kilometers, depending on radio propagation conditions and the nature of the weather system.

2.7   WEATHER RADAR AND THE THEORY OF REFLECTIVITY

Weather radars operate by emitting short (250 m) pulses of coherent microwave energy. When any target object is encountered, for example a precipitation droplet, the emitted energy is scattered. Small amounts of energy, known as backscatter, are returned to the radar where they are detected and recorded. The intensity of the returned signal is then related to the size of the object and analyzed according to the time required for the pulse to reach the target and return. This provides information regarding the range and Doppler velocity of the target relative to the radar. The Doppler radar in the USA is a 10-cm wavelength (S-band) radar. It is designed for long-distance surveillance of precipitation since its wavelength

penetrates rainfall with little attenuation. Shorter wavelength radars suffer attenuation caused by absorption and scattering of the electromagnetic radiation that degrades performance of radar in precipitation estimation as the distance from the radar increases. Under most conditions, the useful range of a 10-cm radar is considered to be approximately equal to 180 km, even though the WSR-88D system produces precipitation estimates out to 230 km. As distance increases, the beam becomes increasingly higher above average ground level because the radar's lowest elevation angle is 0.5 degrees, the Earth is curved, and there is atmospheric refraction. Additional details on radar characteristics related to precipitation measurements may be found in Fulton *et al.* (1998), Doviak and Zrnic (1993), Vieux (2004), Bedient *et al.* (2009), and in Vieux and Bedient (2004). Though the advantages of the use of radar technology are evident, a radar pulse of electromagnetic energy does not give a direct measurement of rainfall.

The transformation of the radar reflectivity into rainfall rates using an empirical power-law function defining a reflectivity–rain (Z–R) rate, called a relationship, is a critical step in radar-based precipitation estimation. Overviews of weather theory and derivation of the radar–rain transformation equation are found in the work of Clement (1995). Further discussions can also be found in Bedient *et al.* (2009). The Z–R relationships may vary from storm to storm but only a few Z–R relationships have been stated to be in use operationally. As radar emits pulses of radiation and a hydrometeor (rain drop) target is encountered, the radiation is absorbed or scattered depending on the wavelength of the incident radiation and the hydrometeor's phase and size. As the radar waves hit the hydrometeors the scattering of waves with a wavelength larger than the hydrometeors is defined by the Rayleigh scattering laws. Waves that are scattered in the exact opposite direction (backscatter) may have hit a rain drop with a cross section, σ, as defined by Rayleigh scattering, of

$$\sigma = \frac{\pi^5}{\lambda^4} |K_w|^2 D^6 \qquad (2.1)$$

where λ is the wavelength of the radar, $|K_w|$ is the complex index of refraction for water, and D is the diameter of an equivalent volume spherical rain drop. Since the radar operation does not know the size of each rain drop or the number of drops of each size, a probability distribution – defined as the drop size distribution (DSD) – which affects the strength of the return signal, is used to reference a volume of space. The DSDs of hydrometeors can be quantified using direct measurements with the help of disdrometers.

Reflectivity values in decibels (dBZ), which is the measure of the backscatter of the rain drop distribution within the volume of space, are defined as:

$$dBZ = 10 \log_{10} \left[\frac{\lambda^4}{\pi^5 |K_w|^2} \int_0^{D_m} D^6 N(D) \, dD \right] \qquad (2.2)$$

where D is the maximum drop diameter, and $N(D)$ is the DSD. The rainfall rate is related to the sixth power of the hydrometeor's distribution.

Another important element in the development of Z–R relationships is the rain rate, which is the product of the drop's mass, DSD, and fall speed, integrated over the range of drop diameters. Rain rate, R (depth of water per given time), is given by:

$$R = \frac{\pi}{6} \int_0^{D_m} D^3 N(D) v(D) \, dD \qquad (2.3)$$

Marshall and Palmer (1948) gave one form of DSD, referred to as the inverse potential:

$$N(D) = N_o e^{-\Lambda D} \qquad (2.4)$$

Fall speed as a function of diameter has been defined by Spilhaus (1948) as:

$$v_t(D) = 14 D^{0.5} \qquad (2.5)$$

When the drop size and fall speed are combined with the reflectivity and rain rate in the equations above, a Z–R relationship can be developed. A number of Z–R relationships have been developed and a listing of them can be found in many textbooks on radar meteorology. The next section discusses this relationship and its variants. There is no single Z–R relationship that works best for all meteorological and storm events. Suitable Z–R relationships are always developed for the local area's precipitation climatology. Systematic variations of drop size and radar–rainfall relations are discussed by Atlas et al. (1999). They provide a possible range of values for the coefficient and exponent for different storm types.

2.7.1 Issues and errors related to reflectivity measurements

Errors in reflectivity (Z) value estimation may be possible for several reasons, and they include (1) ground clutter, (2) anomalous propagation, (3) partial beam filling, and (4) wet radome conditions. Ground clutter refers to high reflectivity values received from ground targets. This generally results in overestimation of rainfall. Ground clutter can be removed by developing a clutter map based on reflectivity measurements on a clear day (without any precipitation) at low angles. The clutter map with reflectivity patterns is removed from the reflectivity measurements during storm events to achieve the required filtering to reduce overestimation of rainfall. The clutter map varies in space and time. For accurate estimation of rainfall amounts, the clutter map should be updated often. Base reflectivity measured at the lowest angle of elevation scan is extremely useful for spatial evaluation of occurrence of precipitation around the radar.

Anomalous propagation is mainly due to radar beams propagating anomalously due to super-refraction. Partial beam filling

refers to the target being partially filled, thereby underestimating the rainfall amounts. Due to a wet radome, less power is transmitted to the targets and fewer reflectivity values are registered, therefore again underestimating rainfall. The attenuation of power is generally a function of the thickness of a water film. Hail in thunderstorms provides good surfaces for reflection of energy. Hailstones, when coated with a thin layer of water and due to their travel through the thunderstorm cloud, enhance the reflectivity values leading to overestimation of the amount of rain. Also, dry hail, without the coat of water, is known to be a poor reflector of energy, thus leading to underestimation of rainfall amounts.

Below-beam effects with the wind blowing towards or away from the rain gages may suggest respectively over- or underestimation of rainfall by the radar beam sampling high in the atmosphere. Evaporation below the beam may result in overestimation of rainfall as the rain gage is not registering the same amount of rainfall as estimated from radar reflectivity. An example of such an evaporation process in arid or semi-arid climates is referred to as virga. Below-beam effects due to coalescence of rain drops can lead to variations in the reflectivity values measured. Measurement of reflectivity close to the radar is limited by the maximum tilt angle, which enables the radar to scan the vertical atmosphere. The radar is generally unable to scan directly overhead due to the tilt angle limit. The area over which the radar is unable to measure is commonly referred to as the cone of silence. Reflectivity returns from non-precipitation objects that are moving aerial targets are common, and these are referred to as unwanted echoes. These echoes can be filtered from reflectivity images by evaluating and editing the images to detect and remove most non-precipitation features.

2.7.2 Reflectivity–rainfall rate relationships

Reflectivity is defined as the sixth power of the diameter of rain drops (mm) per cubic meter of atmosphere. Reflectivity is related to rainfall rate through the designation of a DSD. Commonly, the two parameters are linked through a Z–R relationship of the form:

$$Z = a R^b \qquad (2.6)$$

where Z is the reflectivity (mm^6/m^3), R is the rainfall rate (mm/hr), and a and b are fitting coefficients. The coefficient a typically varies between 100 and 900 and exponent b varies between 1.00 and 2.00. If the Z value is not reported as a dimensionless number and not referenced based on unit reflectivity (1 mm^6/m^3), then the coefficient a will have units as $mm^{6-b} \, m^{-3} \, hr^b$. In practice, at least in the USA, the reflectivity values are reported as decibels of reflectivity (dBZ). The reflectivity can range over many orders of magnitude, and for operational use Z is converted to decibels ($10 \log_{10} Z$).

Table 2.2 *Transformation of reflectivity values to rainfall rates using different Z–R relationships*

Z (mm^6/m^3)	dBZ $10(\log_{10} Z)$	Tropical (mm/hr)	Stratiform (mm/hr)	Standard (mm/hr)
10	10	0.068	0.154	0.088
100	20	0.466	0.648	0.456
1,000	30	3.175	2.734	2.363
3,162	35	8.287	5.615	5.378
10,000	40	21.630	11.531	12.240
31,623	45	56.457	23.679	27.856
100,000	50	147.361	48.625	63.395

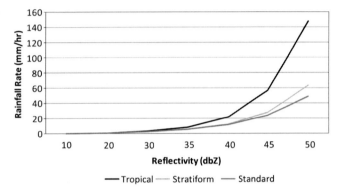

Figure 2.6 Variation of rainfall rate calculated using different Z–R relationships.

Some standard Z–R relationships for different storm types used in some parts of the USA are given below for convective, stratiform, and tropical storms by Equations 2.7, 2.8, and 2.9 respectively.

$$Z = 300R^{1.4} \qquad (2.7)$$

$$Z = 200R^{1.6} \qquad (2.8)$$

$$Z = 250R^{1.2} \qquad (2.9)$$

The three Z–R relationships are used to calculate the rainfall rate for a given Z value and are reported in Table 2.2 and Figure 2.6. It is evident from Figure 2.6 that at lower values of reflectivity the rainfall rates are almost equal, and as the Z values increase the rainfall rates differ substantially once the Z value crosses a specific threshold. This example illustrates the point that use of an appropriate Z–R relationship is critical for accurate estimate of rainfall.

It is important to note that for a given value of R different Z values are possible, and also for a given value of Z different R values are possible. A Z value of 640 mm^6/m^3 is estimated if 640 drops of 1 mm are in 1 m^3 of space or 10 drops of 2 mm are in 1 m^3 of the same space in the atmosphere. This suggests that a volume of atmosphere with different DSDs might provide

the same reflectivity value. Also, if the falling velocities of the drops are different, then the rainfall rates will be different for a given value of Z. Battan (1973) and Raghavan (2003) summarized different Z–R relationships used in various regions of the world for different storm types. These data are important in approximating the coefficient and exponent values based on the storm type and the region. Tables 2.3 and 2.4 provide various Z–R relationships applicable to several regions in the world.

Radar measurement of reflectivity assumes a distribution of drop sizes that are known to vary geographically and during storm evolution. As indicated previously, reflectivity is converted to rainfall rates by means of a Z–R relationship. These relationships are based on the assumed DSDs that may be typical for storm types. However, the evolution of the storm and the processes that produce precipitation may result in DSDs that differ from the assumed Z–R relationship. Radar-centric factors affecting the Z–R relationship include antenna, radar transmitter, and receiver characteristics that diminish or increase signal strength. These radar characteristics can change due to radar equipment replacement and maintenance. For example, two radars observing precipitation at the same location may differ substantially due to differences in the strength of the transmitted microwave signals and the atmosphere traversed by the respective radar signals.

Geographic variation and storm morphology also affect the DSDs. Local enhancement near coastal or inland water bodies can affect precipitation processes and DSDs. Differences between assumed and actual DSDs, and radar signal transmission strength, result in bias that varies with space and time. To correct for radar bias, gage data are used to enhance the accuracy of radar rainfall measurements. Where rain gages are sparsely distributed, radar precipitation sometimes shows accumulations that are more or less in error according to the assumed Z–R relationship. Derivation of Z–R relationships that are specific to a radar setup and for specific geographic regions is expected to improve radar rainfall estimation where gages are lacking or are too sparse. A framework for analysis is needed for assessment of how much improvement can be expected from derived Z–R relationships.

It is well known that stratiform precipitation can be best sampled at the lowest 1 km of the atmosphere that occurs typically within 50 km of the radar. Beyond 50 km from the radar, the sample may be questionable. This range can be limited further by terrain blocking of the radar beam. It is reported in the literature that sampling of stratiform precipitation is generally poor beyond 100 km from the radar. However, the convective precipitation is distributed over a much deeper layer of the atmosphere than stratiform precipitation (UCAR, 2011). In seasons and locations where convection is common and terrain effects are minimal, the radar may be able to sample precipitation at long distances from the radar. Shallow convection may be under-sampled by the radar at ranges closer than 150 km.

Table 2.3 *Z–R relationships obtained for various regions and storm types: data set 1*

a	b	Region	Type of storm	Reference
247	2.4	South Dakota (USA)	Convective storms	Miller (1972a) Climatological Z–R relation
300	1.4	Florida Area Cumulus Experiment (FACE) USA	Convective	Woodley et al. (1977), Gajin et al. (1985)
230	1.3	GATE		Hudlow and Arkell (1978)
185	1.2	GATE		Atlas et al. (1990) Climatological Z–R relation (Atlas et al. give different sets of constants for different ranges and for different radar in GATE. Other constants have been given for GATE by other authors. They are based on drop size measurements onboard ships or aircraft.)
55	1.6	US National Weather Service	Convective (flash flood situations)	Jorgensen and Willis (1982)
300	1.4	US National Weather Service	Hurricane	Jorgensen and Willis (1982)
300	1.4		Default values for US WSR-88D	Fulton et al. (1998)
250	1.2		For tropical environments	
340	1.5	COPT-81 (West Africa)		Hauser et al. (1988) For surface measurements of precipitation from optical spectropluviometer
300	1.5	Swiss Meteorological Institute	For rain	Joss and Lee (1995)
100	1.2	Chennai (India)	Seasonal mean for northeast monsoon	Raghavan and Sivaramakrishnan (1982) By comparison of radar and rain-gage derived areal averages of precipitation
100	1.3	Chennai (India)	Seasonal mean for southwest monsoon	Raghavan and Sivaramakrishnan (1982) By comparison of radar and rain-gage derived areal averages of precipitation
230	1.4	Massachusetts	Average ordinary rain	Austin (1987)
400	1.3	Massachusetts	Intense convective cells	Austin (1987)
100	1.4	Massachusetts	Non-cellular rain	Austin (1987)
650	1.4	Bauru (Brazil)	All rain for near range comparison of probabilities of Z and R in non-simultaneous measurements	
50	1.5	Darwin (Australia)	Convective	Steiner and Houze (1993) Convective and non-convective parts were separated according to reflectivity criteria and then compared with rain gages

Source: Raghavan (2003)

2.7.3 Reflectivity measurements and extraction

Reflectivity measurements of radar sources are generally available as volume scans in a polar grid coordinate system. Spatial analysis needs to be carried out using any geo-spatial analysis tool to collocate the rain gages (i.e., ground truth) and the reflectivity polygons available under a polar coordinate system grid. However, if the values of reflectivity are available in text files with closed polygon information, reflectivity data can be extracted directly without the need for a spatial analysis environment. Polygons in which selected rain gage stations are located can be identified by using a point-in-polygon approach. The height above the radar level at which the reflectivity data are collected is crucial, as is evident from Figure 2.7. Reflectivity data are generally collected at the lowest angle for comparisons with the ground truth and estimation of precipitation at ground level.

Radar reflectivity data are generally available in irregular time intervals ranging from 4 to 10 minutes, which corresponds to a complete radar sweep. These data need to be converted into a rainfall depth using a Z–R relationship and integrated into a total value of depth based on the temporal domain of the rain gage measurement for comparison purposes and bias corrections.

Table 2.4 *Z–R relationships obtained for various regions and storm types: data set 2*

a	b	Region	Type of storm	Reference
150	1.5	Darwin (Australia)	Non-convective	Steiner and Houze (1993) Convective and non-convective parts were separated according to reflectivity criteria and then compared with rain gages
82	1.5	Darwin (Australia)	Convective	Steiner *et al.* (1995) after adjustment with rain gages
143	1.5	Darwin (Australia)	Stratiform	Steiner *et al.* (1995) after adjustment with rain gages
334	1.2	Kapingamarangi	Convective	Atlas *et al.* (1999)
147	1.3	Atoll	Transition	Based on disdrometer and 915 MHz profiler measurements
278	1.4		Stratiform	Based on disdrometer and 915 MHz profiler measurements
865	1.1	Toga Coare	Stratiform	Classification based on variation of median volume diameter and rain rate. Values shown here are for one occasion. Atlas *et al.* obtain widely different values for other occasions.
178	1.5	National MST Radar Facility Gadanki (India)	Convective	Narayana Rao *et al.* (2001) *Z* and *R* from disdrometer data. Classification from MST radar and lower Atmospheric Wind Profiler, seasonal values are irrespective of classification.
162	1.4	National MST Radar Facility Gadanki (India)	Transition	Narayana Rao *et al.* (2001) *Z* and *R* from disdrometer data. Classification from MST radar and lower Atmospheric Wind Profiler, seasonal values are irrespective of classification.
251	1.5	National MST Radar Facility Gadanki (India)	Stratiform	Narayana Rao *et al.* (2001) *Z* and *R* from disdrometer data. Classification from MST radar and lower Atmospheric Wind Profiler, seasonal values are irrespective of classification.
155	1.4	National MST Radar Facility Gadanki (India)	Northeast monsoon	Narayana Rao *et al.* (2001) *Z* and *R* from disdrometer data. Classification from MST radar and lower Atmospheric Wind Profiler, seasonal values are irrespective of classification.
407	1.3	National MST Radar Facility Gadanki (India)	Southwest monsoon	Narayana Rao *et al.* (2001) *Z* and *R* from disdrometer data. Classification from MST radar and lower Atmospheric Wind Profiler, seasonal values are irrespective of classification.

Source: Raghavan (2003)

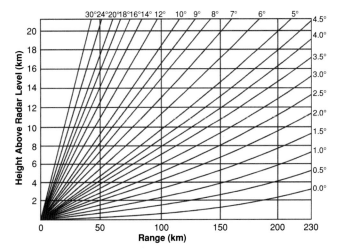

Figure 2.7 Height above the ground of radar sample as a function of range (source: NOAA).

2.7.4 Reflectivity–rainfall rate relationships

The temporal spread of reflectivity measurements based on radar scans in a given time frame is shown in Figure 2.8. The reflectivity values are typically used in a *Z–R* relationship to estimate the rainfall rate (Equation 2.10) and ultimately the rainfall depth (Equation 2.11).

$$R_n = \left[\frac{10^{\frac{dbZ_n}{10}}}{a} \right]^{\left(\frac{1}{b}\right)} \quad \forall n \qquad (2.10)$$

$$\theta_i^m = \sum_{n=1}^{k} R_n \Delta t_n \quad \forall n \qquad (2.11)$$

Equation 2.10 is appropriate when the reflectivity values are expressed in decibels ($dBZ = 10(\log_{10} Z)$) and the power relationship $Z = aR^b$ is used to link reflectivity values and rainfall rates.

2.7.5 Reflectivity data processing: an example from the USA

Radar data can be requested from the National Climatic Data Center (NCDC), which provides data for several radar stations in the USA (NCDC, 2011). The time intervals for which radar data are required can be specified and the data can be retrieved from the HDSS (Holographic Data Storage System) tape archive system. Initially, the data are moved from the tape archive system so

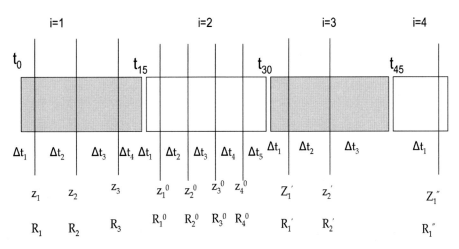

Figure 2.8 Temporal spread of reflectivity measurements.

that these files are available for download from the HDSS Access System (HAS) web interface using a generic FTP (file transfer protocol) client. Pre-processing of the data includes uncompressing the original .tar files into binary format readable files by the NOAA Weather Climate Toolkit (WCT) (WCT, 2011).

2.7.6 NOAA Weather Climate Toolkit

The WCT is a data extraction and processing tool developed in Java by NOAA and it allows for visualizing and exporting radar (NEXRAD) data to different file formats. The toolkit also allows the application of filters and other data manipulation procedures. Each radar (NEXRAD Level II) file contains reflectivity, mean radial velocity, and spectrum width data collected for the entire range of the radar station. When performing the data collection sweep, the elevation angle of the radar also varies, allowing it to obtain information from the lower to the upper levels of the atmosphere. Base reflectivity at the lowest elevation angle, that is, closer to the Earth's surface within the area of interest, should be obtained. The WCT then can be utilized to convert reflectivity data from the binary format to a text format, by cropping the coverage area to a rectangle encompassing the rain gage locations. The latitudes and longitudes of a closed polygon can be defined to specify the region of interest, and reflectivity values in the region alone can be extracted.

2.7.7 Rain Radar Retrieval System

Data collected by NEXRAD are generally expressed in the form of a radial grid, at one degree and one kilometer (radial and in radius respectively) increments. Each polygon in this grid contains the reflectivity value obtained for that area during a specific sweep (scan) of the radar in a given time interval. Recent NEXRAD data are provided at finer resolution, of $0.25°$ and 0.25 km.

To proceed with the optimization of the Z–R relationship, it is required to obtain the reflectivity of the areas that coincide with the rain gages' locations. A tool named the Rain Radar Retrieval (R^3) System (Teegavarapu, 2009), developed by the Department of Civil, Environmental, and Geomatics Engineering at Florida Atlantic University, or a similar radar data extraction tool can be utilized for that task. R^3 encompasses several tools for acquisition and manipulation of rain gage and raw reflectivity data and utilizes text format files previously converted using NOAA WCT (Figure 2.9).

The Reflectivity Extraction tool under R^3 allows the user to specify a set of coordinates of interest (in this case, the location of the rain gages) and retrieve the reflectivity observed at the matching polygons for a given period, generating a matching time-stamped reflectivity file specifically for each given coordinate pair. The tool utilizes a smart search procedure to locate polygons near the point of interest and performs a point-in-polygon check on those. Using the polygon's identification number or the combination of the angle and radial distance for this task is not possible, as the IDs are assigned differently on each file according to the number of non-blank cells, and the grid position shifts randomly, therefore justifying the above-mentioned method.

As the NEXRAD grid is composed of simple convex monotone polygons, the method suggested by Bourke (1987) is used in the R^3 system to define if a specific point lies within a polygon. Using the coordinates of the polygon as vertices, a point is on the interior of the polygon if it is always on the same side of all the line segments making up the path. Given a line segment between $P_0(x_0, y_0)$ and $P_1(x_1, y_1)$, the point $P(x, y)$ has the following relationship to the line segment (Equation 2.12):

$$\delta = (y - y_0)(x_1 - x_0) - (x - x_0)(y_1 - y_0) \quad (2.12)$$

If δ is less than 0 then the point P is to the right of the line segment; if greater than 0 it is to the left; if equal to 0 then it lies on the line

Feature[id=0 , geom=POLYGON ((-80.37374056702825 26.93033750716703,-80.37167071912226   ,26.921505017482485,   -80.39466292258989 26.917357997353996,   -80.3965642350833   26.926220889761467,   -80.37374056702825 26.93033750716703)) , sweep=0 , sweepTime=20080111 19:03:23 , elevAngle=0.48339844 , value=7.5 , radialAng=168.48633 , begGateRan=133995.05 , endGateRan=134995.02 , height=2198.909]

Figure 2.9 Sample text format string of a single polygon of NEXRAD Level II reflectivity data at lowest elevation angle.

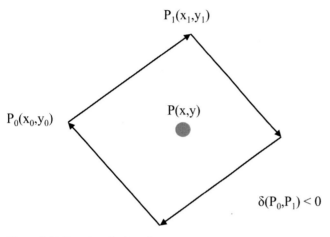

Figure 2.10 Sample point-in-polygon test.

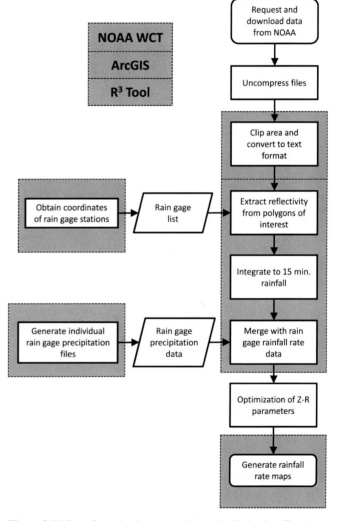

Figure 2.11 Steps for radar data processing and optimization (Ferreira *et al.*, 2009).

segment. This test is repeated for each of the four segments that make up each polygon. If the test indicates that the point is within the polygon, the associated reflectivity value with the polygon is extracted (Figure 2.10).

Rain gage data are expressed in a specific temporal resolution (e.g., 15-minute interval). As the radar data retrieval interval ranges from 4 to 10 minutes, it needs to be combined and integrated so as to allow the comparison of rain totals for the same time window. The total rain from each radar interval is weighted based on the number of occurrences within a given time window (e.g., 15 minutes). The last step performed by R^3 is the merging of time-consistent rain gage data with radar data for a pairwise comparison. A flowchart describing the whole process of radar data extraction is explained in Figure 2.11. A few steps indicated in the flowchart are specific to radar (NEXRAD) data from the USA. However, many of the steps defined in the flowchart from reflectivity data extraction, collocation of rain gages with the radar (in polar coordinate system) grid, use of a *Z–R* relationship to obtain a rainfall estimate, and use of bias correction are general steps carried out in any radar-based rainfall estimation process.

2.7.8 *Z–R* relationships

This section describes a few relevant studies that closely relate to the influence of *Z–R* relationships on radar-based precipitation estimates. In many studies, typical power relationships are used and the sensitivity of relationship parameters in relation to rainfall rate estimation is investigated. Observations from a 16-month field study using two vertically pointing radars and a disdrometer at Wallops Island were analyzed by Tokay *et al.* (2008) to examine the consistency of the multi-instrument observations with respect to reflectivity and Z–R relations. The radar-based precipitation is in better agreement using reflectivity values and observed values at the collocated rain gages compared to those from disdrometer and other radar observations. Baldonado (1996) discusses an approach to determine the Z–R relationship. Calculations of the simultaneous reflectivity factors and rain rates in a volume of air are performed based on the solution of the radar equations by infusion of the different radar parameters and measurement of the return power of the electromagnetic wave. Linear regressions are applied to the transformed data sets and constants and coefficients in the relations that take the form $Z = aR^b$, were determined. A set of empirical Z–R relations for different locations in a river basin, as well as for particular storms, are obtained. A study by Malkapet *et al.* (2007) focuses on uncertainty caused by natural variability in radar reflectivity–rainfall relationships and how it impacts runoff predictions. Rainfall intensity, R, is estimated from one-minute disdrometric reflectivity, Z, using different Z–R relationships. The Z–R relationships are derived at different time scales: climatological, daily, event-based, and physical-process based. The effect of using different estimation methods, such as least squares fitting, bias-corrected, fixed-exponent, and default operational relation, are also examined. The Z–R relationships for physical processes are derived by classifying and separating rain records into convective, transitional, and stratiform phases based on the vertical profile of radar reflectivity, developed from volume-scan WSR-88D data. Lee *et al.* (2006) derived a Z–R relationship with the help of linear regression to convert the radar reflectivity Z into the rainfall rate R. The relationship is expressed in terms of an equation of the form $Z = aR^b$.

Villarini *et al.* (2007) investigated the effects of random errors and used a procedure to convolve the radar–rainfall with a multiplicative random uncertainty factor. They discuss the effects of different magnitudes and spatial dependences. They indicate the basic systematic factors that can affect the results of such analyses, which include the selection of a Z–R relationship, the rain/no rain reflectivity threshold, and the distance from the radar. Ulbrich and Atlas (1977) describe a method by which radar reflectivity and optical extinction are used to determine precipitation parameters, such as liquid water content and rainfall rate, and an exponential approximation to the rain drop size spectrum. The improvement in the accuracy is demonstrated by the authors using empirical relations and also by applying the method to a set of experimentally observed rain drop size spectra. The authors examine these problems using high-resolution radar–rainfall maps generated with different parameters based on the Level II data from

WSR-88 D radar in Kansas, USA. Trafalis *et al.* (2005) used artificial neural networks (ANNs), standard support vector regression (SVR), least squares-support vector regression (LS-SVR), and linear regression to relate reflectivity and rainfall rate. They found that LS-SVR generalizes better than ANNs, linear regression, and a rain rate formula in rainfall estimation and for rainfall detection, and SVR performs better than the other techniques. A simplified probability matching method was introduced by Rosenfeld *et al.* (1994) that relies on matching the unconditional probabilities of R and Z, using data from a C-band radar and rain gage network near Darwin, Australia. This is achieved by matching rain gage intensities to radar reflectivities taken only from small windows centered about the gages in time and space. The windows must be small enough for the gage to represent the rainfall depth within the radar window, yet large enough to encompass the timing and geometrical errors inherent in such coincident observations. Amitai (2000) discusses spatial smoothing by the radar beam as well as post-detection integration that reduces the variability of the distribution of rainfall rate in space. It is shown that when rain gage data are smoothed in time there is an optimum smoothing time interval such that the random error and the bias are reduced to a negligible level. A method is suggested for the optimum comparison of radar and rain gage data and the possibility of a determination of Z–R relationships from such comparisons is discussed. To provide a foundation for other radar studies in the Miami area, 50 comparisons were made by Woodley and Herndon (1969) between shower rainfall recorded by rain gages and observed with radar to evaluate the Z–R relation, $Z = 300R^{1.4}$, referred to as the Miami Z–R relation.

2.8 EVALUATION OF EXPONENTS AND COEFFICIENT VALUES IN A Z–R POWER RELATIONSHIP

The power relationship, $Z = aR^b$, is a standard relationship often used for calculation of rainfall rates based radar reflectivity values. The coefficient a and exponent b values are defined by considering the storm type, rain drop distribution, and other factors. As a first step in understanding the sensitivity of these parameters of the Z–R relationship towards rainfall estimation, incremental values of a and b within a specific numerical range of values can be used to obtain rainfall rates or rainfall depths. This exercise is similar to Monte Carlo simulation where random values of inputs following a specific probability distribution are used to understand the sensitivity of model (i.e., the Z–R relationship) outputs to these inputs. However, in this exercise, systematic variation of a and b values can be attempted as opposed to random value utilization in a Monte Carlo simulation. These experiments are summarized by two schematic diagrams shown in Figure 2.12. Variants of probability distributions of input values (in the current context, a

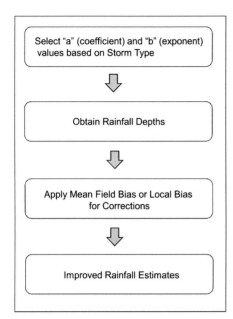
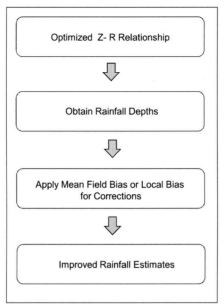

Figure 2.12 Steps for assessing uncertainties in *Z–R* relationships (Teegavarapu, 2010).

and *b* values) can be adopted by two different sampling schemes: uniform random and Latin hypercube. Absolute errors obtained from observed precipitation and *Z–R* relationship-derived precipitation values for different events for variations in the values of coefficients and exponents can be evaluated.

2.9 FORMULATION FOR OPTIMAL COEFFICIENTS AND EXPONENTS

A non-linear optimization formulation (Teegavarapu and Pathak, 2011) can also be developed for obtaining optimal coefficients and exponents for several candidate events selected for the current study. The formulation is provided below.

Minimize

$$\sum_{i=1}^{no}(\hat{\phi}_i^m - \theta_i^m)^2 \qquad (2.13)$$

subject to:

$$R_n = \left[\frac{10^{\frac{dBZ_n}{10}}}{a}\right]^{\left(\frac{1}{b}\right)} \quad \forall n \qquad (2.14)$$

$$\theta_i^m = \sum_{n=1}^{k} R_n \Delta t_n \quad \forall i \qquad (2.15)$$

$$T_i = \sum_{n=1}^{k} \Delta t_n \quad \forall i \qquad (2.16)$$

$$a_l \leq a \leq a_u \qquad (2.17)$$

$$b_l \leq b \leq b_u \qquad (2.18)$$

where θ_i^m is the estimated rainfall rate at a rain gage station; *no* is the number of candidate events or all the events available in the database for which reflectivity and rainfall rate data are available; *a* and *b* are the coefficient and exponent values in the *Z–R* relationship; a_u is the upper bound on the coefficient and b_u is the upper bound on the exponent; Z_i is the reflectivity; and $\hat{\phi}_i^m$ is the observed rainfall rate at a gage. The above formulation can be used to obtain the optimal coefficient and exponent values. An objective function that is based on the summation of absolute deviations (i.e., rain gage–radar estimate) or the exponential function of these deviations can be used. Absolute errors using standard and optimal *Z–R* relationship parameters obtained using an optimization solver with temporal time frame of 5 hours for several stations in Florida, USA, are shown in Figure 2.13.

It is evident from the graph of absolute errors in Figure 2.13 that use of optimized coefficients and exponents improves rainfall estimation.

2.9.1 Multiple *Z–R* relationships: moving temporal window

A varying "temporal window" can be selected for optimizing different *Z–R* relationships over time as the storm event unfolds. Optimization of coefficient and exponent values is achieved by using any mathematical programming formulation (optimization formulation). In this case, as shown in Figure 2.14, the temporal window size for optimization increases as the storm event unfolds, providing more information about the storm in optimizing the power relationship.

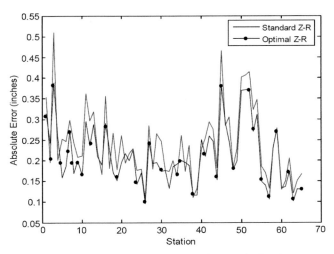

Figure 2.13 Absolute errors calculated based on standard and optimized Z–R relationships.

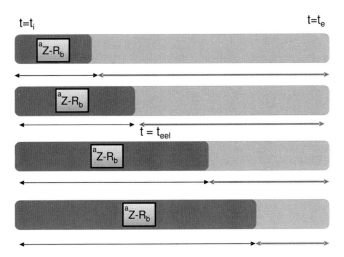

Figure 2.14 Development of changing Z–R relationships based on varying window length of the storm event already elapsed.

The near-time optimal Z–R relationship is more suitable for a storm type that is not changing over a shorter duration of time (stationarity is assumed).

2.9.2 Multiple Z–R relationships: constant temporal window

A constant "temporal window" can be selected for optimizing the Z–R relationships. Again, optimization of coefficient and exponent values is achieved using a mathematical programming formulation (optimization formulation). Again, in this case, as shown in Figure 2.15, the temporal window size for optimization is constant and advances as the storm event unfolds, providing more recent information about the storm in optimizing the power relationship. The near-time optimal Z–R relationship is more suitable for a storm type that is changing over time.

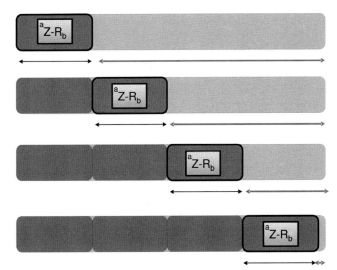

Figure 2.15 Development and use of Z–R relationships based on constant window length of the storm event already elapsed.

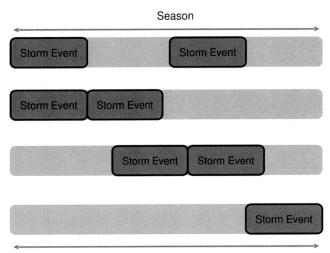

Figure 2.16 Development and use of Z–R relationships based on storm events in a season.

2.9.3 Seasonal Z–R relationships

A select number of storm events (as illustrated in Figure 2.16) can be used for development of optimized Z–R relationships. Again, optimization of coefficient and exponent values is achieved using a mathematical programming formulation (optimization formulation). The objective is to obtain a seasonal Z–R relationship based on select storm events.

2.9.4 Optimization of Z–R relationships using probability matching

Probability distributions of raw reflectivity values can be derived based on observed data for elapsed storm events and optimal

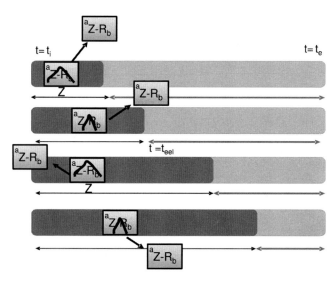

Figure 2.17 Development and use of Z–R relationships based on constant window length of the storm event already elapsed.

Z–R relationships can be developed. The optimized Z–R relationships can be applied to those events that have similar probability distributions of the reflectivity values. The implementation of this approach in real time may be difficult as distributions are not known a priori. The approach is schematically shown in Figure 2.17.

The methods for optimizing Z–R relationships have shown improvements over other methods. Readers are referred to the works of Ferreira *et al.* (2009) and Teegavarapu (2009a) for details of these methods and their applications.

2.9.5   Multiple station analysis

Analysis of uncertainties in Z–R relationships for different storm events and selection of optimal exponents and coefficients requires selection of a number of stations within a region. Multiple stations (rain gages) can be selected based on different criteria, which are described in the next section.

2.9.6   Criteria for selection of stations

Selection of observations at rain gage stations for optimization or analysis of Z–R parameters was based on several criteria. These include (1) rain gage stations reporting positive rainfall; (2) a threshold correlation coefficient estimated based on application of a standard Z–R (e.g., $300R^{1.4}$) relationship to rain gage and radar-based observations; (3) stations located within a given distance from the radar site; and (4) rain gage stations recording positive rainfall in the temporal window selected. The criterion for selection of temporal windows is discussed next.

2.9.7   Criterion for selection of windows

Each event (any temporal frame) with sub-temporal events is analyzed visually to select a window or multiple windows based on occurrence of positive precipitation. Temporal windows in which a large number of stations report rainfall are selected for obtaining optimal Z–R relationships. Generally, two windows are selected: one for optimizing the Z–R parameters and another for testing the optimized Z–R parameter values via performance measures.

2.9.8   Optimal Z–R relationships: key issues

A completely different functional form for the Z–R relationship (i.e., different from the traditional or standard Z–R power relationship) for a given temporal window (e.g., 15 minute interval, day, month, or a season) may be examined for a region for obtaining improved rainfall estimates.

There is a need to fine tune the coefficient and exponent parameter values of the power relationship (i.e., $Z = aR^b$) via optimization considering the realized storm events.

The minimum and maximum values of coefficients and exponents need not be fixed a priori for optimization formulations.

Improved estimates of precipitation can be realized when a temporal window preceding the window in which the optimal relationship of Z–R is applied. In real time, this information about the window will depend on the amount of information available about the elapsed time interval with observed precipitation at rain gages.

Accurate continuous rain gage observations are critical for the success of the Z–R optimal relationships.

Storm type information can be incorporated within the optimization or selection of Z–R relationships indirectly by analyzing the raw reflectivity values.

The number of rain gages used for developing optimal Z–R relationships depends on the performance criteria used for the selection. However, to obtain a robust estimate of coefficients and exponents of Z–R relationships all available rain gages should be used.

Rain gages can be selected based on a performance criterion such as the correlation coefficient derived from radar-based precipitation estimates using the standard Z–R relationship and rain gage observations.

Rain gages with high correlation with preliminary radar-based estimates (based on standard Z–R relationship) in deriving the optimal Z–R relationship can be used.

The constant and varying window approaches are applicable for optimization of Z–R relationships in real time provided an optimization solver is linked to a radar data processing tool and the execution time of the solver is reasonable.

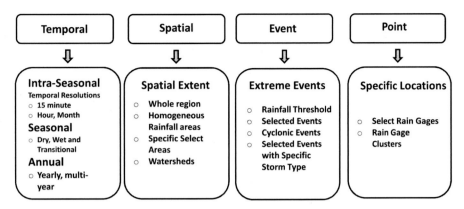

Figure 2.18 Categories for evaluation of biases.

Average coefficient and exponent values for specific seasons can be recommended for further evaluation and applicability in real time.

The probability distribution of raw reflectivity values from the past storms events can be used to obtain optimal coefficient and exponent values for similar events in future.

The optimal relationships are applicable in real time and can also be adopted for post-processing of the radar data. The seasonal average optimal coefficient and exponent values obtained from this study can be used as starting values for rainfall estimation before bias corrections are applied.

2.10 BIAS EVALUATION AND CORRECTIONS

Several indices using radar and rain gage data can be developed that are similar to performance measures used in hydrologic model evaluation and forecast verification of numerical weather prediction models. The indices developed are expected to answer the following major questions:

- What is the overall quality of the radar data compared to rain gage data (i.e., ground truth)?
- Are the radar and rainfall data characteristics statistically similar?
- What is the overall bias?
- What is the error structure and how it is varying in space and time?
- What is the variability of the bias in space and time?
- What is the skill of the method used for radar-based rainfall estimation in the definition of rain or no rain events?
- Do the radar-based rainfall estimates preserve the site (at a rain gage location) and regional (a set of rain gages) statistics?
- Are the rainfall fields generated by radar-based estimates and rain gage-based observations similar?

- Can a specific index or a set of indices help provide directions to improve the radar-based rainfall data?
- How do radar and rain gage data quantitatively compare at different temporal resolutions?

Evaluation of biases uses four categories: (1) time, (2) space, (3) event, and (4) point. Specific indices are developed for each category. Some of the categories (other than time) are also analyzed over time even though they are defined for a region, event, or a specific location. Three sets of indices can be developed that are applicable to each of the categories. These three sets of indices can be classified as (1) visual, (2) quantitative, and (3) statistical. The categories for evaluation and indices are summarized in Figure 2.18 and Figure 2.19 respectively.

Evaluation of bias can be made by comparing rainfall and radar data by collocating rain gages in the fixed tessellations (i.e., grids) of spatial radar rainfall surfaces. Pooled and stratified samples can be obtained and evaluated for bias. Pooled sampling involves pooling of sample pairs over a specific time or space, which is similar to space-based comparison. The stratified version is the same as the space-based comparison in homogeneous or quasi-homogeneous areas. The comparisons can be for a specific temporal scale, within a region, or for specific events. Intra-seasonal comparisons are made at different temporal resolutions (e.g., 15 minutes, hour, day, and month). Seasonal comparison of data can be made by classification of months into different seasons (wet, dry, etc.). Annual evaluations can be carried out to determine changes in the bias values over years. Space-based comparisons can be made using mean-field bias values for the whole region or a specially identified region (e.g., basin, watershed, meteorologically homogeneous area) and spatial variation can be understood. Event-based comparison involves use of selected storm events based on rainfall thresholds and extreme events with high intensity and short duration. Point-based bias evaluation can be addressed by selecting a specific rain gage and cluster of gages.

Several indices can also be devised to evaluate the bias. Visual indices involve development of scatter plots to assess

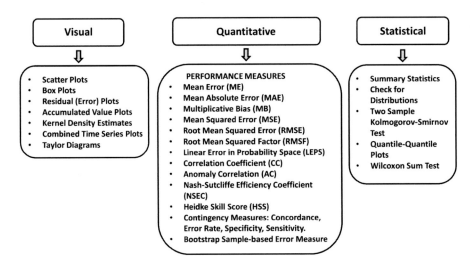

Figure 2.19 Indices and measures for evaluation of biases.

evaluation of under- and overestimation for select gages and regions and for different temporal scales along with box plots of two data sets (i.e., rain and radar data). Histograms and box plots of the data sets provide information about the location, spread, and skewness. Residual (error) plots help assess the error structure, check for any time dependency of errors, autocorrelation aspects and heteroscedasticity. Accumulated rain and radar plots can help evaluate deviations and divergences. Non-parametric distribution procedures (e.g., kernel density estimates) can be used to evaluate the similarity in distributions of rain and radar data. Time series plots of rain and radar data can help identify the biases.

Quantitative indices for bias evaluation involve calculation of summary statistics such as mean, standard deviation, coefficient of variation, skewness and kurtosis, trimmed means, and others. Error performance measures based on rain gage and radar data need estimation of mean error (ME), mean absolute error (MAE), multiplicative bias (MB), root mean squared error (RMSE), root mean squared factor (RMSF), linear error in probability space (LEPS), correlation coefficient (CC), anomaly correlation (AC), Nash–Sutcliffe efficiency coefficient (NSEC), and Heidke skill score (HSS). Linear error in probability space, as the name suggests, measures the error in probability space as opposed to measurement space. Contingency measures such as concordance, error rate, specificity, and sensitivity can also be used. Bootstrap samples (i.e., sampling with replacement) of radar–rainfall pairs can be used to develop confidence intervals for all the error statistics. Error statistics are calculated for each sample pair. All the error measures discussed in this section may be evaluated by combining them into one score. However, having one combined score may not provide details of individual measures and also there is difficulty in deciding weighting factors for each measure in such a combined score. Combining

several measures into one single graph, such as Taylor's diagram (Taylor, 2001), can be beneficial to evaluate RMSE, the correlation coefficient, and the standard deviations based on data sets simultaneously.

Statistical hypothesis tests (e.g., the two-sample Kolmogorov–Smirnov (KS) test) can be used to compare the distributions of the values from the two data sets. The null hypothesis of the KS test is that rainfall and radar data are from the same continuous distribution. The alternative hypothesis is that they are from different continuous distributions. Another non-parametric test that can be used is the Wilcoxon rank sum test, which can help to evaluate if the two data sets (rainfall and radar) are independent samples from identical continuous distributions with equal medians, against the alternative that they do not have equal medians. The Mann–Whitney U test is similar to the Wilcoxon rank sum test. Also, Kuiper's test, which is similar to the KS test, is as sensitive in the tails as it is in the median values and can be used in place of the KS test. This test statistic is considered to be an improvement over the KS test.

A quantile–quantile plot (Q–Q plot) can be used as a graphical procedure to compare two distributions based on rain and radar data sets. A Q–Q plot is essentially a plot of the quantiles of two distributions against each other. Another approach to assess radar–rainfall data sets is to evaluate similarity in wet–dry sequences of the two data sets. First-order Markov-chain probabilities can be developed and assessed. These probabilities are defined as: p_{00} (probability of no precipitation in time interval t, given no precipitation in the previous time interval, $t − 1$), p_{01} (probability of precipitation in time interval t, given no precipitation in the previous time interval, $t − 1$), p_{10} (probability of no precipitation in time interval t, given precipitation in the previous time interval, $t − 1$), and p_{11} (probability of precipitation in time interval t, given precipitation in the previous time interval, $t − 1$).

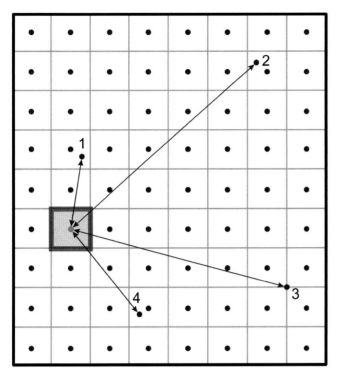

Figure 2.20 Schematic showing the location of rain gages and centroids of radar grids of specific size.

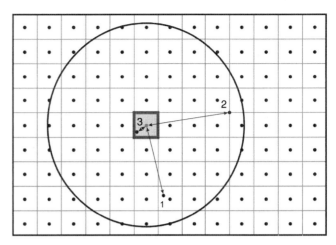

Figure 2.21 Schematic showing the location of rain gages and centroids of radar grids with radius of influence.

Bias corrections refer to corrections applied to radar-based (or satellite-based) precipitation data using the ground truth (i.e., rain gage). Several questions need to be answered before bias correction procedures are attempted. The questions are:

- What are the improvement methods (bias correction) that can be used?
- What are the spatial and temporal resolutions of the rain gage data that can be used?
- What is the best bias correction: local or global?
- If local, how many rain gages should be used?
- If local, can the number of rain gages be optimally selected for a given time period (e.g., season, etc.)?
- If local, what is the spatial extent to be considered?
- Are bias correction methods different for extreme events?

The following sections provide an exhaustive discussion of local and global spatial weighting methods for bias corrections.

2.10.1 Model I: global weighted bias based interpolation

The inverse distance method for spatial interpolation uses distance as weight in a weighted estimate at a point in space. A schematic showing rain gages and the centroids of the radar (grids) pixels is shown in Figure 2.20.

The method for estimation of an observation, θ_m, at a point in space, using the observed values at other cells is given by

$$\theta_m = \emptyset_m \frac{\sum_{i=1}^{n} \beta_i d_{m,i}^{-k}}{\sum_{i=1}^{n} d_{m,i}^{-k}} \quad \forall m \tag{2.19}$$

$$\beta_i = \frac{G_i}{R_i} \quad \forall m \tag{2.20}$$

where θ_m is the adjusted radar-based precipitation value for pixel (grid) m; n is the total number of rain gages; $\emptyset_m$ is the unadjusted value of the radar observation in grid cell n; $d_{m,i}$ is the distance from the location of the centroid of the NEXRAD grid or cell, m, to the ith rain gage; and k is referred to as the friction distance (Vieux, 2001), which ranges from 1.0 to 6.0. A value of 2 is generally used for k. The variable β_i is the gage (G_i) to radar-based precipitation (R_i) ratio. In this method distances used are Euclidean distances.

2.10.2 Model II: local weighted bias based interpolation

The local weighted bias based interpolation method is similar to the global method discussed in the previous section with minor modifications. In this method, all the rain gages within a fixed radius (D_{fr}) from the centroid of the radar grid are considered for spatial interpolation. A schematic showing the rain gages, collocated NEXRAD pixels with centroids, and the limiting distance based on a fixed radius is shown in Figure 2.21. The interpolation is limited to control points within a specified distance as represented by the constraint defined by:

$$d_{m,i} = D_{fr} \quad \forall i \tag{2.21}$$

2.10.3   Model III: inverse exponential weighting method

The inverse exponential weighting method (IEWM) uses a negative exponential function replacing the reciprocal distance as weight in the traditional inverse distance weighting method (IDWM) and is expressed in Equation 2.22. The IEWM is commonly used in the field of quantitative geography for surface generation (Goodchild, 1986; O'Sullivan and Unwin, 2010). The weighting factor in IDWM, d_{mi}^{-k}, is replaced by $e_{mi}^{-\alpha d}$ in Equation 2.19. The factor α is a decay parameter which can be chosen by a variety of methods that include the de-correlation length or the average network spacing (Vieux, 2004). The choice of α based on mean gage spacing leads to the Barnes objective analysis (BOA) (Barnes, 1964) approach, which is applicable to spatial interpolation of meteorological measurements. The value can be chosen after the analysis of inter-gage distances. Vieux (2004) confirms that local bias approaches account for spatial variations of rain drop size distributions and storm evolution over the regions.

$$\theta_m = \emptyset_m \frac{\sum_{i=1}^{n} \beta_i e^{-\alpha d_{m,i}}}{\sum_{i=1}^{n} e^{-\alpha d_{m,i}}} \quad \forall m \qquad (2.22)$$

Two other global standard methods of gage adjustment of radar data can be evaluated. The first method (Method A) uses the average of gage to radar-based precipitation ratio values, and the second method (Method B) uses the cumulative ratio of gage and radar-based precipitation values. These ratios are referred to as mean field bias corrections applied once the radar values have been estimated using one of the spatial weighting methods. These adjustment methods are described in the next section.

2.10.4   Model IV: mean field bias based interpolation

METHOD A
Mean field bias (MFB), φ, is a correction using gage to NEXRAD precipitation estimates, and it can be calculated using:

$$\varphi = \frac{\sum_{i=1}^{n} \frac{G_i}{R_i}}{n} \qquad (2.23)$$

where G_i is the precipitation recorded at the gage, R_i is the NEXRAD based precipitation value for cell i, and n is the number of rain gages.

METHOD B
$$\lambda = \frac{\sum_{i=1}^{n} G_i}{\sum_{i=1}^{n} R_i} \qquad (2.24)$$

Once the mean field bias corrections (φ or λ) have been obtained, they are used to adjust the radar-based precipitation estimates using Equation 2.25 or 2.26, depending on the correction method

Table 2.5 *Yearly rain gage and radar observations and bias corrections*

Rain (G)	Radar (R)	G/R	Bias corrected radar	
			Method A	Method B
1,485	1,668	0.890	1,564	1,562
1,497	1,572	0.952	1,474	1,473
1,515	1,507	1.005	1,413	1,412
1,417	1,707	0.830	1,601	1,599
1,381	1,533	0.901	1,438	1,436
1,412	1,588	0.889	1,490	1,488
1,448	1,631	0.888	1,529	1,528
1,564	1,897	0.825	1,779	1,777
1,784	1,717	1.039	1,611	1,609
1,673	1,592	1.051	1,493	1,491
1,905	1,781	1.070	1,670	1,668
1,621	1,750	0.926	1,641	1,639
1,680	1,816	0.925	1,703	1,701
Total 20,383	Total 21,760	Average 0.938	Total 20,406	Total 20,383

selected.

$$\theta_m = \phi_m \varphi \quad \forall m \qquad (2.25)$$

$$\theta_m = \phi_m \lambda \quad \forall m \qquad (2.26)$$

2.10.5   Example of bias corrections

Bias corrections are usually carried out at temporal resolutions of 15 minutes or less. However, to illustrate the corrections, yearly rain gage and radar totals are used in the first example. Table 2.5 provides the rain gage and radar-based precipitation data in millimeters. Methods A and B are used to calculate the bias correction factors and the radar data are corrected accordingly. Table 2.5 also provides calculations related to corrections.

The bias correction factors for Methods A and B are 0.938 and 0.937 respectively. In the second example, 15-minute rain gage and radar-based precipitation totals are used. Table 2.6 provides the rain gage and radar-based precipitation data in millimeters and the bias corrections using two methods.

Method B provides lower absolute errors compared to Method A in both the examples provided in this section.

2.11   EVALUATION OF METHODS

The performance of the bias correction methods can be compared using widely recognized and commonly used error measures (Kanevski and Maignan, 2004; Ahrens, 2006; Chang, 2009)

Table 2.6 *Fifteen-minute rain gage and radar observations and bias corrections*

Rain (G)	Radar (R)	G/R	Bias corrected radar	
			Method A	Method B
4.3	3.8	1.133	8	5
2.3	0.8	2.986	2	1
8.1	7.2	1.128	15	10
9.1	5.9	1.555	12	8
9.1	13.9	0.660	28	18
4.8	3.4	1.427	7	4
2.3	0.3	7.670	1	0
8.1	5.7	1.434	12	7
1.3	0.5	2.689	1	1
0.0	0.6	0.000	1	1
8.4	4.9	1.720	10	6
8.9	5.1	1.732	10	7
3.8	1.6	2.381	3	2
71	54	2.040	109	71

such as RMSE, absolute error (AE), goodness-of-fit measure criterion, and coefficient of correlation (ρ), based on adjusted radar and rain gage values. The error and performance measures are given by Equations 2.27–2.30. Similar error measures are calculated based on unadjusted radar-based precipitation and rain gage values for comparison purposes.

$$RMSE = \sqrt{\frac{1}{n}\sum_{i=1}^{n}(G_i - R_i)^2} \quad (2.27)$$

$$AE = \sum_{i=1}^{n}|G_i - R_i| \quad (2.28)$$

$$MRE = \frac{1}{n}\sum_{i=1}^{n}\frac{|G_i - R_i|}{R_i} \quad (2.29)$$

$$\rho = \frac{\sum_{i=1}^{n}(G_i - \mu_g)(R_i - \mu_R)}{(n-1)\sigma_g\sigma_R} \quad (2.30)$$

Summary statistics for rain gage and radar data are also compared, including the coefficient of variation (CV) given by:

$$CV = \frac{\sigma}{\mu} \quad (2.31)$$

In Equations 2.27–2.30, n represents the total number of rain gage observations in any time interval, θ_i is the adjusted NEXRAD data value, and R_i is the value of the observation at a rain gage.

2.12 WEIGHTING FUNCTIONS

The error measures described in the previous section are used to select the best method out of the proposed two methods. The

use of several error measures provides advantages as well as few disadvantages in the selection process. The advantages include (1) accurate assessment of performance of methods using different indices; (2) evaluation of error structure and correlations between observed rain gage and adjusted radar data. However, the main disadvantage is that different error statistics have different units and therefore comparison of one method with another is difficult. Therefore, a method by which the error measures are transformed to a common dimensionless parameter and can be used for the selection process is required. The weighting functions that are similar to fuzzy membership functions (Teegavarapu and Elshorbagy, 2005) can be utilized as a way of generating non-dimensional weights from each of the error/performance measures.

The functions can be designed in such a way that the maximum value is always attached to the best performance (W_{max}) based on a specific error measure. Linear and non-linear weighting functions (Teegavarapu et al., 2011) can be developed considering the upper and lower bounds of each performance measure. The functions should be chosen carefully considering the importance attached to each of the error measures.

2.12.1 Difference in means

This function will help assess the similarity in the magnitude of mean precipitation values estimated from radar-based precipitation and observed rainfall at the rain gage. Equation 2.32 provides an expression for calculation of the weight as a function of the absolute difference in mean values.

$$W_\mu = W_{max} - \left(\frac{W_{max}}{\delta_{max}}|\mu_g - \mu_R|\right) \quad (2.32)$$

2.12.2 Difference in standard deviations

This function will help assess the similarity in the magnitude of standard deviations of precipitation values estimated from radar and observed at the rain gage. Equation 2.33 provides the expression for calculation of weight as a function of the absolute difference in standard deviation values.

$$W_\sigma = W_{max} - \left(\frac{W_{max}}{\delta_{max}}|\sigma_g - \sigma_R|\right) \quad (2.33)$$

2.12.3 Difference in coefficient of variation

This function will help assess the similarity in the magnitude of coefficient of variation of precipitation values estimated from radar and observed at the rain gage. Equation 2.34 provides the expression for calculation of weight as a function of absolute difference in the coefficient of variation values.

$$W_{cv} = W_{max} - \left(\frac{W_{max}}{\delta_{max}} \left| cv_g - cv_R \right| \right) \qquad (2.34)$$

2.12.4 Correlation coefficient

Two weighting functions (Equations 2.35 and 2.36) are developed to obtain weights based on the magnitude of correlation coefficients. An exponential relationship that relates weight and the correlation coefficient can be used when the latter is greater than 0.7. This suggests that higher importance is attached to correlation coefficients greater than 0.7.

$$W_c = C\rho \quad \forall \rho \leq 0.7 \qquad (2.35)$$

$$W_c = Ce^{\alpha\rho} \quad \forall \rho \geq 0.7 \qquad (2.36)$$

2.12.5 Absolute error

A functional form for weight relating AE is provided in Equation 2.37. An exponential decay functional form is used to suggest the decreasing preference attached to an increase in the AE.

$$W_{e'} = C_e e^{\left(-\frac{e'}{b} \right)} \qquad (2.37)$$

2.12.6 Root mean squared error

An exponential functional form is chosen for the error measure RMSE as provided in Equation 2.38. RMSE is considered one of the important measures for the assessment of hydrologic simulation models.

$$W_{RMSE} = Z^o e^{\left(-\frac{RMSE}{y} \right)} \qquad (2.38)$$

2.12.7 Mean relative error

A simple linear relationship (Equation 2.39) is used to relate the weight and the error measure MRE.

$$W_{MRE} = W_{max} - \left(\frac{W_{max}}{\delta_{max}} MRE \right) \qquad (2.39)$$

The total weight based on individual weights computed for different components is given by:

$$W_t = W'_\mu + W'_\sigma + W'_{cv} + W'_c + W'_{e'} + W'_{RMSE} + W'_{MRE} \qquad (2.40)$$

The weights, W'_μ, W'_σ, W'_{cv}, W'_c, $W'_{e'}$, W'_{RMSE} and W'_{MRE} are the average values calculated based on the number of gages used in the bias corrections. For example, the average value of weight calculated for differences in mean values of rain gage and the

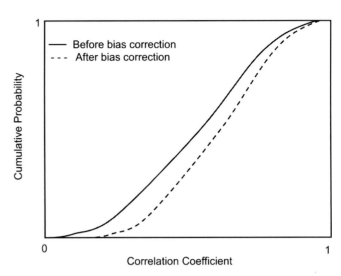

Figure 2.22 Distribution of correlation coefficients before and after bias correction.

radar-based precipitation values for all the rain gages used for bias corrections, *mg*, is given by:

$$W'_\mu = \sum_{m}^{mg} W_{\mu,m} \qquad (2.41)$$

The total weight calculated based on Equation 2.41 for each method is used to compare the methods.

2.13 PERFORMANCE EVALUATIONS WITH MULTIPLE STATIONS

In general, bias estimation and correction procedures are undertaken using a number of rain gage stations. Collocated rain gage and radar pixel (grid) data are used to evaluate the improvements after bias corrections have been completed. Performance evaluations can be compared by using empirical cumulative density functions of the correlation coefficient, AE, or any other performance measure. Figures 2.22 and 2.23 show these functions for the correlation coefficient and AE respectively. These functions indicate improvement in the radar data after bias corrections.

2.14 OPTIMAL PARAMETERS FOR WEIGHTING METHODS

An optimization formulation can be developed and solved for obtaining optimal parameters in the bias corrections methods. The formulation is given by Equations 2.42–2.47. The objective is to minimize the AE based on rain gage observations and bias corrected NEXRAD values. The parameters optimized are (1) distance, (2) friction factor (or exponent used for the distance),

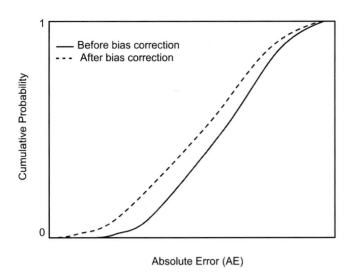

Figure 2.23 Distribution of absolute errors before and after bias correction.

and (3) upper limit value on the gage to radar (*G/R*) observations ratio.

Minimize

$$\sum_{m=1}^{np}\sum_{z=1}^{nt} |\theta'_{m,z} - \theta_{m,z}| \qquad (2.42)$$

subject to:

$$\theta_{m,z} = \emptyset_{m,z} \frac{\sum_{i=1}^{n} \beta_i d_{m,i}^{-k}}{\sum_{i=1}^{n} d_{m,i}^{-k}} \quad \forall m \; \forall k \qquad (2.43)$$

$$\beta_{i,z} = \frac{G_{i,z}}{R_{i,z}} \quad \forall m \; \forall k \qquad (2.44)$$

$$\beta_{i,z} \leq \omega \quad \forall i, \forall z \qquad (2.45)$$

$$k_l \leq k \leq k_u \qquad (2.46)$$

$$d_{m,i} = d_u \qquad (2.47)$$

The variables *np* and *nt* are the number of pixels and time intervals respectively. The decision variables, k, d_u, and ω, are the friction factor in the distance-based method, the maximum distance, and the upper limit on *G/R*, respectively. The formulation can be extended to the exponential weighting method in which the optimal parameter value for α can be estimated.

2.15 BIAS CORRECTIONS WITH LIMITED RAIN GAGE DATA

In bias correction procedures, one of the main problems encountered with the rain gage data is the lack of consistent records of observations. In many instances, the time series data within a time frame (e.g., a month or a year) are not of constant length for all the rain gages. Lack of constant length of rain gage data poses difficulties in implementing the bias correction methods as consistent

record lengths are crucial. Rain gage selection should be carried out and the temporal resolution must be decided before any bias correction methods are applied to the radar data for improvement. Rain gage stations should be selected based on two criteria: (1) number of gages and (2) consistency of rain gage observations.

2.15.1 Bias corrections: temporal resolution issues

Bias correction methods require accurate rain gage data observations without any gaps at a temporal resolution equivalent to that of radar data. Missing data should be filled before the correction procedures are adopted. Optimal selection of parameters for weighting methods is possible and may not always be computationally efficient or tractable for corrections at finer time resolutions, which are less than a day. Mean field bias correction procedures may sometimes yield better performance compared to other local weighting methods. Therefore, an MFB procedure should always be tested along with other methods. A comparative analysis of bias correction methods using different temporal resolutions (e.g., 15 minutes, hour, and day) is recommended whenever rain gage and radar data are available at these resolutions. Absence of rain gage data at the same resolution as radar data (e.g., 15 minute data) forces bias corrections based on a coarser time scale (e.g., day or hour) to be applied to data at a finer time scale (e.g., 15 minute or less). Improvement in bias corrections can be realized only if data at similar temporal resolutions for rain gage and radar are used.

2.15.2 Recent developments in radar meteorology

Dual-polarized (dual-pol) or polarized radar is the most recent development in the area of radar meteorology. This new radar uses horizontally and vertically polarized radar beams for accurate measurement of precipitation along with assessment of shapes of hydrometeors. Dual-pol radar also helps in reducing errors associated with uncertainties in hydrometeor drop size, distribution, and phase. Dual-pol radar is expected to be operational in the year 2012 in the USA.

2.16 SATELLITE-BASED RAINFALL ESTIMATION

Satellite-based measurement of precipitation in the past decade has impacted the field of hydrometeorological observation networks that focus on accurate spatial measurement of precipitation. Comprehensive discussion of methods for satellite-based measurement of precipitation, and instruments and their workings are avoided in this chapter. Descriptions of precipitation estimation by cloud indexing, bispectral and life-history methods, and from passive and active microwave measurements are

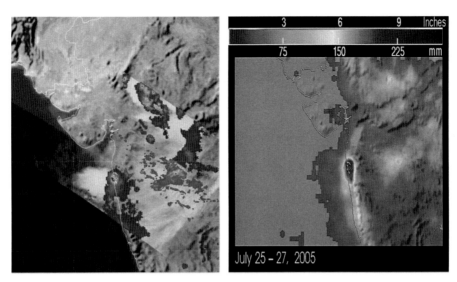

Figure 2.24 TRMM images of extreme precipitation event over the Indian city of Mumbai. (Image courtesy of Hal Pierce, earthobservatory.nasa.gov.)

provided by Strangeways (2007). Mekonnen and Hossain (2009) discuss the most recent advances in satellite-based rainfall estimation and its utility for hydrologic modeling. Brief descriptions of algorithms used to obtain satellite-based precipitation estimations are provided by Sene (2009). In this chapter, precipitation estimates obtained from the Tropical Rainfall Measuring Mission are discussed and compared with rain gage and a multi-sensor precipitation data set.

2.16.1 Precipitation estimation using Tropical Rainfall Measuring Mission

The Tropical Rainfall Measuring Mission (TRMM) is a joint USA–Japan satellite mission to monitor tropical and subtropical precipitation (TRMM, 2011; http://trmm.gsfc.nasa.gov/). The TRMM Multi-satellite Precipitation Analysis (TMPA) provides a calibration-based sequential scheme for combining precipitation estimates from multiple satellites, as well as gage analyses, where feasible, at fine scales ($0.25° \times 0.25°$ and 3 hourly). TMPA is available both after and in real-time, based on calibration by the TRMM combined instrument and TRMM microwave imager precipitation products, respectively. Only the after-real-time product incorporates gage data at present. The data set covers the latitude band 50° N–S for the period from 1998 to the delayed present (Huffman *et al.*, 2007). Figure 2.24 shows the image provided by the TRMM satellite of the rain over Mumbai (formerly Bombay) in India at 3:39 p.m., local time (10:09 UTC), on July 26, 2005.

The left image in Figure 2.24 shows rain intensity as measured by TRMM's sensors, with the heaviest rainfall in dark and lighter rainfall in light gray. A dark spot shown directly on Mumbai reveals a rain rate as high as 50 millimeters per hour. The right image in Figure 2.24 shows record rainfall accumulations between July 25 and July 27. The two images show the extreme precipitation realized in that region of India.

A precipitation data set processed for a region in Florida, USA (refer to Figure 2.25), is used to illustrate the utility of TRMM precipitation data. Comparison of TRMM-based precipitation data, HRAP (Hydrologic Research Analysis Project) multisensor precipitation data, and rain gage data is made. The TRMM data are based on the combined instrument rain calibration algorithm, referred to as 3B42V6. This algorithm uses both passive microwave and infrared information. Figure 2.25 shows the location of the region in Florida for which the precipitation data are obtained for nine grids with each grid size equal to $0.25° \times 0.25°$. The data are available at 3-hour intervals and are compared with multi-sensor precipitation estimates (MPE) available for the same region and also with 58 rain gages in that region.

Results from this comparison are shown in Table 2.7. The values presented in Table 2.7 are yearly rain gage totals (in millimeters) for nine grids in the order shown in Figure 2.25. As expected, the MPE–rain gage correlations are higher than those from TRMM– rain gage based data.

Improvements are noticed when the TRMM precipitation data are bias corrected. The bias corrections for TRMM data are made using the MFB correction procedure using 58 rain gage stations. The correlations are low for grids 3, 6, and 9, where there are fewer rain gages than in other grids.

2.16.2 Global precipitation measurement

The main mission of global precipitation measurement (GPM) (GPM, 2011) in the future is to unify precipitation measurements from an array of satellites, which include GPM Core Observatory

Table 2.7 *Precipitation totals based on gage, MPE, and TRMM, and performance measures*

Precipitation	Year					
Product	2000	2001	2000	2001	2000	2001
Gage	564.64	793.09	994.45	1027.48	902.83	961.46
HRAP (MPE)	704.30	966.94	1057.68	1087.06	1030.85	1417.22
ρ	0.76	0.70	0.71	0.79	0.57	0.57
TRMM	977.09	1409.46	1088.63	1415.76	1205.94	1535.30
ρ	0.53	0.49	0.36	0.38	0.32	0.39
TRMM'	632.70	813.52	675.31	849.31	723.79	1107.88
ρ'	0.65	0.68	0.69	0.71	0.67	0.60
Gage	803.70	848.76	839.89	936.60	892.00	1111.09
HRAP (MPE)	1025.05	1238.42	1088.01	1268.31	1170.38	1513.45
ρ	0.81	0.63	0.74	0.68	0.80	0.71
TRMM	1156.61	1503.71	1184.61	1541.90	1220.36	1463.65
ρ	0.54	0.48	0.58	0.41	0.54	0.42
TRMM'	693.20	826.97	728.60	828.96	925.57	928.04
ρ'	0.66	0.68	0.74	0.75	0.56	0.65
Gage	1056.73	1027.02	1077.35	1305.59	912.01	1243.31
HRAP (MPE)	1194.53	1297.57	1134.15	1373.42	1225.76	1506.13
ρ	0.90	0.72	0.74	0.80	0.81	0.79
TRMM	1244.63	1665.89	1305.80	1578.29	1344.03	1582.16
ρ	0.61	0.42	0.50	0.46	0.44	0.45
TRMM'	832.02	1062.47	908.75	970.28	932.88	1089.88
ρ'	0.73	0.56	0.63	0.66	0.58	0.57

TRMM': Gage adjusted

ρ: Correlation coefficient

ρ': Correlation coefficient (after adjustment)

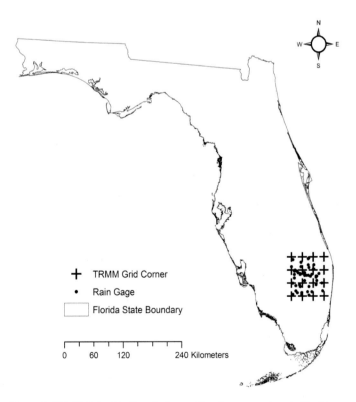

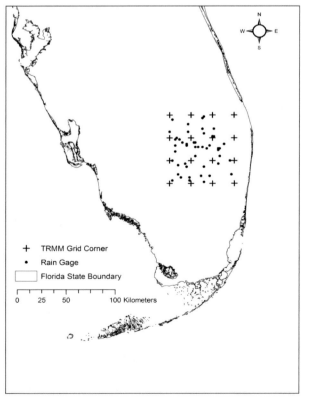

Figure 2.25 TRMM grids (3×3) along with rain gage stations in Florida.

(NASA/JAXA), GPM (low-inclination observatory, NASA), GCOM-W1(JAXA), Megha-Tropiques (CNES, ISRO), DMSP-F Series (DOD), NOAA-19 (NOAA), METOP Series (EUMET-SAT), JPSS Series (NASA/NOAA), and DWSS (DOD). The Megha-Tropiques satellite was launched by the Indian Space Research Organization (ISRO) and was put into orbit on October 12, 2011. It is expected that NASA and JAXA will deploy the Core satellite, which will operate as a precipitation physics observatory, in 2014. Complete details about GPM are available at http://pmm.nasa.gov/GPM. The GPM's array of satellites is expected to provide precipitation measurements every 2–4 hours throughout the globe. The measurements are based on advanced active/passive microwave sensors and provide greater sensitivity to light rain and falling snow. GPM will provide exhaustive description of the space–time variability of global precipitation, and insights into storm structure and large-scale atmospheric processes.

2.17 PRECIPITATION MONITORING NETWORKS

A precipitation monitoring network refers to a network of rainfall gaging stations located in space. According to WMO (2004), analysis of precipitation data is much easier and more reliable if the same gages and siting criteria are used throughout the networks. This should be a major consideration in designing networks. However, these conditions are not applicable all the time as the objectives of the monitoring network differ from region to region.

2.17.1 Precipitation monitoring network density

Precipitation monitoring network density is simply the number of rain gages per given area of a specific region. The WMO recommendations (WMO, 1974) for minimum network density of precipitation stations for hydrometeorological purposes are given in Table 2.8. Dingman (2008) provides details of rain gage densities from different countries throughout the world to highlight the issue of lack of sufficient density in many developing countries. The rain gage density is an important aspect of the monitoring network design. However, placement of rain gages within a basin or region is critical for hydrologic monitoring and modeling purposes.

2.18 CLUSTERING OF RAIN GAGES

Clustering and non-clustering of point patterns in space can be assessed using spatial statistical techniques. Considering rain

Table 2.8 *Minimum density of precipitation stations in different regions*

Type of region	Area (km^2) per station
Flat regions of temperate Mediterranean and tropical zones	600–900
Mountainous regions of temperate Mediterranean and tropical zones	100–250
Small mountainous islands with very irregular precipitation	25
Arid and polar zones	1,500–10,000

gages as points in space, these techniques can be extended to assess the rain gage monitoring networks. Spatial statistical tools provide several indices to identify clusters and dispersed point patterns. Two widely known indices are discussed in this chapter and they are the average nearest neighbor distance tool and the G statistic (Getis–Ord statistic defined by Getis and Ord, 1992). A sample set of rain gage stations is used to illustrate the use of these indices. Figure 2.26 shows the location of several rain gages in Florida that have lengthy historical data.

It is evident from Figure 2.26 that clustering of rain gages exists, and needs to be confirmed by one or more spatial statistical tests. In the case of the average nearest neighbor index, the distance between each rain gage (i.e., point) and its nearest neighbor's (rain gage) location is calculated. The average of all nearest neighbor distances is calculated. If this average distance is less than the average for a hypothetical random distribution, the distribution of the rain gages is considered to be clustered. If the average distance is greater than a hypothetical random distribution of points (rain gages), then the rain gages are considered to be dispersed. The index is the ratio of the observed distance divided by the expected distance (expected distance is based on a hypothetical random distribution of the same number of rain gages within an area). Figure 2.27 shows an example of average nearest neighbor index calculated using the ArcGIS software. The conclusion is that the pattern of rain gage locations is in fact clustered.

The general G-index is another indicator of cluster and dispersion of point features in an area. This index was defined by Getis and Ord (1992). Figure 2.28 shows an example of a G-index calculated using the ArcGIS software. The conclusion is that the pattern of rain gage locations is highly clustered.

Clustering of rain gages is very common and is often caused by lack of adequate planning and design of monitoring networks. In many situations, rain gages are installed to fulfill project objectives driven by the requirements of hydrologic and water quality modeling and management studies.

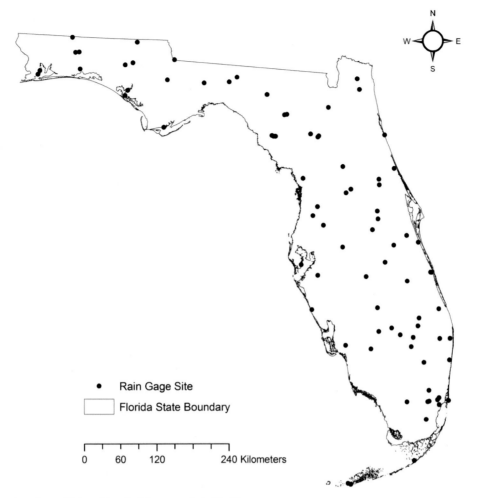

Figure 2.26 Rain gage locations with long historical data records in Florida.

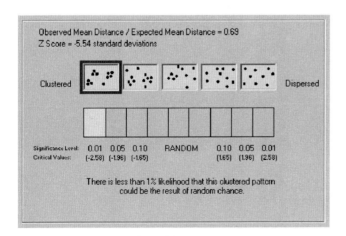

Figure 2.27 Average nearest neighbor index using ArcGIS spatial analysis to evaluate the high/low clustering of points.

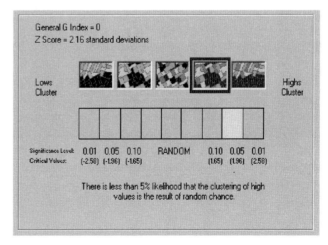

Figure 2.28 General G-index using ArcGIS spatial analysis to evaluate the high/low clustering of points.

2.19 OPTIMAL DENSITY

Optimal density refers to the optimal number of rain gages within a specific region. This density is defined by first fixing an analysis block (a sub-region of constant size in a region) and then determining the optimal number of rain gages in this block. The optimal number is obtained by pre-specified objectives that are discussed in the next section.

2.20 OPTIMAL MONITORING NETWORKS

Design of optimal monitoring networks requires a critical assessment of information to be gained about the heterogeneity of the precipitation process between distances among the sampling locations. Fortin and Dale (2005) indicate that the design of an optimal spatial sampling scheme requires a careful balance between sampling locations that are too close to one another, thus not providing enough new information (data highly autocorrelated), and sampling locations that are too sparse, so that processes at other spatial scales introduce too much variability (Haining, 1990).

2.20.1 Objectives for optimal monitoring network design

This section and the following sections describe rain gage monitoring network design based on variance and geostatistical methods. The geostatistical method of rain gage optimization is one of the methods that are available for network design. It is important to note that the design of a monitoring network is aimed at capturing and characterizing the spatial variability of the precipitation across a specific region. This generally should be the primary objective. However, other secondary objectives related to placement of gages can be due to reasons such as (1) validation of modeling efforts; (2) precipitation measurements required to derive inputs for a hydrologic simulation model; and (3) requirement of a precipitation monitoring effort for a specific hydrologic or hydraulic project. The design of the network depends on several factors that include (1) placement of rain gages to maximize the information obtained from the rain gages; (2) existing network of rain gages; (3) monetary cost involved in the placement (relocation and installation) of the rain gages; and (4) feasibility of installation of a rain gage at a location within a specific region.

2.21 METHODS FOR DESIGN

Two types of methods are discussed in the following sections for design of rainfall monitoring networks.

2.21.1 Variance-based methods

The number of stations required in a specific region (e.g., watershed, basin) can be determined by conceptually simple methods based on variance of rainfall in space or its variants. The variance of rainfall is calculated based on the existing number of rain gages in the region. Rakhecha and Singh (2009) provide discussion of three methods based on variance of rainfall. Methods by Rycroft (1949), Ganguli *et al.* (1951), and Ahuja (1960) are reported by Rakhecha and Singh (2009).

Rycroft (1949) uses variance in space and allowable variance in estimate of mean rainfall to determine the optimum number of rain gages. The optimum number is given by:

$$N = \left[\frac{2v}{\omega}\right]^2 \qquad (2.48)$$

where N is the optimum number of rain gages, v is the variance, and ω is the allowable variance in the estimate of mean rainfall.

A method by Ganguli *et al.* (1951) described by Rakhecha and Singh (2009) uses the mean monthly coefficient of variation (CV_m), and the tolerance value of CV_m, τ, and the existing gages, n, in the following equation to obtain the optimum number of rain gages, N:

$$N = \left[\frac{CV_m}{\tau}\right]^2 n \qquad (2.49)$$

A minor variant of the above equation was proposed by Ahuja (1960) and is given by:

$$N = \left[\frac{CV}{x}\right]^2 \qquad (2.50)$$

where x is the desired percentage error in the estimate of mean rainfall.

All the above methods provide the optimum number of rain gages. This number is compared with the number of already existing gages, and additional rain gages are added based on pre-determined density.

2.21.2 Geostatistics-based method

This section provides an example of optimal monitoring network design for precipitation. Geostatistics and spatial interpolation methods are combined in this illustrative example to obtain an optimal precipitation monitoring network. Details of this precipitation monitoring network design can be found in Vieux (2004). To devise a monitoring network, long-term precipitation data are required. The precipitation data are initially analyzed for variability across a region and regions of fixed area (or analysis blocks are established). These analysis blocks of pre-specified sizes are further evaluated using geostatistical methods to model the variability of rainfall within each block through fitted semi-variograms for the precipitation data. The semi-variogram that is used to model the covariance structure of the precipitation data provides the de-correlation lengths as range parameters of the model.

The standard error (Vieux, 2004) of mean is then used, as a metric, to identify the optimal number of ground rain gages in this example. The standard error can be considered as one measure of accuracy of the monitoring network. The required number of rain gages in a specific analysis block can be obtained by improving the accuracy of the mean precipitation measurement. The effective number of rain gages is obtained from the correlation distance calculated from precipitation data and point variance from gage data for each analysis block. Haan (2002), and Matalas and Langbein (1962) described the information contained in a number of correlated stations. The information contained in data from n stations in a region having a correlation of ρ is equivalent to the information contained in n' uncorrelated stations (rain gages) in the region. The value of n' is given by:

$$n' = \frac{n}{1 + \rho_{ave}(n - 1)} \qquad (2.51)$$

where ρ_{ave} is the average correlation coefficient among n stations. Equation 2.51 suggests that as n increases, n' approaches $1/\rho_{ave}$. The effective number of rain gages n' was related to separation distance h. The covariance structure defined by the spatial correlation coefficient can be modeled by an exponential semi-variogram or any empirical semi-variogram that can be fitted to precipitation data in each analysis block. The relationship between the spatial correlation coefficient ρ_{ave}, the separation distance h, and the range parameter R is given by:

$$\rho_{ave} = e^{\left(-\frac{3h}{R}\right)} \qquad (2.52)$$

The standard error of the mean (SE) is calculated for all the blocks using:

$$SE = \sqrt{\frac{\sigma^2}{n'}} \qquad (2.53)$$

where σ^2 is point variance and n' is again the effective number of stations.

A scaled standard error (Vieux, 2004) SE' is defined by:

$$SE' = \frac{\sqrt{\frac{\sigma^2}{n'}}}{SE_{n'=1}} \qquad (2.54)$$

The standard error is calculated based on effective number of rain gages (n'). Depending on the value of the average correlation coefficient, range parameter, and inter-rain-gage distance, the values of n' will have an increasing trend up to a threshold value, and then decrease as the number of rain gages increases.

2.21.3 Optimal network design

The optimal network design is based on the standard error and the number of gages in each analysis block. Based on the covariance structure of the precipitation data, uniform or non-uniform analysis block sizes may be identified. Vieux (2004) outlined two

methods of optimal network design in a study related to rain gage network optimization. The network designed is optimal considering the stipulations imposed on the block size and other constraints. However, it should be noted that a strict mathematical programming model (i.e., an optimization model) was not used in the network design example presented here.

2.21.4 Variable density analysis block approach

In this approach, the point variance (σ^2) for each block is used to identify the number of rain gages (n) that have a correlation coefficient (ρ) to achieve a desired level of constant accuracy defined by the magnitude of the standard error (SE). The relationship is given by:

$$n_k(SE)^2 = 1 + (\rho_k\,(n_k - 1)\,\sigma_k^2) \quad \forall k \qquad (2.55)$$

where k identifies each analysis block. Once the SE value is fixed, the above equation is solved for n_k. This equation can be solved for different values of SE for each of the analysis blocks using any optimization solver. The value obtained for n needs to be rounded off to the nearest integer, as the number of rain gages. The approach considers only the number of rain gages within a block. Further study would be needed (using land-cover data, etc.) to assess exact placements of rain gages within a given block.

Uniform standard error can be adopted for all the analysis blocks and the optimum number of rain gages can be obtained for each block. Once the optimum number is obtained for each block, this number is compared with the existing number of rain gages in each analysis block. If the existing number of gages in a block is higher than the optimum number, then a recommendation is made to remove the additional gages. When the existing number of gages is less than the optimum, additional gages can be recommended for that block.

2.22 RECOMMENDATIONS FOR RAIN GAGE PLACEMENTS

The placement of rain gages should be based on those existing, those proposed for addition, and removal or relocation of existing rain gages, with the following considerations.

- The proposed rain gages to be added may include both new and spatially relocated rain gages.
- The relocation of rain gages within a specific analysis block needs to be undertaken considering the inter-rain-gage placement distance.
- If an analysis block already has the required number of rain gages, relocation should only be considered within the block when the spacing between the rain gages is lower than the optimal separation distance.
- Cost structure (installation, maintenance, etc.) needs to be incorporated into decision making regarding the installation

of new and relocation of existing rain gages within and across the analysis blocks.

- The excess rain gages from one or more analysis blocks can be utilized in analysis blocks where the number of rain gages required is greater than the existing number.
- The location of the rain gages in analysis blocks should be based on hydrometeorological considerations.

A separate study needs to be carried out to clearly define the implementation strategies for the recommended network design. A cost-based assessment of the different implementation strategies is generally recommended as the next step in this process. Also, a feasibility study of using optimization models in a multi-objective framework that maximizes the information obtained from the rain gages and minimizes the cost structure for installation and maintenance needs to be evaluated.

2.23 GLOBAL PRECIPITATION DATA SETS

Global precipitation data sets are available as point or areal measurements. The areal measurements are driven by radar and satellite observations and interpolation of point measurements to grid-based estimates. There are two types of organizations that deal with precipitation and other essential climate variable data sets. The objectives and functions of these organizations are different. These organizations can be broadly classified as either data providers or data coordinators. Data providers (e.g., Climatic Research Unit (CRU), University of East Anglia, UK) conduct research and assemble data sets that can help in climate variability and change studies. The data are generally provided free of cost; a few data sets are available only on request. Data coordinators (e.g., WMO, Global Climate Observing System) develop guidelines for collection, storage, and dissemination of data. They also provide support to data providers in organizing the data for public use, development of monitoring networks and decision making, and understanding the climate variability and change at regional and global scales. The next few sections deal with historical precipitation data sets provided by different global organizations.

2.23.1 World Meteorological Organization

The WMO is the leading organization that acts as a data organizer and coordinator and is a specialized agency of the United Nations (UN). The organization describes itself as the UN system's authoritative voice on the state and behavior of the Earth's atmosphere and its interaction with the oceans and climate. It advocates and promotes cooperation among all nations in the establishment of networks for meteorological, climatological, hydrological, and geophysical observations. It also works for free exchange, processing, and standardization of related data,

and assists technology transfer, training, and research. As data organizer and coordinator, it facilitates the development of standards for national monitoring agencies and improvement of operational hydrology infrastructure. The WMO was established in 1950, and the organization has more than 189 members.

2.23.2 Intergovernmental Panel on Climate Change

The data distribution center (DDC) of the IPCC provides data used in the AR4 assessment and global climatology (high resolution) data sets from CRU. The IPCC DDC website provides links to the data sets as well as an option to use a data visualizer to assess the global trends in precipitation.

2.23.3 Global Precipitation Climatology Center

The Global Precipitation Climatology Center (GPCC) (Rudolf *et al.*, 2003; Rudolf and Schneider, 2005) is an international data provider operated and supported by the meteorological service of Germany. The GPCC full data reanalysis data set uses the complete GPCC station database (approximately 64,400 stations with at least 10 years of data) available at the time of analysis. The GPCC recommends that these data can be used for global and regional water balance studies, calibration/validation of remote sensing-based rainfall estimations and also for verification of numerical models. It is important to note that applications and assessments based on GPCC data (Rudolf *et al.*, 1994) should understand the implications of the varying number of stations available for a specific analysis grid in a specific month.

The precipitation data sets from GPCC (Rudolf *et al.*, 2005) are available for $0.5° \times 0.5°$ and $1° \times 1°$ grid resolutions as monthly totals. The users of these data are cautioned against the variable number of rain gages (stations) per grid over time, which might lead to homogeneity issues. Abundant caution is essential when GPCC reanalysis data are used for studies dealing with climate variability and trend analyses in data-sparse areas. The GPCC product adjusted to support climate variability and trend analyses in the best possible way is the GPCC VASClimO 50-year data set (GPCC website). A web-based GPCC visualizer is available to evaluate and assess the precipitation data set. A recent snapshot of precipitation totals on a global scale for the month of October 2010 is shown in Figure 2.29.

2.23.4 Global Historical Climatology Network

The Global Historical Climatology Network (GHCN) is jointly operated by NOAA and NCDC. The precipitation anomaly values provided by GHCN are based on data sets of land surface stations with a base period of 1961–90. Precipitation anomalies on a month-to-month basis are provided and often used to

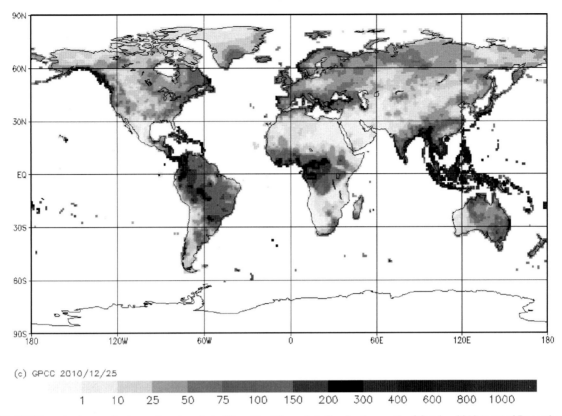

Figure 2.29 GPCC monitoring product gage-based analysis (1° × 1° grid) precipitation for the month of October 2010 (created December 2010).

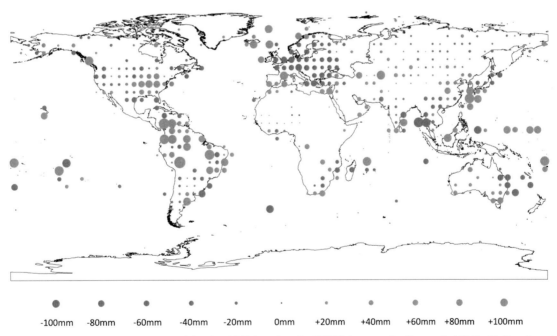

Figure 2.30 Precipitation anomalies for the month of October 2010, based on the GHCN data set using base period 1961–90. See also color plates.

evaluate the precipitation variability across the globe. A sample map of precipitation anomalies provided by GHCN for global precipitation for the month of October 2010, is shown in Figure 2.30.

2.23.5 Climatic Research Unit

The CRU is one of the leading organizations in the world working towards organizing and collecting essential climate variables for climate change studies. The CRU is part of the School of

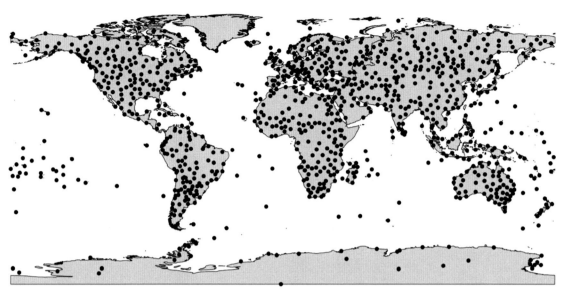

Figure 2.31 GCOS surface network of stations (station location data source courtesy of GCOS).

Environmental Sciences, University of East Anglia, UK. Global monthly precipitation data sets from 1900 to 1998 are available from this organization. The precipitation data sets are available for $5° \times 5°$ and $2.5° \times 3.75°$ grid resolutions as monthly totals. The data are provided through CRU and also through the Tyndall Centre for Climate Change Research (http://www.tyndall.ac.uk/) (TYC, 2011). High-resolution data sets are also available at 10-minute grid resolution for Europe.

2.23.6 Global Terrestrial Network – Hydrology

The Global Terrestrial Network – Hydrology (GTN-H) (GTN, 2011) connects the existing networks and systems for integrated observations of the global water cycle. Established in 2001 and referred to as a "network of networks" the GTN-H supports a range of climate and water resources modeling activities. Based on the existing networks and data centers, the GTN-H is a joint project of GCOS (GCOS, 2011), the WMO Climate and Water Department (WMO/CLW), and the Global Terrestrial Observing System (GTOS). The GTN-H website is a gateway to a great number of global observing systems for hydrologic data. The GTN-H currently provides precipitation data from CRU through its data viewer.

2.23.7 World Data Center for Meteorology

The World Data Center (WDC) for Meteorology (WDC, 2011) is a global climate data clearing house, which was mainly established to serve as a data exchange center. The WDC for Meteorology, Asheville, is maintained by the US Department of Commerce, NOAA, and is co-located and operated by the NCDC in Asheville, NC, USA (WDC, 2011). The US center partners

with other WDCs in Russia and China. It provides historical data and current data by country or region, daily global surface summaries, and monthly data.

2.23.8 Global Climate Observing System

The GCOS (GCOS, 2011) is one of the leading data organizing centers. It is co-sponsored by the WMO, the Intergovernmental Oceanographic Commission of UNESCO, the United Nations Environment Programme (UNEP), and the International Council of Scientific Unions (ICSU) (Strangeways, 2007). The GCOS surface network of stations is shown in Figure 2.31.

GCOS develops several policy documents and guidelines for monitoring of essential climate variables. GCOS (2010) suggests that precipitation is one of the most difficult quantities to observe to the extent needed to meet climate needs, because (1) its physical nature makes reliable point measurement challenging; (2) amounts can vary widely in space and time; and (3) the impacts themselves can depend critically on location, timing, and precipitation type. There is thus a pressing need to develop and implement improved observation and estimation of precipitation from local to global scales.

GCOS (2010) has also identified the following issues related to collection, dissemination, and contribution to the development of global data portals for providing monitoring data for essential climate variables. Parties need to (1) submit all national precipitation data, including hourly totals and radar-derived products where available, to the international data centers; (2) develop and implement improved methods for ground-based measurement of precipitation; (3) develop and sustain operation of a constellation of satellites providing data on precipitation, building on the system to be implemented in the GPM mission; and (4) support

Table 2.9 *Historical global precipitation data sets available from different agencies*

Organization	Grid (spatial resolution)/site	Region	Period	Temporal frame
CRU	2.5° × 3.75°, 5° × 5°	Global, Europe	1900–1908	Monthly
	0.5° × 0.5° (Global)		1961–1990	
	10′ × 10′ (Global)		1961–1990	
	10′ × 10′ (Europe)		1961–1990	
	0.5° × 0.5° (Global)		1901–1995	
	10′ × 10′ (Europe)		1901–2000	
	0.5° × 0.5° (Global)		1901–2000	
	0.5° × 0.5° (Global)		1901–2002	
	0.5° × 0.5° (Global)		1901–2006	
GCOS	3.75° × 2.5°, Station	Global	1961–1990 (Daily)	Daily, monthly
	Station		1900–1978 (Monthly)	
	Station		1950–1985 (Monthly)	
	Station		1990–2007 (Monthly)	
GHCN	Station	Global	Varying	Daily
GPCC	1° × 1°	Global	1901–2009	Monthly
	0.5° × 0.5°			
	0.25° × 0.25°			
GPCP	0.5° × 0.5°	Global	1979–Present	Monthly, daily
	1° × 1°		1997–Present	
GTN-H	From CRU (0.5° × 0.5°; 1° × 1°)	Global	1901–2002	Monthly
Hadley Centre, UK	0.5° × 0.5°	UK, England and Wales	1931–Present	Daily
	1° × 1°		1931–Present	
IPCC	CRU grids	Global	1901–2000	Decadal
NARR (NCEP)	32 km (0.3°)	USA	1979–2009	3-hourly, daily, monthly
	32 km (0.3°)		1979–2000	
NCDC-NOAA	Station	USA	1971–Present (15-min)	15-minutes, hourly
	Station		1948–Present (Hourly)	
TRMM	0.25° × 0.25°	Latitude range 50° N–S	1998–Present	3-hourly
TYC	CRU	Global	1961–1990	Monthly
	Country		1901–1998	
	Country		1901–2000	
WDC	Station	Country, geographical region	Varying	Monthly, daily

the continued development of improved global precipitation products.

2.24 GLOBAL PRECIPITATION DATA SETS: AVAILABILITY AND FORMATS

2.24.1 Observed gridded data sets

Global precipitation data sets developed by different international agencies are not completely free of problems. The utility of data for long-term trend analysis to assess the climate variability in any region of the globe is questionable for several reasons. These reasons include (1) limited precipitation measurements over oceans; (2) inaccuracies in interpolation of point measurements to grids; (3) non-stationary monitoring networks; (4) variations of instruments; (5) coarse resolution of available gridded precipitation data sets; (6) limited availability of radar- and satellite-based data sets; and (7) uncertainty associated with interpolated precipitation data sets from sparse rain gage networks. It is important to note that the format and availability of these data sets is continuously changing and users are advised to always check for recent data. Table 2.9 provides a list of organizations that provide historical global precipitation data sets. Some of these data sets can be used to, among others, define baseline climatic conditions and validate the GCM skills for reproducing spatial (gridded data sets) and temporal (time series data) patterns of key climate variables under current climate (Lu, 2006). The data list provided in Table 2.9 is accurate as of October 2011.

The availability of data sets for a specific period of record, resolution, and global or regional extent is continuously changing as new organizations are emerging and existing agencies are continuously updating their data sets as when new observations are available. Many of these data sets are updated on a yearly basis. Data from the different agencies (details provided in Table 2.9) are available in several formats. The two common formats are the ASCII and Netcdf (Network Common Data Form). ASCII data sets are easy to read, while NetCDF files require special readers. These files are self-describing, machine-independent data formats that support the creation, access, and sharing of array-oriented scientific data.

The 10-minute data sets provided by the Tyndall Centre for Climate Change Research are useful for spatial applications, such as vulnerability zoning or mapping over a large area/region (Lu, 2006). The 30-year average climatology (1961–90), is not conducive to any analysis pertaining to inter-annual or intra-annual variability. Lu (2006) also indicates that 0.5° data sets for the period 1901–2002 may not be adequate to capture the influence of complex topographic features (e.g., mountains) or landscapes (e.g., lakes, coastlines) on a small-scale climate due to their coarse resolution.

2.24.2 Reanalysis data sets

Reanalysis data sets are fine-resolution gridded data which combine observations with simulated data from numerical models (Lu, 2006; IPCC-TGICA, 2007). Sparsely available observational data, with the aid of data from satellites, and information from a previous model forecast are input into a short-range weather forecast model using a data assimilation process. This is integrated forward by one time step (typically 6 hours) and combined with observational data for the corresponding period. The result is a comprehensive and dynamically consistent three-dimensional gridded data set (the "analysis") that represents the best estimate of the state of the atmosphere at that time. The assimilation process fills data voids with model predictions and provides a suite of constrained estimates of unobserved quantities such as vertical motion, radiative fluxes, and precipitation (IPCC-TGICA, 2007).

A publicly available reanalysis data source is the Physical Sciences Division of the Earth System Research Laboratory, NOAA, USA. Details of this data set are provided by Kalnaya et al. (1996). These data sets include 6-hourly observations of daily and monthly averages for numerous meteorological parameters. They are available from year 1948 to present. Widmann and Bretherton (2000) caution that precipitation data based on reanalysis data are unreliable and should not normally be used as proxies for observed climate data.

2.24.3 Outputs from GCM control runs

The data sets from GCM control runs are outputs from multi-century, unforced GCM simulations that can be used to analyze natural multi-decadal climate variations (Lu, 2006, IPCC-TGICA, 2007). Outputs from unforced GCM experiments (i.e., control runs) assessed by the IPCC AR4 are available from the IPCC.

2.24.4 Other data resources

Additional data sources for precipitation documented by Puma and Gold (2011) are:

- WorldClim interpolated climate surfaces for global land areas at a spatial resolution of 30 arc seconds (1 km spatial resolution). Monthly precipitation data are available from this source: http://www.worldclim.org/.
- Global 50-year (1948–2000) data set of meteorological forcings derived by combining reanalysis with observations. Available at 2.0° and 1.0° spatial resolution and daily and 3-hourly temporal resolution: http://hydrology.princeton.edu/data.php.
- APHRODITE daily precipitation (0.5° and 0.25°) for portions of Asia: http://www.chikyu.ac.jp/precip/cgi-bin/aphrodite/script/aphrodite_cgi.cgi/register.

These data sources are useful for regions where data scarcity exists due to lack of dense monitoring networks.

2.25 EVALUATION OF OBSERVED GRIDDED PRECIPITATION DATA SETS

Gridded precipitation data sets from two data sources GPCC (GPCC, 2011) and CRU (CRU, 2011) are evaluated for use in future analysis for Florida. The evaluation is aided by the available rain gage data. It is essential to note that the gridded data are developed based on long data from rain gages that are available in the region. The three gridded data sets used for illustration are:

- GPCC 0.5° × 0.5° grid
- GPCC 1° × 1° grid
- CRU 2.5° × 3.75° grid

The GPCC analyses are focused on developing the monthly precipitation based on rain gage station data. The GPCC supports global and regional climate monitoring and research and is a German contribution to the World Climate Research Program (WCRP) and to GCOS (Schneider et al., 2010). GPCC gridded data with 0.5° and 1° spatial resolutions are available for the period 1951 to 2004. The spatial extent of the 0.5° GPCC grid for Florida is shown in Figure 2.32, and for the 1° GPCC grid is shown in Figure 2.33. A variety of data sets are available from CRU at different spatial and temporal resolutions. Historical monthly precipitation data sets for global land areas are available from 1900 to 1998 gridded at 2.5° latitude by 3.75° longitude resolution (see Figure 2.34). The average monthly and yearly correlations for

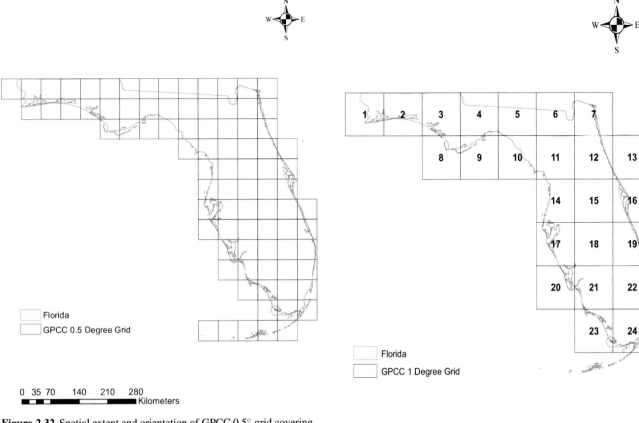

Figure 2.32 Spatial extent and orientation of GPCC 0.5° grid covering Florida.

GPCC 0.5° are 0.916 and 0.867 respectively. Similarly for CRU, the average monthly and yearly correlations are 0.813 and 0.736. Strong correlations with rain gage data sets suggest that these data can be used in lieu of gage data for spatial trend analysis. It is obvious that improved correlations are observed when finer grid resolutions are used for analyses.

Tables 2.10 and 2.11 provide monthly and yearly correlations based on observed rain gage data and gridded precipitation data from Florida. The location of these stations and further analysis of these data sets can be found in the study by Teegavarapu and Goly (2011).

Figures 2.35 and 2.36 show cumulative density plots of monthly and yearly correlations for GPCC and CRU data sets respectively. It is evident from these figures that gridded data provide higher correlations for monthly data compared to yearly data. The monthly and yearly correlations are closer in magnitude for GPCC compared to those from CRU data sets. The main reason may be related to the coarser grid size of the CRU data set compared to that of GPCC.

Figures 2.37 and 2.38 show cumulative density plots for common stations for the two data sets from Tables 2.10 and 2.11. The density plots of correlations indicate that GPCC data sets are performing better than those from CRU.

Figure 2.33 Spatial extent and orientation of GPCC 1° grid covering Florida.

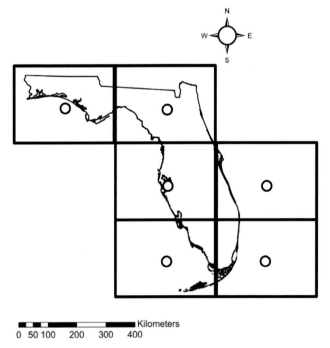

Figure 2.34 Spatial extent and orientation of CRU grid covering Florida.

Table 2.10 *Monthly and yearly correlations for NOAA rain gages based on GPCC 0.5°
gridded precipitation data sets*

Station	Correlations Monthly	Yearly	Station	Correlations Monthly	Yearly
Arcadia	0.942	0.898	Kissimmee2	0.964	0.965
Archbold Biological	0.931	0.890	La Belle	0.938	0.872
Avon Park	0.902	0.900	Lake Alfred	0.934	0.880
Bartow	0.921	0.921	Melbourne WSO	0.928	0.886
Belle Glade	0.936	0.870	Miami Beach	0.897	0.861
Bradenton	0.964	0.917	Miami WSO City	0.917	0.884
Brooksville	0.925	0.876	Moore Haven	0.917	0.867
Bushnell NWS	0.914	0.873	Mountain Lake	0.909	0.829
Clermont	0.924	0.883	Myakka River Station	0.954	0.921
Crescent City	0.898	0.862	Naples	0.957	0.923
Desoto City NWS	0.894	0.746	Ocala	0.958	0.936
Fellsmere 7 SSW	0.925	0.869	Okeechobee	0.935	0.894
Fort Lauderdale	0.931	0.891	Plant City	0.915	0.878
Fort Myers Page Field	0.941	0.869	Pompano Beach	0.936	0.915
Fort Pierce	0.886	0.793	St. Leo	0.937	0.923
Gainesville U. of Florida	0.904	0.797	St. Petersburg	0.950	0.898
Hialeah	0.914	0.845	Tamiami Trail 40	0.952	0.910
High Springs	0.925	0.835	Tampa Intl Airport	0.925	0.868
Hillsborough River Park	0.920	0.802	Tarpon Springs	0.946	0.906
Homestead Station	0.899	0.835	Tavernier	0.805	0.786
Immokalee	0.936	0.894	Venice NWS	0.907	0.873
Inverness	0.955	0.934	Wauchula	0.924	0.885
Isleworth	0.915	0.874	West Palm Beach	0.928	0.871
Key West City	0.497	0.395	Winter Haven	0.913	0.820
Kissimmee	0.935	0.944			

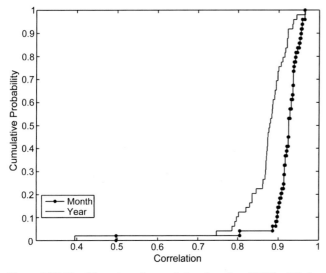

Figure 2.35 Monthly and yearly correlations based on GPCC gridded datasets.

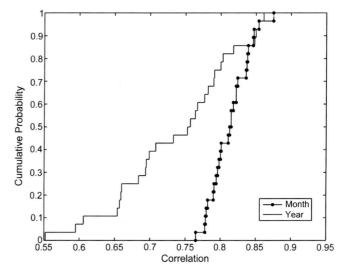

Figure 2.36 Monthly and yearly correlations based on CRU gridded datasets.

Table 2.11 *Monthly and yearly correlations of NOAA stations based on CRU 2.5° × 3.75° gridded precipitation data sets*

Station	Correlations	
	Monthly	Yearly
Arcadia	0.854	0.791
Archbold Biological	0.790	0.767
Avon Park	0.794	0.818
Bartow	0.838	0.850
Bradenton	0.837	0.757
Brooksville	0.796	0.660
Bushnell NWS	0.782	0.733
Cedar Key WSW	0.765	0.699
Clermont	0.779	0.552
Crescent City	0.791	0.777
Desoto City NWS	0.798	0.659
Fort Myers Page Field	0.875	0.708
Gainesville U. of Florida	0.813	0.684
High Springs	0.780	0.606
Hillsborough River Park	0.822	0.595
Inverness	0.778	0.657
Lake Alfred	0.824	0.764
Mountain Lake	0.815	0.694
Myakka River Park	0.839	0.782
Ocala	0.811	0.695
Plant City	0.846	0.854
St. Leo	0.801	0.753
St. Petersburg	0.815	0.654
Tampa Intl Airport	0.836	0.800
Tarpon Springs	0.847	0.861
Venice NWS	0.800	0.803
Wauchula	0.822	0.790
Winter Haven	0.818	0.848

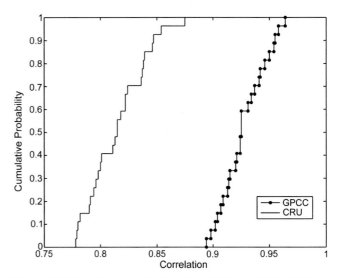

Figure 2.37 Monthly correlations based on GPCC and CRU gridded datasets.

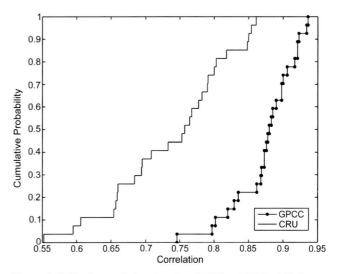

Figure 2.38 Yearly correlations based on GPCC and CRU gridded datasets.

A comprehensive assessment of gridded data sets should involve other evaluations using different performance measures discussed previously in this chapter before these data sets can be accepted for climate change or variability studies.

2.26 MONITORING NETWORKS FOR EXTREME EVENTS

Networks for extreme precipitation events are no different from regular monitoring networks. Extreme precipitation events are not always captured using point measurements (i.e., rain gages). Often radar- and satellite-based measurements help to register the extreme events that are missed by rain gages. Reliable satellite- and radar-based observations have been available since the 1990s (Strangeways, 2007; Sene, 2009). Evaluation of extreme events

using radar- and satellite-based estimates has now become easier. However, there is a need for improvements in the algorithms used for processing radar- and satellite-based data.

2.27 PRECIPITATION MEASUREMENTS IN THE FUTURE

Precipitation measurements in the future will be driven and improved by advances in weather radar- and satellite-based techniques. Improvements in satellite-based rainfall estimation algorithms for precipitation measurement (Sene, 2009) and enhancements in weather radar via dual-polarization techniques (Meischner, 2005) will clearly benefit the distributed hydrologic modeling and weather forecasting communities. The availability of satellite-based data in many parts of the globe with the lowest rain gage densities is of immense help to hydrometeorological studies and the enhancement of our understanding of precipitation variability in different regions of the world. The TRMM satellite may be terminated in 2012, leading the way to the GPM mission that is expected to be operational in 2014. Radar- and satellite-based precipitation estimates can be used for infilling precipitation data at rain gages. Strangeways (2007) provides a comprehensive review of available precipitation data sets that provide a strong foundation for climate change and variability studies.

2.28 SUMMARY AND CONCLUSIONS

Rain gages still remain the quintessential observation devices for measurement of terrestrial precipitation. Unlike their recent and emerging counterparts such as radar and satellites, which rely on surrogate measurement of precipitation amounts, rain gages always serve as the ground truth. From humble beginnings as pre-historic measuring devices, which aided agriculture and other water resources-based sectors, they have come a long way in helping hydrologic and meteorological science communities in understanding the complexities of the hydrologic cycle. Systematic and random errors often plague the measurements, and approaches to limit them (if not completely eliminate them) have always received attention. Recent advances in weather radars and satellite-based measurements and continued understanding of cloud microphysics and precipitation processes are now helping hydrologic studies. Radar estimates of precipitation provide the much-needed spatial domain of measurement and provide the impetus for improved distributed hydrologic modeling. Monitoring network optimization based on precipitation processes should be implemented and refinements of these networks should be undertaken using other rain gages available for estimation. Future precipitation measurements will rely heavily on radar- and satellite-based estimations.

EXERCISES

Data sets available on the website and the specialized codes can be used for solving some problems in this section.

2.1 Differentiate between systematic and random errors in precipitation measurements. Identify the systematic errors that can be easily eliminated.

2.2 Use the precipitation data provided on the website for four spatial locations and identify the station (rain gage) with inconsistent records.

2.3 Reflectivity values in decibels ($dBZ = 10(\log_{10} Z)$), time intervals in which these reflectivity values are observed at a location (i.e., at a rain gage), and observed rainfall values are provided in Table 2.12. Use different Z–R relationships for estimating radar-based rainfall. The value -999 suggests no observation was recorded for reflectivity in the associated time interval. The total time duration for which reflectivity values are available is 15 minutes.

2.4 Evaluate the radar–rainfall biases for the estimated radar-based rainfall using different Z–R relationships using the data in Exercise 2.3.

2.5 Radar (NEXRAD) data are provided on the website for a Doppler radar station from Florida, USA. Use the standard Z–R (reflectivity (Z) – rainfall rate (R)) to obtain rainfall estimates for 15-minute intervals. Evaluate the sensitivity of the coefficient and exponent parameters utilized in obtaining rainfall estimates.

2.6 Use any optimization solver to obtain optimal coefficient and exponent parameters for the data in Exercise 2.3.

2.7 Use the radar reflectivity data provided on the website and develop a non-linear functional form other than a power relationship linking reflectivity and rainfall rate and compare the performance of this functional form to the standard power relationships discussed in this chapter.

Table 2.12 *Reflectivity values, observed rainfall values, and time durations*

Rain (mm)	dBZ_1	dBZ_2	dBZ_3	dBZ_4	t_1	t_2	t_3	t_4
5.334	28.5	53	48.5	−999	3.78	5.07	5.05	1.1
8.382	48.5	44	41.5	−999	3.95	5.05	5.07	0.93
1.778	35	30.5	25.5	−999	4.1	5.08	5.07	0.75
12.446	45	55.5	52.5	50.5	3.87	5.06	5.07	1
28.702	50.5	46	50	49	4.07	5.05	5.05	0.83
22.86	49	44.5	46	40.5	4.27	5.05	5.08	0.6
4.826	40.5	36.5	27.5	30.5	4.45	5.07	5.08	0.4
2.032	30.5	29.5	29.5	−999	5.53	5.05	4.42	−999

2.8 Rain gage and radar-based rainfall for 10 rain gage station locations are provided on the website. Use these data to develop different bias correction methods to improve the radar data.

2.9 Evaluate the performance of mean-field bias correction in comparison to that of local bias correction.

2.10 Use the 15-minute rain gage and radar-based precipitation data in millimeters provided in Table 2.13 to evaluate Method A and Method B bias correction procedures.

Table 2.13 *Rain gage and radar-based precipitation data*

Rain gage (*G*)	Radar (*R*)
3.05	2.36
7.62	4.86
2.79	0.07
0.25	0.01
3.81	0.78
2.29	0.12
1.52	0.23
0.76	0.38
8.89	1.61
5.33	1.05
1.02	0.31
0.25	0.22
0.25	0.05

2.11 Use the executable code (TRMMcode) provided on the website to extract satellite precipitation data from the TRMM website and compare with any rain gages that can be collocated in the TRMM grid.

2.12 Clustering or non-clustering (randomness) of rain gage placement in a region can be assessed as a problem of point pattern analysis. Use a geospatial or spatial analysis tool available under any GIS platform to assess whether the rain gage locations in any given region are clustered or not, using measures for spatial autocorrelation.

2.13 Download the GPCC and CRU data sets from the websites provided at the end of the chapter. Evaluate the GPCC and CRU precipitation data available for your region.

WEBSITES FOR DATA ACQUISITION AND RESOURCES

- TRMM: http://trmm.gsfc.nasa.gov/ (accessed December, 2011).
- NCDC: http://www.ncdc.noaa.gov/oa/climate/climatedata.html (accessed December, 2011).
- NOAA: http://www.ncdc.noaa.gov/oa/wct/ (accessed December, 2011).
- GCOS: http://www.wmo.int/pages/prog/gcos/ (accessed December, 2011).
- GPCC: http://gpcc.dwd.de/ (accessed December, 2011).
- GTN: http://gtn-h.unh.edu/ (accessed December, 2011).
- CRU: http://www.cru.uea.ac.uk/ (accessed December, 2011).
- IPCC: http://www.ipcc-data.org/ddc_about.html (accessed December, 2011).
- WDC: http://www.ncdc.noaa.gov/oa/wdc/ (accessed December, 2011).
- TYC: http://www.tyndall.ac.uk/ (accessed December, 2011).
- GPM: http://gpm.nasa.gov (accessed September, 2011).

3 Spatial analysis of precipitation

3.1 SPATIAL ANALYSIS OF PRECIPITATION DATA

Understanding and quantifying the spatial and temporal variability of precipitation in a watershed are crucial tasks for hydrologic modeling, analysis, and design of water resources systems. Availability of continuous precipitation data at different spatial and temporal scales is essential for hydrologic simulation models that use precipitation as input for accurate prediction of watershed response to different precipitation events. Hydrologists often encounter the problem of missing data for a variety of reasons. Also, measurement of hydrologic variables (rainfall, streamflows, etc.) is prone to systematic and random errors (Larson and Peck, 1974; Vieux, 2001). Complete lack of data is also possible in many situations wherein the rain gage malfunctions for a specific period of time. In this chapter, missing rainfall data and their estimation techniques are of interest. The need for serially complete time series of hydrologic data, especially precipitation records, is becoming increasingly evident from the studies that require reconstruction of past climate, hydrologic history of a site, and modeling and management of water resources systems. Studies relating to frequency analysis of extremes (Hosking and Wallis, 1998; Wallis *et al.*, 1991), climate change, and precipitation variability (Strangeways, 2007) require complete data series without any gaps. These data sets are essential for assessment of short- and long-term trends. Simulation of hydrologic processes requires extension of point measurements to a surface representing the spatial distribution of a specific parameter or input (Vieux, 2001). Spatial interpolation is often necessary due to limited sampling of these variables in space, and the need for understanding a specific process over larger spatial domains, where observations are not available at the scale required for modeling due to prohibitive cost associated with exhaustive sampling or monitoring.

Recent availability of improved quality radar- and satellite-based precipitation data is providing an impetus to many studies that involve infilling missing precipitation records. Infilling of historical precipitation data that are missing for various reasons is an important task that relies on availability of data from monitoring networks and computationally tractable interpolation methods that are conceptually sound and robust. Spatial interpolation methods ranging from conceptually simple weighting techniques to methods using artificial intelligence paradigms (Teegavarapu, 2007, 2008) are now available for estimation of missing precipitation data.

Deterministic weighting and stochastic interpolation methods have been used for the spatial construction of rainfall fields or estimating missing rainfall data at points in space. Traditional weighting and data-driven methods generally are used for estimating missing precipitation. The reader is directed to several research studies in this chapter that provide an exhaustive survey of the literature discussing a spectrum of methods that are available and can be used for infilling historical precipitation or estimating missing data. However, several fundamental questions relevant to the performance of these methods and the rainfall process itself still remain unanswered, leading to several potential research directions. These questions are: (1) What is the influence of spatio-temporal precipitation patterns on estimation methods? (2) How do we quantify and delineate the observation space (number of points, clusters, etc.)? (3) How do we capture the correlation structure of the observations from a monitoring network for estimation of missing data? (4) How do we use auxiliary data (other than traditional rain gage observations) to improve the estimation? While computationally efficient algorithms are required for implementation of improved estimation methods based on auxiliary data, surrogate measures of correlation among observations can be used for conceptually simple weighting techniques (Teegavarapu and Chandramouli, 2005) and statistical distance-based methods (Ahrens, 2006). Universal functional approximation methods such as artificial neural networks (ANN) are viable techniques for estimation of missing precipitation data. However, over-fitting, lack of generalization, unrealistic numeric results, and repeatability are some of the main concerns of these techniques. Reliable data-driven, artificial intelligence including soft computing or statistical techniques for prediction of "rain" or "no-rain" conditions are essential. These techniques may sometimes precede the application of spatial interpolation techniques for estimating missing precipitation records. Rain gage–radar relationships or rain gage–satellite observations need to be characterized via functional forms to effectively utilize them in estimating missing precipitation data. However, reliability of radar-based estimation of precipitation is a still a contentious issue. Optimum clustering of gages to address any redundancy issues for efficient and robust estimation of missing precipitation records should be investigated. Mathematical programming formulations with binary variables can be used for this purpose. Spatial and temporal variability of rainfall processes, and local and regional meteorology and climatology should form essential considerations for infilling or estimation of missing rainfall data.

This chapter presents a comprehensive review of several spatial interpolation methods useful for estimation of missing precipitation data. The methods described in this chapter are applicable for estimation of missing data at a gage (i.e., a point) or creation of a surface (i.e., rainfall field or surface). In this chapter and in other chapters of this book, infilling or imputation refers to filling in of missing data. As this chapter and the rest of the book are focused on precipitation observations, phrases such as control points, observations points, stations, and gages refer to spatial locations where measurement of precipitation is made. Weather radar-based precipitation data are often used for infilling missing precipitation data. In the USA, radar-based precipitation data are referred to as NEXt generation RADar (NEXRAD) data.

3.2 MISSING DATA ESTIMATION

3.2.1 Gaps in data

Missing data exist whenever data time series exist based on observations of any process. Data gaps are a major problem in hydrologic time series, and filling the data time series is essential for a number of reasons. Valid statistical analysis cannot be conducted without a complete data series. Trends in data cannot be assessed if the data set has gaps. Understanding the occurrence of missing data and their nature are essential for the data filling process. Continuous data estimation in this chapter refers to filling of data without any gaps in a data time series. O'Sullivan and Unwin (2010) appropriately suggest the phrase "knowing the unknowables" for spatial estimation of missing data. Missing data mechanisms are important for the development of estimation methods and these are:

- MAR (Missing At Random)
  - Data for a given variable (e.g., Y) are said to be MAR if the probability of missing data on Y is unrelated to the value of Y, after accounting for other variables (X).

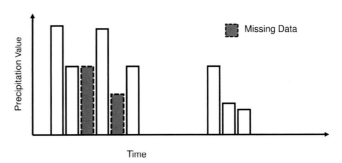

Figure 3.1 Missing data in a precipitation time series as a process of missing completely at random (MCAR).

- MCAR (Missing Completely At Random)
  - Data on Y are said to be MCAR if the probability of missing data on Y is unrelated to the value of Y or any values of other variables (X) in a data set.
- MNAR (Missing Not At Random)
  - Data on Y are said to MNAR if the probability of missing data on Y is related to value of Y or any values of other variables in a data set.

In many precipitation studies, gaps are attributed as MCAR data, as defined by Little and Rubin (1987). The amount and temporal occurrence of gaps (missing) in precipitation data at a site (i.e., rain gage) are not dependent on the data at that site or any other sites.

3.2.2 Interpolation in time

Interpolation in time refers to estimation of missing precipitation data in a precipitation time series. Missing data in a time series, as shown in Figure 3.1, are estimated generally by methods ranging from simple pattern-based approaches to autoregressive time series models.

3.2.3 Interpolation in space and time

Estimation of missing point measurements at a point in space and multiple points in space and also in a temporal domain is referred to as interpolation in space and time. Refer to Figure 3.2 and observe the missing data in different time intervals and at different points in space. For all practical purposes, missing precipitation data are considered to be MCAR. Estimation of missing precipitation data, or any data, requires a stationary monitoring network and time-consistent spatial observations without any systematic errors. Also, the monitoring network used in the model calibration stage (parameter estimation for spatial interpolation methods) and the estimation stage should be constant. Figure 3.3 shows the missing data at control points that are not time-consistent. Observations are said to be time-consistent at any two control points if

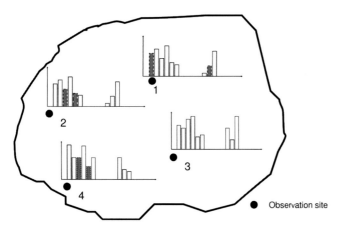

Figure 3.2 Missing data in a precipitation time series with data sets at different spatial observation points.

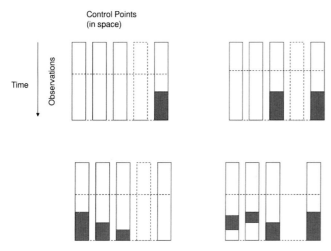

Figure 3.3 Missing data in a precipitation time series with data sets at different spatial observation points in the calibration and estimation phases.

those observations are not missing at any one of the points within the time frame considered.

The calibration phase and estimation phases in Figure 3.3 refer to parameter estimation and validation phases in a typical model building exercise.

3.2.4 Point estimation

Point estimation refers to the process of estimating a missing value at a point in space based on observations from neighboring control points. In this process we are not interested in variation of the process variable over the entire space but at one control point of interest. Therefore, estimation mainly confirms the establishment of models that link the past observations at the point of interest and the observations available at all other locations.

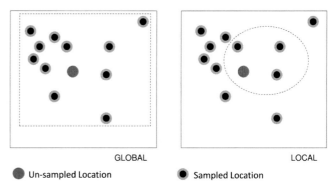

Figure 3.4 Selection of control points in global and local interpolation schemes.

3.2.5 Surface generation

Surface generation is the process of developing a model to understand and evaluate the spatial variation of a process variable. Surface generation is the most common process of spatial interpolation. Later on in this chapter, spatial interpolation methods for surface generation will be discussed.

3.3 SPATIAL INTERPOLATION

3.3.1 Local and global interpolation

The definitions of local and global interpolations depend on the control space and the number of interpolators (observed points) used for interpolation. Local interpolation schemes use only a subset of a complete set of observation points for estimation. Global interpolation schemes use all the available data observation points in the process of estimation. Figure 3.4 shows the difference between local and global interpolation. In the case of local interpolation, the observations can be selected based on a specific criterion. Observations in local interpolation schemes can be selected by different methods that include (1) region of influence (defined by a region in space with the observation point in the center of the regular or irregular polygon); (2) number of nearest observation points; (3) selection of points based on the division of space by a specific geometric pattern; and (4) selection of neighbors based on an optimization approach with pre-specified numbers. Figure 3.5 shows the division of observation space into different equal closed polygons to define circular, elliptical, quadrant, and triangular neighborhoods to select stations from each of these neighborhoods for estimation of missing data.

3.3.2 Exact and inexact interpolation

Exact and inexact interpolations refer to how well interpolation methods closely estimate the values at known observation points in space. As the name suggests, exact interpolation refers to a

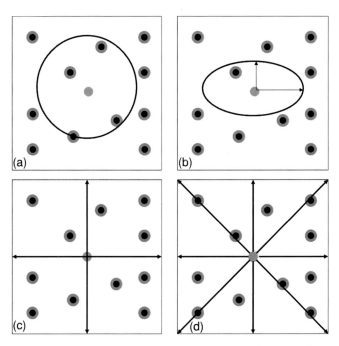

Figure 3.5 Division of monitoring space for selection of control points in local interpolation schemes.

process by which the value estimated by a scheme or model is equal to the known observation value at a point.

3.4 DETERMINISTIC AND STOCHASTIC INTERPOLATION METHODS

3.4.1 Point estimation methods

The following sections describe methods for estimation of missing precipitation data at a point based on observations at all other control points. The first few sections deal with naïve or conceptually simple interpolation methods. These methods are generally used as benchmark methods for comparison purposes.

3.4.2 Single best estimator

The single best estimator (SBE) procedure is one of the simplest methods for estimating missing data. Data from the gage "closest" to the gage with missing data are used for infilling. This closest station can be selected by Euclidean distance and also by using information about the strongest positive correlation. The SBE can also be identified using an optimal gage selection procedure. The SBE is illustrated in Figure 3.6. The availability of historical data again is essential to use this estimator. Eischeid *et al.* (2000) have described the use of the SBE method for estimating time series of temperature and precipitation. Teegavarapu (2009b) used it for correcting interpolated precipitation records.

The estimate of missing data at a station m is given by Equation 3.1. The value from the SBE is the same as the value

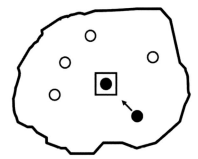

Figure 3.6 Schematic representation of SBE for estimation of missing data.

from a station $\theta_{s,n}()$ selected from $ns - 1$ stations based on criteria discussed earlier.

$$\theta_{m,n} = \theta_{s,n} s \in ns - 1 \quad \forall n \qquad (3.1)$$

Selection of SBE from a set of gages can be done by evaluating the probability density functions of the precipitation data at the gage with the missing data and gage to be used as SBE.

3.4.3 Gage mean estimator

Estimating missing precipitation values by the gage mean estimator (GME) method uses an arithmetic average of all the gages reporting observed rainfall. The method is a special case of inverse distance weighting and correlation coefficient weighting where all the weights are raised to an exponent of zero. It is similar to the average precipitation method recommended by McCuen (1998).

$$\theta_{m,n} = \frac{\sum_{j=1}^{ns-1} \theta_{j,n}}{ns - 1} \quad \forall n \qquad (3.2)$$

This method, or any other interpolation method, fails at estimating missing precipitation in two situations: (1) when precipitation is measured at all or a few other stations but no precipitation actually occurs at the base station; and (2) when precipitation occurs at the base station and no precipitation is measured or occurs at all the other stations. In case 1 (shown in Figure 3.7), the method provides a positive estimate, while in reality zero precipitation occurred at the base station (refer to Figure 3.7a). It is impossible to estimate non-zero precipitation values in the second case (refer to Figure 3.7b) since the observations at all the other stations are zero. Data from other sources (e.g., radar-based precipitation estimates) could be used in this situation to estimate missing values.

In case II (shown in Figure 3.8), the method provides a positive estimate, while in reality zero precipitation occurred at the base station (refer to Figure 3.8a). It is possible to estimate non-zero precipitation values in the second case (refer to Figure 3.8b) since the observation at the nearest station is non-zero even though observations at all the other stations are zero.

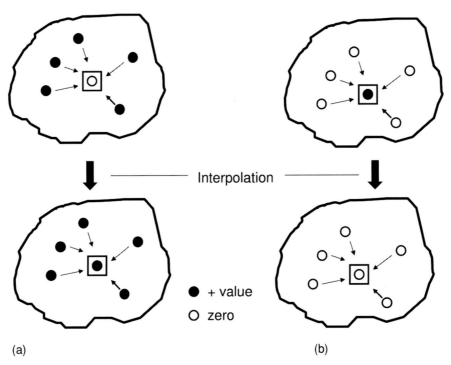

Figure 3.7 Schematic representation of SBE for estimation of missing data (case I).

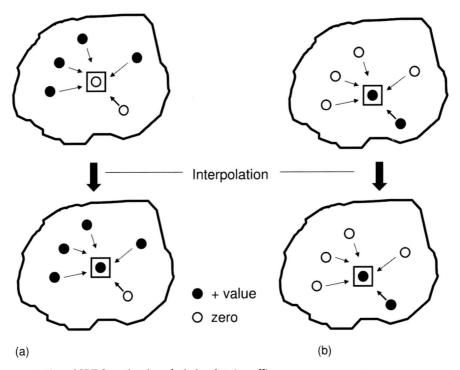

Figure 3.8 Schematic representation of SBE for estimation of missing data (case II).

3.4.4   Climatological mean estimator

The climatological mean estimator (CME) method is similar to GME but the estimate is an average based on the climatological mean for a time interval (e.g., day) based on the available historical record at the gage. This method requires at least one available measurement for each time interval of the year. When information in a particular time interval is not available, it is possible to use the following or previous time interval's information or an average

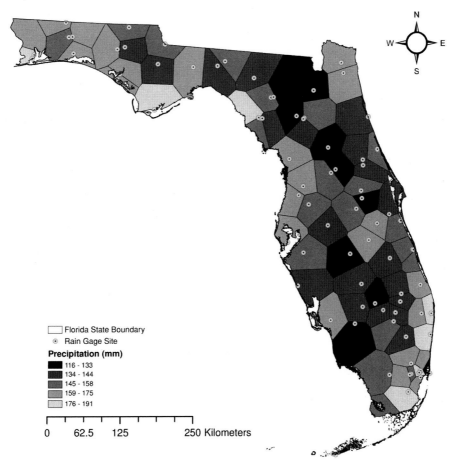

Figure 3.9 Thiessen polygons created based on extreme precipitation data in Florida.

estimated value centered along the time interval in which the estimate is sought.

$$\theta_{m,n} = \frac{\sum_{ny=1}^{nyo} \theta_{m,ny}}{no} \quad \forall n \qquad (3.3)$$

Climatological mean is the simplest and one of the naïve estimation methods. However, information used for this method should account for multi-year or decadal oscillations of climate variability (e.g., El Niño, La Niña, AMO). These oscillations, described and discussed in Chapter 6, may have an impact on the precipitation extremes and totals in many regions across the globe.

3.4.5 Thiessen polygons

Thiessen polygons are often used for spatial interpolation or generation of surfaces in geographical and hydrologic sciences. These polygons are also referred to as Dirichlet cells (tessellations) and Voronoi polygons. These polygons define the area of influences around each point in such a way that the polygon boundaries are equidistant from neighboring points, and each location within a polygon is closer to its contained point than to any other point (Maggio and Long, 1991). Thiessen polygons are constructed

using the Delaunay triangulation approach, which ensures that each known point is connected to its nearest neighbors, and that triangles are as equilateral as possible (Chang, 2010). Given a set of points in the plane, there exists an associated set of regions surrounding these points such that all locations within any given region are closer to one of the points than to any other point. These regions may be regarded as the *dual* of the point set, and are known as proximity polygons or Thiessen polygons. An example of Thiessen polygons created based on a set of rain gages within the state of Florida, USA, is shown in Figure 3.9.

Exhaustive descriptions and analyses of Thiessen polygons are given in a monograph by Boots (1985). Thiessen polygons are built from a set of points and these generated polygon surfaces are used to assign values to regions or cells (grids) based on the attributes of the point set. Distributed hydrologic modeling requires precipitation data as a gridded input and this input is obtained from different sources, which include radar, satellite, and rain gage. As rain gage observations are point measurements, these measurements can be transformed into a surface to assign values to any specific grid or tessellation. This process requires spatial interpolation that helps in generation of surfaces. Once the Thiessen polygons are constructed, the value of an observation

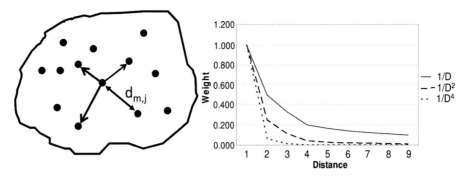

Figure 3.10 (Left) Definition of distance. (Right) Variation of weights with the distance for different exponent values.

at any point in space within a polygon is considered to be the same value at the point that is used to construct the polygon. The surface created (see Figure 3.9) consists of a series of steps with regions of different values separated by lines. The method does not provide an estimate of the error, and neighboring points or any other sampled observations are not used in definition of the surface (Webster and Oliver, 2007).

Thiessen polygons can be used for estimation of missing precipitation data using three steps: (1) use all the rain gages (including the rain gage that has missing observations) to create the Thiessen polygons; (2) remove the station that is missing observations and then create the new set of Thiessen polygons; (3) overlay these two sets of polygons from step 1 and step 2 to identify intersecting areas to estimate the missing values at a particular station by area weighting the observations from all the Thiessen polygons.

Thiessen polygons belong to the class of exact interpolators. The Thiessen polygon approach has the major limitation of not providing a continuous field of estimates when used for spatial interpolation (O'Sullivan and Unwin, 2010). The interpolation can be regarded as abrupt as there is a sudden change in the values across the boundaries of the polygons (Li and Heap, 2008).

3.4.6 Inverse distance weighting method

The inverse distance (reciprocal distance) weighting method (IDWM) is most commonly used for estimation of missing data. The method for estimation of the missing value of an observation at a base station, $\theta'_{\alpha,n}$, using the observed values at other stations is given by:

$$\theta'_{m,n} = \sum_{j=1}^{ns-1} w_{m,j}\theta_{j,n} \quad \forall n \tag{3.4}$$

$$w_{m,j} = d_{m,j}^{-f} \left\{ \sum_{j=1}^{ns-1} d_{m,j}^{-f} \right\}^{-1} \tag{3.5}$$

where $ns - 1$ is the number of neighboring stations; $\theta_{j,n}$ is the observation at station j; $d_{\alpha,j}$ is the distance from the location of station j to the base station α; and f is the exponent referred

to as the friction distance (Vieux, 2001), which ranges from 1.0 to 6.0. The most commonly used value for f is 2. The $d_{m,j}^{-f}$ is the weighting factor and varies with the exponent value. As the exponent value increases, the rate at which associated weights decrease is higher, as shown in Figure 3.10. IDWM is referred to as the National Weather Service (NWS) method in the USA. The IDWM is an exact, convex, and continuous spatial interpolation method (Hengl, 2009). The convex nature of the method suggests that all the predictions (or estimations) are within the range of the values observed in the region.

The IDWM can be applied for estimation of missing precipitation at a point or generation of precipitation fields (or surfaces). Application of IDWM requires selection of two important parameters: (1) power (exponent or friction distance) and (2) number of neighbors. The selection of appropriate values for these parameters is arbitrary; however, optimal values of these parameters can be obtained by optimization formulations. These formulations are discussed later in this chapter. Inverse distance interpolation is an exact interpolator in one way; the interpolated values are equated to observed values at sites where observations are available. This requirement is enforced so that the inverse distance never becomes indeterminate when zero values are encountered for zero distances. Also, two observation sites located at equal distances from a point will have similar weights, resulting in no distinction between two observation sites located in a region. Therefore, the method is not sensitive to sampling network or organization in space. The IDWM is referred to as a weighted average method when the power value is greater than 1 and a simple average when the value is equal to zero (Li and Heap, 2008; Burrough and McDonnell, 1998).

Vieux (2001) points out several limitations of the IDWM, with a major one being the "tent-pole effect," which leads to greater estimates closer to the point of interest. This effect creates "bull's eyes" around the observation points. Grayson and Bloschl (2000) also list several limitations of inverse distance methods. They suggest that this method and the Thiessen polygon method should not be recommended for spatial interpolation considering their limitations. However, they recommend thin splines and kriging for interpolating hydrologic variables. Brimicombe (2003) indicated

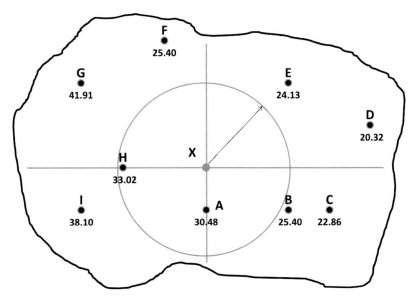

Figure 3.11 Location of rain gages and precipitation amounts (in millimeters) for the example problem.

that the main point of contention in applying the IDWM to spatial interpolation is the selection of the number of relevant observation points.

The success of the IDWM depends primarily on the existence of positive spatial autocorrelation, which is summarized by the following statement. Data from locations near one another in space are more likely to be similar than data from locations remote from one another. In general, distance-based weighting methods suffer from one major conceptual limitation based on the fact that Euclidean distance is not always a definitive measure of the correlation among spatial point measurements. This also negates Tobler's first law of geography (1970): "Everything is related to everything else, but near things are more related than distant things," which forms the basis for many interpolation techniques. Also, the interpolation methods fail to estimate missing values correctly if errors are introduced into the measurement process of rainfall at one or more rainfall stations. These are artefacts of interpolation techniques that cannot be avoided or eliminated altogether in many situations. Also, the existence of negative spatial autocorrelation may be a major limitation in the application of IDWM for estimation of missing data. Distance, a possible surrogate for strength of correlation, often works well when all the rain gaging stations are located in one topographic region.

Variants of the IDWM will involve (1) calculations with a fixed radius from a specific point; (2) application with k (a specific number) nearest neighbors; and (3) introduction of barriers (to avoid specific observation points). The first two variants are considered local, and for the first one restriction to a fixed radius may result in situations where no observation points are available for interpolation. Variable distance can be used to accommodate a specific number of observation points in the interpolation. An IDWM with barriers is required when interpolation is carried across different topographic areas that might have a high variability in precipitation patterns. Recent studies by Teegavarapu and Chandramouli (2005) and Tomczak (1998) provided several conceptual variants of IDWM. Lu and Wong (2008) modified weights in IDWM by a distance-based decay parameter to adjust the diminishing strength in relation to increasing distance. Teegavarapu (2009b) used an association rule mining (ARM) approach to improve estimates of missing precipitation data.

3.4.7 Application of inverse distance weighting method: example

Application of the IDWM is illustrated using a hypothetical example in which a few precipitation values are available at different observation points in space. Figure 3.11 shows the location of the observation points along with the values of precipitation depth measured in millimeters. The observation at point X is assumed to be missing. The value of observation is 35.56 mm. A total of nine observation points are used for estimation of missing precipitation data at point X. Different interpolation schemes with all the observations, different numbers of nearest neighbors, different exponents (i.e., friction distances), the quadrant method, optimal exponents, and optimal weights are evaluated in this section.

The results based on all observations with different exponents and different nearest neighbors used are given in Tables 3.1–3.4. Different estimates of missing precipitation values at point X are also given in these tables. It is evident from the results

Table 3.1 *Estimates using all observations*

Point	Distance (D)	θ (mm)	$\dfrac{1}{D}$	$\dfrac{1}{D^2}$	$\dfrac{1}{D^3}$
A	1.000	30.480	1.000	1.000	1.000
H	2.000	33.020	0.500	0.250	0.125
B	2.236	25.400	0.447	0.200	0.089
E	2.828	24.130	0.354	0.125	0.044
C	3.162	22.860	0.316	0.100	0.032
F	3.162	25.400	0.316	0.100	0.032
I	3.162	38.100	0.316	0.100	0.032
G	3.606	41.910	0.277	0.077	0.021
D	4.123	20.320	0.243	0.059	0.014
		Estimates:	29.380	29.783	30.135

$$\frac{\sum \theta \times \frac{1}{D^k}}{\sum \frac{1}{D^k}}$$

Table 3.2 *Estimates using four nearest neighbors*

Point	Distance (D)	θ (mm)	$\dfrac{1}{D}$	$\dfrac{1}{D^2}$	$\dfrac{1}{D^3}$
A	1.000	30.480	1.000	1.000	1.000
H	2.000	33.020	0.500	0.250	0.125
B	2.236	25.400	0.447	0.200	0.089
E	2.828	24.130	0.354	0.125	0.044
		Estimates:	29.069	29.734	30.148

$$\frac{\sum \theta \times \frac{1}{D^k}}{\sum \frac{1}{D^k}}$$

Table 3.3 *Estimates using two nearest neighbors*

Point	Distance (D)	θ (mm)	$\dfrac{1}{D}$	$\dfrac{1}{D^2}$	$\dfrac{1}{D^3}$
A	1.000	30.480	1.000	1.000	1.000
H	2.000	33.020	0.500	0.250	0.125
		Estimates:	31.327	30.988	30.762

$$\frac{\sum \theta \times \frac{1}{D^k}}{\sum \frac{1}{D^k}}$$

that the local interpolation schemes are providing improved estimates compared to those from global interpolation where all the observations are included. The two nearest neighbors included in the interpolation provide the closest estimate of missing precipitation value.

Table 3.4 *Estimates using seven nearest neighbors*

Point	Distance (D)	θ (mm)	$\dfrac{1}{D}$	$\dfrac{1}{D^2}$	$\dfrac{1}{D^3}$
A	1.000	30.480	1.000	1.000	1.000
H	2.000	33.020	0.500	0.250	0.125
B	2.236	25.400	0.447	0.200	0.089
E	2.828	24.130	0.354	0.125	0.044
C	3.162	22.860	0.316	0.100	0.032
F	3.162	25.400	0.316	0.100	0.032
I	3.162	38.100	0.316	0.100	0.032
		Estimates:	28.986	29.583	30.053

$$\frac{\sum \theta \times \frac{1}{D^k}}{\sum \frac{1}{D^k}}$$

Table 3.5 *Estimates using the quadrant method*

Point	Distance (D)	θ (mm)	$\dfrac{1}{D^2}$
B	2.236	25.400	0.200
E	2.828	24.130	0.125
F	3.162	25.400	0.100
I	3.162	38.100	0.100
		Estimate:	25.738

$$\frac{\sum \theta \times \frac{1}{D^2}}{\sum \frac{1}{D^2}}$$

		Absolute error	8.042

$$\left| \theta_X - \frac{\sum \theta \times \frac{1}{D^2}}{\sum \frac{1}{D^2}} \right|$$

Results from the quadrant method, the optimal exponent and different exponents are presented in Table 3.5. In the case of the quadrant method, the observation at the point nearest (using Euclidean distance) to the point X in each quadrant is used for interpolation.

Results from the quadrant method are no better than those from other schemes using different exponents and optimum exponent (as shown in Table 3.6 and Table 3.7). The optimal exponent value is obtained by using a non-linear programming formulation using a simple gradient-based optimization solver.

The optimal weights obtained using a non-linear optimization solver are provided in Table 3.8. The estimate is equal to the observed value at the location X. This is a trivial result as the objective is to minimize the deviation of the estimated value from observed value at X. The solver provides an optimal set of weights that makes the deviation equal to zero.

Table 3.6 *Estimates using different exponents*

Point	Distance (D)	θ (mm)	$\dfrac{1}{D^2}$	$\dfrac{1}{D^4}$	$\dfrac{1}{D^6}$
A	1.000	30.480	1.000	1.000	1.0000
H	2.000	33.020	0.250	0.063	0.0160
B	2.236	25.400	0.200	0.040	0.0080
E	2.828	24.130	0.125	0.016	0.0020
C	3.162	22.860	0.100	0.010	0.0010
F	3.162	25.400	0.100	0.010	0.0010
I	3.162	38.100	0.100	0.010	0.0010
G	3.606	41.910	0.077	0.006	0.0004
D	4.123	20.320	0.059	0.003	0.0002
		Estimates:	29.783	30.340	30.4650

$$\frac{\sum \theta \times \dfrac{1}{D^k}}{\sum \dfrac{1}{D^k}}$$

Table 3.7 *Estimates using optimal exponent*

Point	Distance (D)	θ (mm)	$\dfrac{1}{D^{kopt}}$
A	1.000	30.480	1.000000
H	2.000	33.020	0.016271
B	2.236	25.400	0.000578
E	2.828	24.130	6.563E-05
C	3.162	22.860	2.335E-05
F	3.162	25.400	2.338E-05
I	3.162	38.100	2.338E-05
G	3.606	41.910	6.927E-06
D	4.123	20.320	1.995E-06
		Estimate:	30.481

$$\frac{\sum \theta \times \dfrac{1}{D^{opt}}}{\sum \dfrac{1}{D^{opt}}}$$

Absolute error	5.079

$$\left| \theta_X - \frac{\sum \theta \times \dfrac{1}{D^{opt}}}{\sum \dfrac{1}{D^{opt}}} \right|$$

3.5 REVISIONS TO THE INVERSE DISTANCE WEIGHTING METHOD

Modifications can be incorporated in the IDWM for estimation of missing data. These modifications are mainly related to distance and weight calculations. In a few variants of the IDWM, the distances are replaced by new parameters that can help in better estimation of missing data.

Table 3.8 *Estimates using optimized weights*

Point	Distance (D)	θ (mm)	Weight, w
A	1.000	30.480	1.208
H	2.000	33.020	1.579
B	2.236	25.400	0.467
E	2.828	24.130	0.282
C	3.162	22.860	0.097
F	3.162	25.400	0.467
I	3.162	38.100	2.320
G	3.606	41.910	2.876
D	4.123	20.320	0.004
		Estimate:	35.560

$$\frac{\sum \theta \times w}{\sum w}$$

Absolute error	0.000

$$\left| \theta_X - \frac{\sum \theta \times w}{\sum w} \right|$$

3.6 INTEGRATION OF THE THIESSEN POLYGON APPROACH AND INVERSE DISTANCE METHOD

Thiessen or proximity polygons (Boots, 1985; Smith, 1993) establish regions of neighborhood around a point of observation. They are generally used to estimate areal averages of precipitation values (Tabios and Salas, 1985; Salas, 1993; Smith, 1993; ASCE, 1996). The property of the proximity polygon also suggests that any point within the polygon is closer to the point used for developing the polygon than any other point in any adjacent polygon. Based on this property, the points located within any polygon and on the boundary separating two polygons will have the same observation values as are obtained from the point of measurement that is used to construct the polygon. A new distance, $d_{m,j}^o$, defined by the property of the proximity polygons, is defined for the integrated estimation method (Teegavarapu and Chandramouli, 2005) and is shown in Figure 3.12.

It should be noted that the new distance ($d_{m,j}^o$) is smaller than the actual distance measured from the base station; the weights in this method, the modified inverse distance weighting method (MIDWM) will now be higher for some of the stations than those used in the IDWM. Estimation of the missing data value, $\theta_{\alpha,n}'$, in MIDWM (Teegavarapu and Chandramouli, 2005) is given by following expression:

$$\theta_{m,n}' = \frac{\sum_{j=1}^{ns-1} \theta_{j,n}(d_{m,j}^o)^{-k}}{\sum_{j=1}^{ns-1}(d_{m,j}^o)^{-k}} \quad \forall n \qquad (3.6)$$

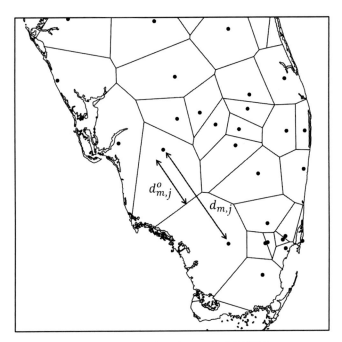

Figure 3.12 Definition of modified distance for MIDWM.

3.7 CORRELATION COEFFICIENT WEIGHTING METHOD

The success of the IDWM strongly depends on the existence of strong positive spatial autocorrelation. One way of establishing the existence of this correlation is to obtain the correlation coefficient between any two data sets obtained from two locations. Since the correlation coefficient is one way of quantifying the strength of spatial autocorrelation, it is meaningful to replace the weighting factor by the correlation coefficient. In the case of the correlation coefficient weighting method (CCWM), the weighting factors are replaced by the correlation coefficients and the estimation is given by:

$$\theta'_{m,n} = \frac{\sum_{j=1}^{ns-1} \theta_{j,n}(\rho_{m,j})^k}{\sum_{j=1}^{ns-1}(\rho_{m,j})^k} \quad \forall n \tag{3.7}$$

where $\rho_{m,j}$ is the coefficient of correlation, which is the ratio of the covariance of two data sets to the product of the standard deviations of the data sets. The coefficient $\rho_{m,j}$ is obtained by using the data at station m and any other station j. In applying this method, available historical data are used for deriving the values of $\rho_{m,j}$. The exponent k in Equation 3.7 can be optimized. The CCWM was first proposed and implemented by Teegavarapu and Chandramouli (2005) for estimation of missing precipitation data. The use, applicability, and superiority of the CCWM over other deterministic methods were discussed in several recent studies (e.g., Kim *et al.*, 2008; Westerberg *et al.*, 2009; Roudier *et al.*, 2011; Vergni and Todisco, 2011). An exponent can be attached to the CC and can be optimized to improve interpolation (Aly *et al.*,

2009). Also, the number of neighbors can be optimally selected for this method (Teegavarapu, 2012b) using binary variables in optimization formulations.

The CCWM (Teegavarapu and Chandramouli, 2005) is a conceptually superior deterministic approach compared to the traditional IDWM and its variants. This superiority is due to its ability to identify anomalies in the observations that are used to estimate the missing data. The CCWM disregards erroneous or uncorrelated observations by assigning reduced influences to these observations via smaller CCs derived from data.

3.8 INVERSE EXPONENTIAL WEIGHTING METHOD

The inverse exponential weighting method (IEWM) uses a negative exponential function to replace the reciprocal distance as the weight in the traditional IDWM. The IEWM is commonly used in the field of quantitative geography for surface generation (Goodchild, 1986; O'Sullivan and Unwin, 2010). The weighting factor in IDWM, $d_{m,j}^{-f}$, is replaced by $e^{-(kd_{m,j})}$ in Equation 3.8. The most commonly used value for k is 2, and usually several values are tested before arriving at a final acceptable value that improves the performance of the method for estimation. Generally a value of 2 is found to be appropriate for many applications.

$$\theta'_{m,n} = \frac{\sum_{j=1}^{ns-1} e^{-(kd_{m,j})}\theta_j}{\sum_{j=1}^{ns-1} e^{-(kd_{m,j})}} \tag{3.8}$$

3.9 REGRESSION MODELS

Regression models link dependent and independent variables via a pre-defined linear or non-linear functional form. In the current context of estimating missing precipitation data at a point, the dependent variable and the observed values at all other control points are designated as independent variables.

3.9.1 Multiple linear regression

A multiple linear regression (MLR) model is developed using observations at stations (as predictors) and missing data to be estimated as a dependent variable. The MLR model is given by:

$$\theta'_{m,n} = \varrho_o + \sum_{j=1}^{ns-1} \theta_{j,n}\varrho_j \quad \forall n \tag{3.9}$$

where, ϱ_j is the coefficient (or weight) for station j and ϱ_o is the constant term. There are disadvantages of using MLR in the context of estimating missing precipitation data without any stipulated conditions: (1) no constant (ϱ_o) and (2) non-negative coefficients

(ϱ_j). A constant positive value of precipitation is always realized when the MLR is used for estimation when the constant term is included and if all the coefficients (or weights) are non-negative and precipitation values are zero at all other stations. Negative values of precipitation may be provided by this method if some of the coefficients are negative.

Variants of local regression models incorporating meteorological variables and variables that change in the spatial domain (i.e., elevation) are used for a real interpolation of precipitation (Daly et al., 1994, 2002); PRISM (precipitation-elevation regression on independent slopes model) (Daly et al., 1994) is one such model. Models incorporating locally weighted polynomials reported by Loader (1999) and Regonda et al. (2006) are conceptual improvements over traditional weighting methods in which the number of neighbors and polynomial functions are objectively chosen. Xia et al. (1999) used the inverse distance, normal ratio, single best estimator, and multiple regression methods for estimation of missing climatological data. The regression method was proved to be the best of all the methods investigated in their study.

Akaike's information criterion (AIC), developed by Akaike (1974), can be a useful statistic to obtain the optimal number of gages to be used in the MLR for interpolation. The statistic or criterion provides a quantitative basis for the tradeoff between bias and variance in model construction. The AIC is defined by:

$$AIC = 2N' + no \left[\ln \left(2\pi \frac{RSS}{no} \right) + 1 \right] \quad (3.10)$$

where no is the number of records used to fit the model, RSS is the residual sum of squares, and N' is the number of parameters used in the linear models. The number of parameters excluding any constant in the current context is the number of gages. The Bayesian information criterion (BIC) (Schwarz, 1978), which is similar to AIC, can also be used to make a decision on the number of gages to be used for interpolation.

3.9.2 Non-negative least squares

The non-negative constraints requirement to obtain positive weights can be enforced using the non-linear least squares constraint formulation defined by:

Minimize

$$\| G.x - H \|_2^2 \quad (3.11)$$

subject to:

$$v_{\alpha,j} \geq 0 \quad (3.12)$$

G is the $no \times j$ matrix of $\theta_{j,n}$ values; x is the matrix $j \times 1$ of $v_{\alpha,j}$ weight values; and H is the matrix of $no \times 1$ values of observed precipitation data at the base station ($\theta_{\alpha,n}$). The formulation minimizes the norm given by Equation 3.10 with constraint on the weights (inequality 3.12). This formulation provides non-negative optimal coefficients when solved. The solution obtained from NLS is obviously better than that of the MLR model as negative precipitation values are not possible using this model. However both MLR and NLS lack the conceptual superiority of models that allow the use of any objective function in any functional form along with the flexibility to use binary variables to select or de-select the stations.

3.9.3 Evaluation of residuals

Evaluation of residuals based on observed and estimated values by interpolation methods is essential. If linear regression models are used, several assumptions need to be checked. The required assumptions of regression include (1) linearity, (2) independence of residuals, (3) normality of residuals, and (4) homoscedasticity. The linear relationship between independent and dependent variables can be assessed by plotting observed values and estimated values and also by evaluation of residuals with respect to estimated values. Independence of residuals can be assessed by autocorrelation plots. Normality of residuals is checked by evaluating the histogram of the residuals and probability plots. Homoscedasticity (i.e., constant error variance) or heteroscedasticity (i.e., nonconstant error variance) can be assessed by plotting standardized residuals and estimated values. A check for heteroscedasticity of residuals is essential and can be done using different tests: the Breush–Pagan test or the White test (Woodridge, 2003). Visual observation of residuals and their patterns is the first step. Visual observations of residuals can be used for matching with the patterns that clearly identify with those with heteroscedasticity (Hair et al., 2010). Also, it is important to note that heteroscedasticity has to be severe before it can lead to serious bias in the standard errors (Allison, 1998). Checks using statistical tests for heteroscedasticity of residuals are recommended for all the interpolation models explained in this chapter if these methods are to be developed and applied to any other regions.

3.10 TREND SURFACE MODELS USING LOCAL AND GLOBAL POLYNOMIAL FUNCTIONS

Local and global polynomial functions, as the names indicate, deal with local and global interpolations. Instead of using observations at control points as variables, the coordinates of space are used for the development of the trend surfaces. These surfaces will ultimately lead to estimation at the point of interest. A trend surface model generating the surface of a precipitation field can be specified as:

$$\theta = f(x, y) \quad (3.13)$$

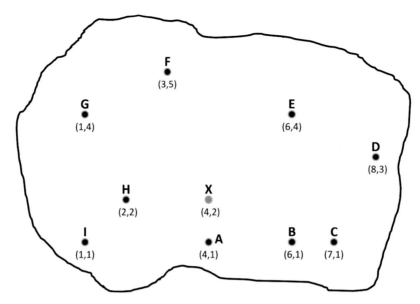

Figure 3.13 Location of rain gage stations defined by Cartesian coordinates.

where θ is a surface generated using x and y coordinates to model the local and global variation of precipitation. The estimate of the missing precipitation value at a point in space can be obtained using the x and y coordinates of the point.

Trend surface models use polynomial functions of different degrees (orders) to fit the surfaces to observation points in space. Smooth and irregular surfaces may result depending on the nature of the polynomial or the degree of the polynomial adopted for the surface. A cubic trend surface model is described by the following equation, with parameters ($\varepsilon_{i,n}$ $\{i = 0 \ldots 9\}$):

$$
\begin{aligned}
\theta'_{m,n} = {} & \varepsilon_{o,n} + \varepsilon_{1,n}\,(x) + \varepsilon_{2,n}\,(y) + \varepsilon_{3,n}\left(x^2\right) + \varepsilon_{4,n}\,(xy) \\
& + \varepsilon_{5,n}\left(y^2\right) + \varepsilon_{6,n}\left(x^3\right) + \varepsilon_{7,n}\left(x^2 y\right) + \varepsilon_{8,n}\left(xy^2\right) \\
& + \varepsilon_{9,n}\left(y^3\right) \quad \forall n
\end{aligned} \tag{3.14}
$$

The applicability of trend surface models for estimation of missing precipitation data for different time intervals requires generation of a surface for each time interval. This is not feasible as it is computationally intensive. Therefore, trend surface models have limited use in the estimation of missing precipitation data at a point. Generalization is a problem when higher order models are used to fit the surface representing the precipitation process in space. Trend surface models are considered to be of limited use as they do not include information about local variables that might affect the estimation. The variables are referred to as geographic predictors (Hengl, 2009), such as distance from the coastline, water body, etc. Use of these variables is beneficial in multiple regression models and can be beneficial in providing more accurate estimates.

3.11 EXAMPLE FOR TREND SURFACE MODELS

The hypothetical example of estimation of missing precipitation data discussed previously (Section 3.4.7) is now used to illustrate the development and application of trend surface models. The models are used for estimation of missing precipitation data at a location X shown in Figure 3.13. The locations of the observation points are specified by x and y coordinates (in the Cartesian coordinate system) and trend surface models described by Equations 3.15 and 3.16 (first and second order respectively) are developed using non-linear regression. The non-linear regression used for this example adopts the Levenberg–Marquardt algorithm (Seber and Wild, 2003) for non-linear least squares to compute fits for the functional form defined in the standard equations 3.15 and 3.16.

$$
\theta'_{m,n} = \varepsilon_{o,n} + \varepsilon_{1,n}\,(x) + \varepsilon_{2,n}\,(y) \quad \forall n \tag{3.15}
$$

$$
\begin{aligned}
\theta'_{m,n} = {} & \varepsilon_{o,n} + \varepsilon_{1,n}\,(x) + \varepsilon_{2,n}\,(y) + \varepsilon_{3,n}\left(x^2\right) \\
& + \varepsilon_{4,n}\,(xy) + \varepsilon_{5,n}\left(y^2\right) \quad \forall n
\end{aligned} \tag{3.16}
$$

Equation 3.14 provided the functional form for the trend surface model using a third-order (cubic) polynomial.

The coefficients estimated for linear, second- and third-order trend surface models are provided in Table 3.9. The observed and estimated values of precipitation at different points using a first-order model are given in Table 3.10.

The error values are also provided in Table 3.10. Figure 3.14 shows the scatter plot of observed and estimated values of precipitation. Tables 3.11 and 3.12 provide results based on

Table 3.9 *Coefficients for different trend surface models*

Coefficients	Linear	Second order	Third order (cubic)
ε_o	41.734	31.416	89.198
ε_1	−2.583	−1.923	−13.693
ε_2	−0.72	9.169	−70.261
ε_3	—	0.007	0.639
ε_4	—	−0.377	12.908
ε_5	—	−1.556	24.233
ε_6	—	—	0
ε_7	—	—	−0.639
ε_8	—	—	−1.755
ε_9	—	—	−2.53

Table 3.10 *Observed and estimated precipitation values from a first-order (linear) trend surface model using all observations*

Point	Observed precipitation (mm)	Estimate (mm)	Error (mm)
A	30.480	30.683	0.203
H	33.020	35.128	2.108
B	25.400	25.517	0.117
E	24.130	23.357	− 0.773
C	22.860	22.934	0.074
F	25.400	30.386	4.986
I	38.100	38.431	0.331
G	41.910	36.271	− 5.639
D	20.320	18.912	− 1.408
X	35.560	29.963	− 5.597

Correlation coefficient (ρ): 0.896
$\theta'_m = 41.734 - 2.583(x) - 0.720(y)$

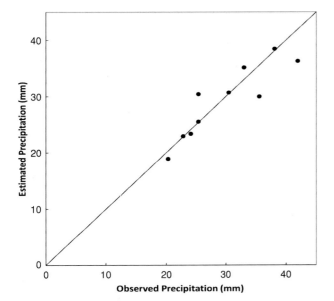

Figure 3.14 Observed and estimated precipitation values from a first-order (linear) trend surface model.

Table 3.11 *Observed and estimated precipitation values from a first-order (linear) trend surface model with six nearest neighbors*

Point	Observed precipitation (mm)	Estimate (mm)	Error (mm)
A	30.480	29.788	− 0.692
H	33.020	32.489	− 0.531
B	25.400	25.883	0.483
E	24.130	22.272	− 1.858
C	22.860	23.931	1.071
F	25.400	26.926	1.526
X	35.560	28.584	− 6.976

Correlation coefficient (ρ): 0.815
$\theta'_m = 38.801 - 1.952(x) - 1.204(y)$

Table 3.12 *Observed and estimated precipitation values from a first-order (linear) trend surface model using four nearest neighbors*

Point	Observed precipitation (mm)	Estimate (mm)	Error (mm)
A	30.480	29.865	− 0.615
H	33.020	33.327	0.307
B	25.400	25.810	0.410
E	24.130	24.028	− 0.102
X	35.560	29.271	− 6.289

Correlation coefficient (ρ): 0.817
$\theta'_m = 38.571 - 2.028(x) - 0.594(y)$

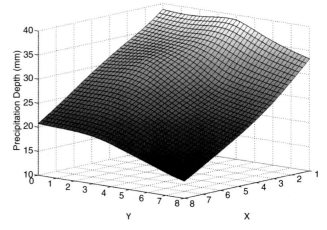

Figure 3.15 Trend surface generated based on second-degree polynomial.

models using six and four nearest neighbors respectively. The errors are smaller when local models using nearest neighbors are used.

The trend surfaces using second- and third-order polynomials are shown in Figures 3.15 and 3.16 respectively.

Table 3.13 *Coefficients of trend surface models using all observations except the observation from point X*

Coefficients	Second order	Third order (cubic)
ε_o	36.704	39.929
ε_1	− 3.606	3.194
ε_2	5.585	− 2.471
ε_3	0.199	− 0.390
ε_4	− 0.325	− 6.399
ε_5	− 0.927	3.775
ε_6	—	0.000
ε_7	—	0.390
ε_8	—	0.666
ε_9	—	− 0.593

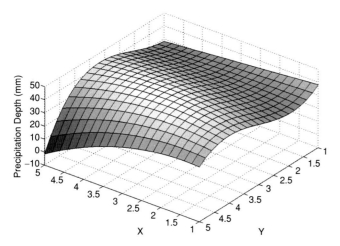

Figure 3.16 Trend surface generated based on third-degree polynomial.

Table 3.13 provides the coefficients estimated based on a non-linear regression method using all observations except those at point X.

Trend surface polynomials may provide good estimates of data when the spatially referenced variables have a spatial structure (Lloyd, 2007). The polynomials are also referred to as moving surface interpolators. O'Sullivan and Unwin (2010) point to several limitations of trend surface models including their dependence on spatial coordinates to explain variation of a process, and autocorrelation in residuals that leads to mis-specification of models. They consider the trend surface methodology to be "dumb" with weak theoretical underpinnings.

3.12 THIN-PLATE SPLINES

Thin-plate splines as exact spatial interpolators can be used to create surfaces that can help estimate values at a location in space.

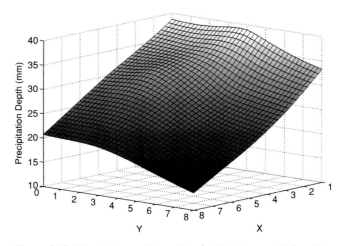

Figure 3.17 Thin-plate smoothing spline fitted to the precipitation data.

The expression for the thin-plate spline surface (Wang, 2006) is given by

$$\theta'_{m,n} = \varepsilon_{o,n} + \varepsilon_{1,n}(x) + \varepsilon_{2,n}(y) + \sum_{j=1}^{ns-1} k_j d_{m,j}^2 \quad \forall n \quad (3.17)$$

where $d_{m,j}$ is the distance from the point where the estimate is required, $\varepsilon_{o,n}$, $\varepsilon_{1,n}$, $\varepsilon_{2,n}$, and k_j are the parameters to be estimated based on data from the $ns - 1$ control points. The parameters are estimated using three equations and are ultimately used for generation of the surface. Wang (2006) indicates that thin-plate splines have problems in interpolation in data-poor areas and suggests that improved variants such as thin-plate splines with tension and regularized splines can be used to avoid the limitations.

Thin-plate splines have the same disadvantages as trend surface methods and are not suitable for multi-time-period estimation of missing precipitation data if the thin-splines are fitted multiple times. Hutchinson (1995) used thin splines for interpolation of precipitation data. Xia *et al.* (2001) reported the use of thin splines, closest station, MLR techniques, and Shepard's method (Shepard, 1968) for estimation of daily climatological data. They indicated that the thin-splines method was the best among all the methods investigated. While thin-plate spline methods have several advantages over others, they tend to generate steep gradients in data-poor areas leading to compounded errors in the estimation process (Wang, 2006). Figure 3.17 shows the surface created using all the observations in the thin-plate spline method.

3.12.1 Thin-plate splines with tension

Thin-plate splines with tension (Mitas and Mitasova, 1988; Chang, 2009) can be used for spatial interpolation or estimation of missing data. The splines belong to the group of radial basis functions and are useful for fitting surface to values that vary smoothly in a spatial domain. Radial basis functions are deterministic interpolators that are exact. Different radial basis functions are possible and they are (1) thin plates, (2) thin plates with splines, (3) regularized splines, (4) multi-quadratic functions, and (5) inverse multi-quadratic splines. Thin-plate splines are also referred to as Laplacian smoothing splines.

There are variants of thin-plate splines that incorporate a tension parameter that allows for controlling the shapes of membranes passing through control points (Franke, 1982). The equation for estimation of a missing precipitation value ($\theta'_{m,n}$) is given by:

$$\theta'_{m,n} = \vartheta_n + \sum_{j=1}^{ns-1} A_{j,n} R(d_j)_n \quad \forall n \qquad (3.18)$$

where ϑ_n represents the trend function, and $R(d_j)_n$ is the basis function. The basis function value is obtained by using

$$R\left(d_j\right)_n = \frac{1}{2\pi\eta^2}\left[\ln\left(\frac{d_j\eta}{2}\right) + c + k_o(d_j\eta)\right] \quad \forall n \qquad (3.19)$$

The variables ϑ_n and $A_{j,n}$ need to be estimated; d_j is the distance between the station at which missing data are prevalent and any other station. The variable η is the tension (or weight) parameter, c is a constant (Euler's constant equal to 0.577215) and k_o is the modified Bessel function (Abramowitz and Stegun, 1965). When the tension parameter is set close to zero, the results from this method approximate those from a thin-plate spline method.

3.13 NATURAL NEIGHBOR INTERPOLATION

Natural neighbor interpolation, also referred to as area-stealing or Sibson interpolation (Sibson, 1981), is evaluated for estimating missing precipitation data. The natural neighbors of any point are those associated with neighboring Thiessen polygons or Dirichlet tessellations. These polygons are constructed using two steps. In the first step, a diagram is initially constructed based on all rain gage stations excluding the station with missing precipitation data. The construction is carried out using the procedure referred to as Delaunay's triangulation. In the second step, the location (point) of the rain gage with missing precipitation data is then added to all other points (rain gages) and a new Thiessen polygon is created. The two sets of polygons from two steps are overlaid on top of the each other. The proportions of overlap among these new polygons and the initial polygons are then used as the weights (Equation 3.20).

$$w_{m,j} = \frac{A_m}{A_j} \quad \forall j \qquad (3.20)$$

$$\theta'_{m,n} = \sum_{j=1}^{ns-1} w_{m,j}\theta_{j,n} \quad \forall n \qquad (3.21)$$

The Thiessen polygons with and without the station with missing precipitation data are shown in Figure 3.18. The overlapping areas of Thiessen polygons are used to obtain the weights as shown in

Figure 3.19. The Thiessen polygons can be constructed using any spatial analysis environment.

3.14 NORMAL RATIO METHOD

The traditional normal ratio (McCuen, 2005) method is based on ratios of observation and long-term annual values at a number of stations. The normal ratio method for estimating missing data at a base station is given by:

$$\theta'_{m,n} = \frac{\theta^a_\alpha}{ns-1}\sum_{j=1}^{ns-1}\frac{\theta_{j,n}}{\theta^a_j} \quad \forall n \qquad (3.22)$$

where $\theta'_{m,n}$ is the estimated value of the observation at the base station m; $ns-1$ is the number of stations used for interpolation; $\theta_{j,n}$ is the observation at station j; θ^a_α and θ^a_j are average annual precipitation values at the base station and at station j, respectively. The average annual precipitation values are obtained from long-term historical data in a region.

3.15 NEAREST NEIGHBOR WEIGHTING METHOD

The nearest neighbor weighting method (NNWM) (Teegavarapu and Chandramouli, 2005) is based on the need to define the most nearby stations that can be used in the IDWM. The nearest neighboring stations are identified by assessing the similarity in the geometric (e.g., trapezoid, rectangle) patterns of the observed rainfall time series. The underlying hypothesis is that similarity in the observations suggests the existence of positive spatial autocorrelation. The existence of high correlation between the data at any two stations should be tested to confirm the hypothesis before the stations are used in calculations. Grayson and Gunter (2001) provide an excellent treatise on the existence of patterns in hydrologic time series and processes. Their work provides a strong motivation to use patterns to establish similarities in observed rainfall time series. Similarity in the observed rainfall values at two stations is established by analyzing simple geometric shapes in the rainfall time series. Deriving simple geometrical shapes (patterns) in rainfall is a straightforward procedure as discussed by Teegavarapu and Chandramouli (2005). The time series of rainfall values is used to define the geometric patterns.

The number of patterns (P_n) obtained depends on the number of rainfall values (nv) considered as one entity. The relationship $P_n = 3^{nv-1} + 1$ provides the number of patterns based on the values considered. Figure 3.20 shows the patterns (i.e., P_1, P_2, P_3) when two consecutive time series values are considered.

Similarly Figure 3.21 provides the patterns (i.e., P_1, P_2, P_3, P_4, P_5, P_6, P_7, P_8, and P_9) when three consecutive values are

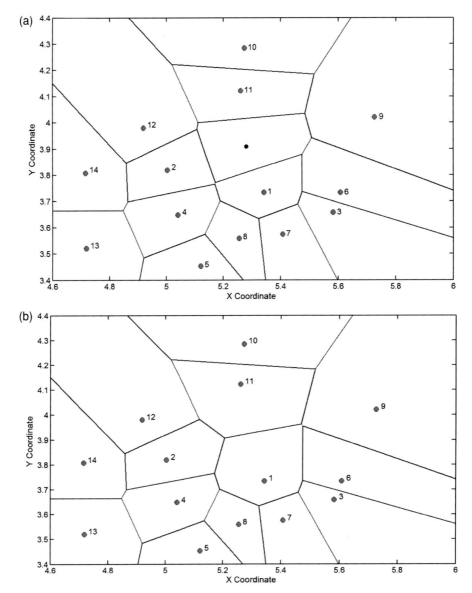

Figure 3.18 Thiessen polygons (a) with base station and (b) without base station.

considered. In the current context, lack of rain constitutes an additional pattern that should be included along with the other patterns described.

Once the patterns are obtained at all the stations, the percentages of these patterns are compared to establish the similarity in the rainfall events. Stations having similar values of percentages of each pattern can be considered highly correlated and the existence of positive spatial autocorrelation can be confirmed. Also, stations with similar values of percentages are considered neighbors. Once the nearest neighbors are identified, distances are calculated from base station to these stations and the traditional IDWM is applied. The method has limitations and should be used only as a method substituting any naïve or conceptually simple method.

3.15.1 Revised nearest neighbor weighting method

A method of finding a surrogate measure for distance using patterns can be proposed:

$$d_{m,j}^s = \sum_{z=1}^{tp} \left| p_m^z - p_j^z \right| \qquad (3.23)$$

where $d_{m,j}^s$ is considered to be a surrogate measure of distance; tp is the total number of patterns; m and j refer to indices for the base station and any other station under consideration; and p_m^z and p_j^z are percentages of patterns at the base station and any other station, j. The distance parameter in the traditional IDWM is replaced by the surrogate measure of distance and the values of missing rainfall are estimated. The

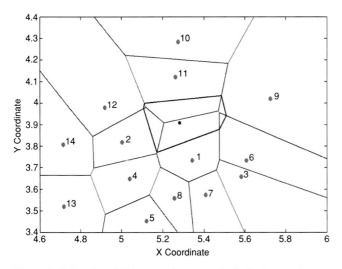

Figure 3.19 Overlay of Thiessen polygons to obtain the intersecting areas defined by nearest neighbors.

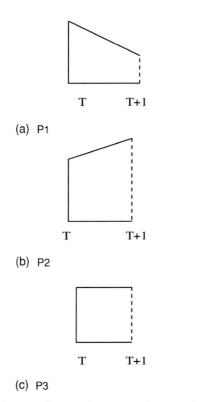

(a) P1

(b) P2

(c) P3

Figure 3.20 Development of geometric patterns using two values of precipitation time series as one entity.

modified NNWM is referred to as the revised nearest neighbor weighting method (RNNWM). While the surrogate measure provided by Equation 3.23 was used successfully for estimation of missing precipitation in the study by Teegavarapu and Chandramouli (2005), the measure is neither robust nor recommended.

3.16 VARIANTS OF MULTIPLE LINEAR REGRESSION METHODS

Two variants of MLR methods are also useful for estimation of missing precipitation data and they are step-wise regression and robust regression. These variants have not been reported earlier in the literature for use in estimation of missing precipitation data. Step-wise regression is a systematic method for adding and removing variables from a MLR model based on their statistical significance tested through the F-statistic (Draper and Smith, 1998). The robust regression uses an iteratively re-weighted least squares method with a bi-square weighting function (Holland and Welsch, 1977). The former regression method helps in selecting the optimal number of variables in the multiple regression, and the latter improves the regression by reducing the influence of the outliers when compared with the original least squares method.

3.17 REGRESSION MODELS USING AUXILIARY INFORMATION

The MLR models previously discussed in this chapter have dealt only with precipitation observations at different sites. Information about auxiliary or ancillary variables (e.g., elevation) can be incorporated into regression models for generation of interpolated surfaces or to obtain values at unobserved locations. Use of these models for temporal interpolation might not be beneficial unless some of the variables used in the model are non-spatial and are time dependent. One example of regression models using auxiliary information is provided by Chang and Li (2000). They describe a model for snow water equivalent estimation using linear regression linking elevation and spatial coordinates and also a parameter referred to as the surface curvature measure. Precipitation patterns are strongly influenced by regional topography. A few methods, such as analysis using relief for hydrometeorology (AURELHY) and PRISM, incorporate the topography into an interpolation of climatic data by combining principal components analysis, linear multiple regression, and kriging (WMO, 2009a). Topography is often described by the elevation, slope, and slope direction, generally averaged over an area in several models.

Genton and Furrer (1998a) question the necessity of robust spatial statistics when a simple good sense prediction approach that utilizes local rainfall observations and knowledge about topography can be used for estimation of missing precipitation data. They conclude that robust spatial statistics are worth the trouble when their comparison of a good sense method with different deterministic and stochastic interpolation methods yields inferior results to the former approach.

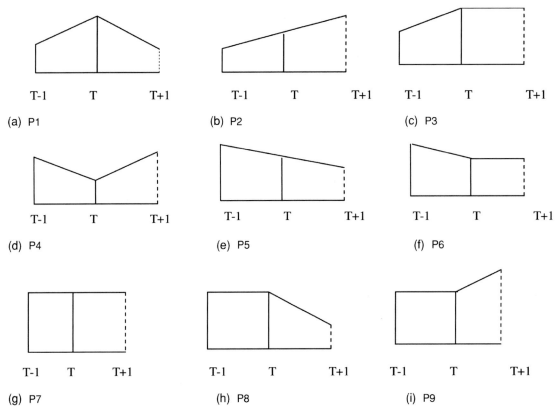

Figure 3.21 Geometric patterns from precipitation time series using three values of precipitation as one entity.

3.18 GEOSTATISTICAL SPATIAL INTERPOLATION

Ordinary kriging (Journel and Huijbregts, 1978; Isaaks and Srivastava, 1989; Vieux, 2004; Webster and Oliver, 2007) is widely recognized as a stochastic interpolation method for surface interpolation based on scalar measurements at different locations. Kriging is an optimal surface interpolation method based on spatially dependent variance (Vieux, 2004).

3.18.1 Semi-variogram

The degree of spatial dependence in the kriging approach is generally expressed using a semi-variogram given by:

$$\gamma(d) = \frac{1}{2n(d)} \sum_{d_{ij}} \left(\theta_i - \theta_j\right)^2 \quad (3.24)$$

where $\gamma(d)$ is the semi-variance, which is defined over observations θ_i and θ_j lagged successively by distance d. Surface interpolation using kriging depends on the semi-variogram model that is selected, which must be fitted with a theoretical form that can be used to estimate the semi-variogram values at arbitrary separation distances. Depending on the shape of the semi-variogram, several mathematical models are possible, including linear, spherical,

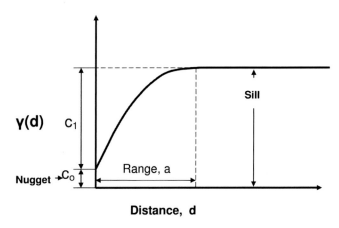

Figure 3.22 Parameters of a typical semi-variogram.

circular, exponential, and Gaussian functional forms. Pentaspherical, cubic, Whittle's elementary correlation, and the Matern function are a few other models documented by Webster and Oliver (2007). A typical spherical semi-variogram is shown in Figure 3.22.

3.18.2 Semi-variogram models

Generally, several semi-variogram models are tested before selecting a particular one. The four most widely used semi-variogram

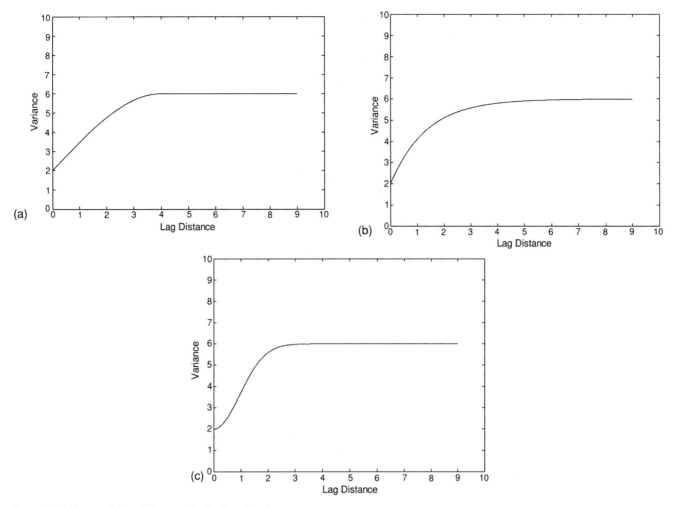

Figure 3.23 Shapes of three different authorized semi-variograms.

models (spherical, exponential, Gaussian, and circular) are given by Equations 3.25, 3.26, 3.27, and 3.28 respectively:

$$\gamma(d)_1 = C_o + C_1 \left[\frac{1.5d}{a} - 0.5 \left(\frac{d}{a} \right)^3 \right] \qquad (3.25)$$

$$\gamma(d)_2 = C_o + C_1 \left[1 - \exp \left(-\frac{3d}{a} \right) \right] \qquad (3.26)$$

$$\gamma(d)_3 = C_o + C_1 \left[1 - \exp \left(-\frac{(3d)^2}{a^2} \right) \right] \qquad (3.27)$$

$$\gamma(d)_4 = C_o + C_1 \left[1 - \frac{2}{\pi} \cos^{-1} \left(\frac{d}{a} \right) + \frac{2d}{\pi a} \sqrt{1 - \frac{d^2}{a^2}} \right] \qquad (3.28)$$

The parameters C_o, d, and a are referred to as nugget, distance, and range, as shown in Figure 3.22. The shapes of three semi-variograms are shown in Figure 3.23.

The summation of C_o and C_1 is referred to as the sill, and the semi-variance at range a equals the sill value. The values of C_o and C_1 are obtained using trial and error or an optimization formulation. Visual observation of the variogram cloud based on

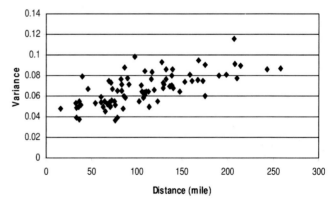

Figure 3.24 Typical variogram cloud created based on daily precipitation data (Teegavarapu, 2007).

the available spatial data (Figure 3.24) should be carried out, and the possible fit of the theoretical variogram model to the cloud should also be evaluated. A pure nugget model will result when the C_o value is equal to zero when $d = 0$ and $C_o = C_1$ when d is not equal to zero.

The spherical semi-variogram model provides the specified sill value, at range value a. The exponential and Gaussian semi-variogram models approach the sill values asymptotically. Weights in the kriging approach are based not only on the distance between the measured points and the prediction location, but also on the overall spatial arrangement among the measured points and their values. The weights mainly depend on the model fitted (i.e., the semi-variogram) to the measured points. The general equation for estimating the missing value, $\theta'_{m,n}$, is given by:

$$\theta'_{m,n} = \sum_{j=1}^{ns-1} \theta_{j,n} \lambda_j^* \quad \forall n \tag{3.29}$$

where, λ_j^* is the weight obtained from the fitted semi-variogram, and $\theta_{j,n}$ is the value of the observation at location j. The observed data are used twice, once to estimate the semi-variogram and then to interpolate the values. To avoid any systematic bias in the kriging estimates, a constraint on weights is enforced that is given by:

$$\sum_{j=1}^{ns-1} \lambda_j^* = 1 \tag{3.30}$$

The estimation variance at a specific point in space is given by:

$$\sigma_m^2 = \sum_{i=1}^{N} \lambda_j^* \gamma_{m,i} + \psi \tag{3.31}$$

where σ_m^2 is the variance, $\gamma_{m,i}$ is the variance between the points m and i, and ψ is the Lagrangian multiplier used in solving kriging equations to minimize estimation variance. It is essential to evaluate an adequate number of semi-variogram forms that will provide the best possible fit for the empirical semi-variogram values obtained through the analysis.

3.18.3   Positive kriging

Kriging weights depend on the shape of the semi-variogram (i.e., the relative nugget effect and anisotropy, and correlation range) and not on the sill (Gooverts, 1997; Li and Heap, 2008). The existence of negative weights can lead to estimated values that are outside the range of data. This is referred to as non-convexity of the estimator. In the case of kriging estimation, negative weights may cause negative precipitation values at the base station where estimates are needed when these weights are attached to high precipitation values. Chiles and Delfiner (1999) indicate that even a fairly crudely determined set of weights can give excellent results when applied to data. However, caution should be exercised if negative weights exist as part of a solution. In general, positive weights are not a requirement for the sufficient and necessary condition for positive estimates in spatial interpolation using ordinary kriging. Negative weights can be eliminated by using a procedure referred to as "positive kriging" (Barnes and Johnson, 1984).

A variant of positive kriging (Teegavarapu, 2007) can be used to restrict the kriging weights to positive values. The approach described by a mathematical programming (optimization) formulation (Teegavarapu, 2007) is given by:

Minimize

$$\sum_{j=1}^{n} \left(\sum_{i=1}^{Ns-1} \left(\lambda_i \theta_i^j \right) - \theta_m^j \right)^2 \tag{3.32}$$

subject to

$$\sum_{i=1}^{Ns-1} \lambda_i = 1 \tag{3.33}$$

$$\lambda_i \geq 0 \quad \forall i \tag{3.34}$$

where θ_m^j is the observed value of the precipitation at the base station; θ_i^j is the observed precipitation value at a specific station, j; $Ns - 1$ is the number of stations excluding the base station; and n is the number of time intervals. The objective function (Equation 3.32) is used to minimize the difference between the kriging estimate and the observed value of precipitation over a period of n time intervals. The constraint defined by inequality 3.34 will ensure that all the weights are positive. There are other ways of limiting the weights to positive values. A constant value that is equal to the absolute value of the largest negative value can be added to all weights to ensure that the condition defined by Equation 3.34 is met. Also, the weights in Equation 3.32 can be restricted to positive weights already obtained and the equation solved for the remaining weights using the optimization formulation.

3.18.4   Limitations of kriging

The success and applicability of kriging for spatial interpolation depend on several factors. These factors are (1) the number of spatial data; (2) location of sampling points in space; (3) selection of fitted or experimental semi-variogram model; and (4) success in the cross-validation process. Voltz and Webster (1990) suggest that to compute a variogram with reasonable confidence, 100 data points for isotropic cases and 200 to 400 data points for anisotropic cases are needed. O'Sullivan and Unwin (2010) indicate that the preferred functional forms for the semi-variogram are non-linear and cannot be estimated easily using standard regression software. The robustness of kriging relies heavily on the proper selection of a semi-variogram model (Grayson and Gunter, 2001).

Choosing semi-variogram models and fitting them to data remain amongst the most controversial topics in geostatistics (Webster and Oliver, 2007). One reason for the difficulty is due to the number of authorized models that are available to choose from. O'Sullivan and Unwin (2010) indicate that the selection of the experimental semi-variogram and a mathematical model is to some extent a "black art" that needs careful analysis informed by good knowledge of the variables being analyzed. Cross-validation is often used as a procedure for selecting semi-variogram models. The most widely used and authorized semi-variogram models in ordinary kriging provide unique solutions for weights required for the interpolation. However, selection of parameters for these models is a tedious trial-and-error task. Work by Jarvis *et al.* (2003) points to the use of an artificial intelligence system for the selection of the appropriate interpolator based on task-related knowledge and data characteristics. Functional forms provided by artificial neural network models can be used to replace the semi-variograms.

3.18.5 Authorized semi-variogram models

Experimental semi-variogram models are generally used to summarize the variance associated with spatial data. These variogram models are referred to as authorized models (Webster and Oliver, 2007), which are bounded models in which the variance has a maximum value known as the sill. One condition should be fulfilled before any functional form can be used for fitting the semi-variogram for the data in the kriging process. The condition is given by:

$$W^T C W \geq 0 \qquad (3.35)$$

where W is the weight matrix, C is the covariance matrix that includes covariances between the observation points and those between the observation points and the location at which an estimate is required. The condition in Equation 3.35 confirms the existence of one and only one unique and stable solution, and suggests that the matrix C satisfies a mathematical condition known as positive definiteness. Only authorized variogram models (e.g., spherical, exponential, or Gaussian) satisfy the condition of positive definiteness and make the kriging equations non-singular. Selection of a semi-variogram model other than an authorized one may result in a non-unique and unstable solution (i.e., weights). However, Isaaks and Srivastava (1989) indicate that it is possible to concoct a new mathematical model or functional form for semi-variogram models and to verify its positive definiteness, and finally use it within kriging for interpolation. The condition specified in Equation 3.35 needs to be satisfied before any semi-variogram can be adopted for use.

3.18.6 Isotropy and neighborhood

Development of experimental semi-variograms should consider two important factors: isotropy and neighborhood. If the precipitation patterns and variability are the same in all directions, then it is appropriate to say that the processes that are being modeled using a semi-variogram are isotropic. Also, if the semi-variograms are varying in different directions, anisotropic conditions need to

be considered. Webster and Oliver (2007) indicate that if the variograms have the same form and values in all directions, then an average semi-variogram can be used for estimation. The selection of neighborhood is important in all local spatial interpolation models and there is no exception for kriging. The estimated values using kriging also depend on the number of observation sites selected in the neighborhood and the orientation of the neighborhood with respect to the base station where missing data are known to exist.

3.18.7 Development of variogams

Development of experimental semi-variograms with available data is one of the major issues when dealing with temporal data. Development of a variogram for a set of data available in space is easy. However, precipitation data at a point have high variability, with several zero values. Variograms can be developed using a specific statistic of spatial data that will help model the spatial variability of the process. If the data characteristics change from time to time, time-specific (or seasonal) variograms can be developed. Also, if the process depicts variability in different directions in space, anisotropy of the process should be considered in the development of several variograms. Estimation at a point in space should be made using a specific variogram applicable to the orientation of the point in space. The estimation using kriging is optimal if the data are normal. Data transformation and a check for normality is often required before a semi-variogram is developed. Also, de-trending of the data is required if a spatial trend is depicted in the process. Webster and Oliver (2007) discuss a number of steps to carry out geostatistical analysis. These steps include (1) screening of data to eliminate observations influenced by instrumental and human errors; (2) exploratory data analysis to evaluate the summary statistics of the data and check for outliers; (3) data transformation and normality checks; (4) spatial distribution of data for trends and patches and isotropic or anisotropic conditions; (5) development of experimental variogram; (6) modeling and fitting of the variogram; (7) spatial estimation or prediction including cross-validation checks; and finally (8) mapping of data.

3.18.8 Precipitation-specific kriging applications in hydrology

Several applications of kriging exist in the fields of hydrology and geosciences. The success of kriging in surface interpolation and estimation of missing precipitation data has been reported in many studies. Ashraf *et al.* (1997) compare interpolation methods (kriging, inverse distance, and co-kriging) to estimate missing values of precipitation. They indicate that the kriging interpolation method provided the lowest RMSE. Co-kriging of radar and rain gage data studies were performed by Krajewski

(1987), Seo *et al.* (1990a, 1990b), Seo and Smith (1993), and Seo (1996) to estimate mean areal precipitation and interpolation of rainfall data. Seo (1998) studied real-time estimation of rainfall fields using radar and rain gage data. Chang (2010) used three methods, namely, ordinary kriging, inverse distance weighting, and universal kriging, and compared them for generating an annual precipitation surface in the state of Idaho, USA. Deraisme *et al.* (2001) used kriging with external drift and collocated co-kriging for interpolation of rainfall in mountainous areas. Goovaerts (2000) used information from a digital elevation model within kriging for spatial interpolation of rainfall.

A study of precipitation data in Austria by Holawe and Dutter (1999) illustrates the utility of kriging and variogram parameters in understanding of the spatial and temporal structure of the precipitation. Kriging within a Bayesian framework was used by Handcock and Wallis (1994) to model the spatial and temporal variations of mean temperature in the northern USA. Kyriakidis *et al.* (2004) used kriging for simulation of daily precipitation time series at a regional scale. Genton and Furrer (1998b) used robust geostatistical methods on a data set of rainfall measurements in Switzerland by fitting the variogram using generalized least squares and taking into account the variance–covariance structure of the variogram estimates. All these studies suggest that kriging with variations is a viable stochastic technique for spatial and temporal interpolation of precipitation data.

3.19 OPTIMAL FUNCTIONAL FORMS

Optimal functional forms (Teegavarapu, 2007) for linking precipitation data at one observation point with others are possible for use in estimation of missing precipitation data. These functions establish an optimal relationship among the observations via a mathematical relationship. Three types of mathematical functional forms (Teegavarapu *et al.*, 2008) are possible: (1) fixed mathematical forms, (2) flexible mathematical forms, and (3) fixed function-free forms. The following sections describe optimal functions developed using a method similar to item 2 above. The method is referred to as the fixed function set genetic algorithm method (Teegavarapu *et al.*, 2008). Function-free forms are universal function approximators such as neural networks.

3.19.1 Fixed function set genetic algorithm method

The fixed function set genetic algorithm method (FFSGAM) of inductive model development was recently proposed by Tufail and Ormsbee (2006) and Teegavarapu *et al.* (2009) to obtain functional forms for a given data set. FFSGAM is particularly suited for cases in which a simple and easy to use functional form is sought to represent the response function that is being modeled.

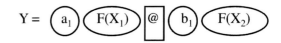

$$Y = a_1 \; F(X_1) \; @ \; b_1 \; F(X_2)$$

Coefficients a1, b1 = {real numbers}
Mathematical Operator @ = {+, -, *, /, ^}
F( ) = 0, 1, X, log(X), eX, Sin(X), X0.5, 1/X, etc.

Figure 3.25 General pre-defined functional forms for response function Y relating independent variables X_1 and X_2.

The method starts with pre-defined (general) functional forms that are combinations of numeric coefficients, functions of decision variables (or model inputs), and mathematical operators. The coefficients of these functional forms are then obtained by optimization formulations. Use of optimization eliminates any potential convergence problems associated with estimation of numeric coefficients (constants) and the search for optimal functional forms (sub-functions of independent variables and mathematical operators) simultaneously. Genetic algorithms (GA) are used for obtaining optimal functional forms and therefore the method is referred to as the fixed function set genetic algorithm method. This static GA approach results in the optimal selection of a functional form or expression based on training on a given data set. The GA optimization involves selection, crossover, and mutation processes by selecting functions of decision variables (model inputs) and mathematical operators that improve the fitness of the function.

The FFSGAM is explained using an example of obtaining a functional relationship between two independent variables (X_1 and X_2) and a dependent variable (Y). For a given function Y for which an empirical functional form is sought, a set of pre-defined functional forms is first selected. Such a formulation can be based on any prior knowledge about Y or can be formulated without such knowledge. The pre-defined functional form is a function of the independent variables (say X_1 and X_2) and one such formulation is given in Figure 3.25.

The objective function in the search process is an explicit functional form sought (Figure 3.25) for Y (model output). The fitness function is based on the mean squared error (MSE) performance measure of the objective function in predicting the value of Y as a function of the two decision variables X_1 and X_2 (model inputs). More than one functional form can be formulated for Y by varying the parameters and functional structures. Also, the number of available sub-functions of the independent variables can be increased for optimal selection in the GA process. The basic genetic operations of GA (selection, crossover, and mutation) are used to select the best sub-functions. The FFSGAM is illustrated by the flow chart shown in Figure 3.26.

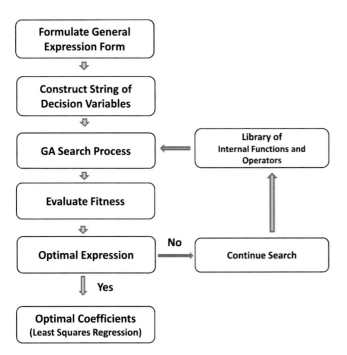

Figure 3.26 Steps used in FFSGAM method (Teegavarapu *et al.*, 2009).

Table 3.14 *List of potential functions for decision variables: distance (d_{mi}) and correlation coefficient (R_{mi}).*

Function	Function f (d_{mi}) or Function f (R_{mi})
1	1
2	d_{mi} or R_{mi} or Sqrt(d_{mi}) or Sqrt(R_{mi})
3	$1/d_{mi}$ or $1/R_{mi}$
4	Exp(d_{mi}) or Exp(R_{mi})
5	Log_e(d_{mi}) or Log_e(R_{mi})
6	Log_{10}(d_{mi}) or Log_{10}(R_{mi})
7	Exp($1/d_{mi}$) or Exp($1/R_{mi}$)
8	Log_e($1/d_{mi}$) or Log_e($1/R_{mi}$)
9	Log_{10}($1/d_{mi}$) or Log_{10}($1/R_{mi}$)
10	d_{mi}*Exp(d_{mi}) or R_{mi}*Exp(R_{mi})
11	d_{mi}*Log_e(d_{mi}) or R_{mi}*Log_e(R_{mi})
12	d_{mi}*Log_{10}(d_{mi}) or R_{mi}*Log_{10}(R_{mi})
13	($1/d_{mi}$)*Exp(d_{mi}) or ($1/R_{mi}$)*Exp(R_{mi})
14	($1/d_{mi}$)* Log_e(d_{mi}) or ($1/R_{mi}$)*Log_e(R_{mi})
15	($1/d_{mi}$)*Log_{10}(d_{mi}) or ($1/R_{mi}$)*Log_{10}(R_{mi})

3.19.2 FFSGAM for estimating missing precipitation data

The optimal functional form for estimating missing precipitation data using the model inputs (i.e., distance and correlation coefficient) and the fitness function based on the MSE as a performance measure is optimized in FFSGAM. As a demonstration of the proposed technique, a fixed functional form was formulated to represent the expression for missing precipitation data and is defined by Equation 3.36. The two decision variables contained in the functional form are (1) the distance between any two stations (d_{mi}) and (2) the correlation coefficient derived based on historical precipitation data between any two stations (R_{mi}).

$$\text{Functional Form} = \{C_1 \ [\text{function\_1} \ (R_{mi}) \ \text{operator\_1}$$
$$\times \ \text{function\_2} \ (R_{mi})]\} \ \text{operator\_3}$$
$$\times \ \{C_2 \ [\text{function\_3} \ (d_{mi}) \ \text{operator\_2}$$
$$\times \ \text{function\_4} \ (d_{mi})]\} \quad (3.36)$$

Table 3.14 provides a list of 15 elementary functions for each of the decision variables (model inputs) that are available for selection by the FFSGAM. This table is adopted from the work by Teegavarapu *et al.* (2009). The number of such elementary functions can be expanded further by introducing additional functions or combination of functions. A larger set of such functions will facilitate the GA search process in finding the optimal functional form that fits a given data set. The mathematical operators that are available for selection by the FFSGAM are also shown in Table 3.14. The

fixed functional formulation given in Equation 3.36 is defined in terms of some function of the decision variables (selected from the set of 15 elementary functions identified in Table 3.14). For example, "function_1 (R_{mi})" in Equation 3.36 can be selected to be any of the 15 internal functions defined in Table 3.14. Similarly, "operator_1" in Equation 3.36 can take on any of the five mathematical operators shown in Figure 3.25. The coefficients C_1 and C_2 are real numbers with appropriate pre-set upper and lower bounds.

Once the functional form (as given in Equation 3.36) has been obtained from the FFSGAM and through subsequent optimization, the estimated precipitation at the base station is given by:

$$\theta'_{m,n} = \frac{\sum_{j=1}^{ns-1} \theta_{j,n} \ (\text{FFSGAMfunctionform})}{\sum_{j=1}^{ns-1} (\text{FFSGAMfunctionform})} \quad \forall n \quad (3.37)$$

In addition to the numeric coefficients included in the fixed functional form given by Equation 3.36, the functional form may also include additional coefficients for each individual precipitation station used in the analysis. Such numeric coefficients can be estimated using optimization by minimizing the MSE derived from the observed and estimated precipitation values for the station for which the precipitation records are sought. Accordingly, the equation to estimate precipitation values at a station, given precipitation for all other stations, can be expressed using:

$$\theta'_{m,n} = \frac{\sum_{j=1}^{ns-1} \theta_{j,n} C_i \ (\text{FFSGAMfunctionform})}{\sum_{j=1}^{ns-1} C_i (\text{FFSGAMfunctionform})} \quad \forall n \quad (3.38)$$

3.19.3 Mathematical programming formulation for optimal coefficients

The approach to obtain optimal coefficient values described by a mathematical programming formulation is given by:

Minimize

$$\frac{1}{no}\left[\sum_{n=1}^{no}(\theta'_{m,n}-\theta_{m,n})^2\right] \tag{3.39}$$

subject to:

$$C_l \le C_i \le C_u \quad \forall i \tag{3.40}$$

where $\theta'_{m,n}$ is the estimated value of the precipitation at the base station; $\theta_{m,n}$ is the observed precipitation value at the base station for a given time interval, n; and no is the number of time intervals. The objective function (Equation 3.38) is used to minimize the mean of the squared difference between the estimated and observed precipitation values over a training period of no time intervals. A constraint defined by inequality 3.40 will ensure that all coefficients are within their allowable upper (C_u) and lower (C_l) bounds. The upper and lower bounds can be modified for improved solutions. The formulation is solved using the Microsoft Excel solver that uses a generalized reduced gradient (GRG) non-linear optimization algorithm for solution. However, there is no restriction on the type of approach or solver used for solution of this mathematical programming formulation.

The FFSGAM relies on two fundamental parameters (distance and correlation coefficient) that capture the spatial correlation of the data in indirect and direct ways. The fact that the approach relies on lengthy historical daily data to develop optimal functional forms provides a more generalized weighting scheme in the interpolation process. The Euclidean distances in the IDWM are fixed as constant weights for training and test data. Also, the correlation coefficients used in the CCWM and FFSGAM are derived based on training data. Since these values are used as weights or constant parameters in the interpolation methods for the test data set, the improvement in the performance measure (correlation coefficient, ρ) can be attributed to the characteristics of the spatial data (point data) during the testing period. The spatial data characteristics provide insight into the reasons for improvement in the performance measure, ρ, for test data compared to that from training data. The spatial data structure (referred to as spatial autocorrelation) (Griffith, 1987; Vasiliev, 1996; O'Sullivan and Unwin, 2010) of the data from rain gages in space may be evaluated using spatial correlations among the data sets at the base station and at other stations. In a conceptually simple weight-based interpolation method the estimated value at a station with missing values is a weighted combination of values at all other stations. The estimate improves if the correlations among data at the base station and any other stations improve when the weights are held constant. Since the estimate is equal to the sum of weighted observations, observations that are weighted high will affect the estimate more than the others. The closer the observations are in magnitude at all other stations to those at the base station, the better the estimate. The training and testing data should be selected by a split-sample method using available chronological or non-chronological data time series. No preferential bias should be introduced into the process of selecting the data time series for testing or training.

The functional form ultimately derived from FFSGAM may not always be as simple in structure (mathematical form) as any functional form used in traditional weighting methods. Also, the functional form may not represent a parsimonious model with simplicity in the form. However, it provides an advantage in developing functional forms with select parameters. The pre-defined set of functions and operators will define and affect the functional form that is obtained using FFSGAM and ultimately used for estimating missing data. An example of a complex functional form (Teegavarapu et al., 2008) derived based on FFSGAM is given by:

$$\theta_m = \frac{\sum_{i=1}^{ns-1}\theta_i C_i\left[C_1^0\left[R_{mi}\log_{10}\left(\frac{1}{R_{mi}}\right)-\left(\frac{1}{R_{mi}}\right)\log_{10}\left(R_{mi}\right)\right]^{C_2^0\left[\frac{\log_{10}\left(\frac{1}{d_{mi}}\right)}{\log_{10}(d_{mi})}\right]}\right]}{\sum_{i=1}^{ns-1}C_i\left[C_1^0\left[R_{mi}\log_{10}\left(\frac{1}{R_{mi}}\right)-\left(\frac{1}{R_{mi}}\right)\log_{10}\left(R_{mi}\right)\right]^{C_2^0\left[\frac{\log_{10}\left(\frac{1}{d_{mi}}\right)}{\log_{10}(d_{mi})}\right]}\right]} \tag{3.41}$$

A parsimonious model using FFSGAM can be obtained by limiting the number of operators and functional forms. The parsimonious model selection can be guided by the principle of parsimony (Occam's razor) to select the simplest functional form. The mathematical programming formulation uses the available data to obtain the optimal coefficients. The use of lengthy data time series ensures generalization and reduces the risk of overfitting non-linear functions to the spatial correlation structure of the observations.

The nature of functional forms obtained using FFSGAM depends on the length of the training data. The existence of data that are representative of all possible spatial and temporal precipitation patterns within the training data set is essential for the success of FFSGAM in the testing phase. The method is data intensive and sensitive, and can be referred to as a data-driven inductive model for functional approximation. The optimized values of local coefficients used in these functional forms will be influenced by the capabilities of the optimization solver used for the mathematical programming formulation.

The functional forms obtained by FFSGAM and discussed in this chapter were forced to resemble those of traditional weighting approaches except that the weight is not one single parameter but a non-linear function of two parameters (i.e., distance and correlation coefficient) that define the strength of autocorrelation. However, there is no conceptual limitation to adopting a completely different form of functional expression with no such resemblance. A close look at the functional forms reveals that distance and correlation coefficient (the former a surrogate of the spatial correlation structure and the latter a direct parameter used for evaluation of similarity in observations among stations) are adopted in the optimal functional forms. Exhaustive studies need to be conducted to test the efficacy of FFSGAM before any conclusive recommendations can be made about the method presented in this chapter. FFSGAM can perform better in situations when rain gages are located in two or more entirely different topographic locations, as it combines information about spatial correlations and Euclidean distances. The former parameter will help FFSGAM to identify existence or lack of spatial correlation among observations separated by locations that are topographically different.

3.20 STRUCTURE OF OPTIMIZATION FORMULATIONS

Optimization formulations that use binary variables for estimation of missing precipitation were proposed by Teegavarapu (2012a). The binary variables are useful in the selection of gages that are used in the spatial interpolation methods. The optimum weighting formulations using binary variables described in this chapter adopt a common structure of mixed-integer non-linear programming (MINLP) formulation. The structure is referred to as a superstructure (Floudas, 1995) and is given by the set of equations 3.42–3.46. The superstructure form is adopted from Floudas (1995) and Teegavarapu and Simonovic (2000) and is given by:

Maximize or minimize

$$f(x, y) \tag{3.42}$$

subject to:

$$h(x, y) = 0 \tag{3.43}$$

$$g(x, y) \leq 0 \tag{3.44}$$

$$x \in X \subseteq \Re^n \tag{3.45}$$

$$y \in Y = \{0, 1\}^\tau \tag{3.46}$$

where x is a vector of n continuous variables; y is a vector of τ 0–1 (integer) variables that denote the existence or non-existence of a process; $h(x, y) = 0$ are m equality constraints; $g(x, y) \leq 0$ are p inequality constraints; and $f(x, y)$ is the objective function, which is a performance evaluation function that needs to be minimized. If the integer variables are 0–1 variables (i.e., binary variables), then the formulation is referred to as a MINLP model with binary variables. In the case of general mathematical

programming formulations, the constraints are functional relationships that relate variables in the formulation. However, for the problem formulations defined here, historical data need to be used simultaneously with the evaluation of the objective function.

3.20.1 Optimum weighting models

In all the optimal weighting methods the optimal station weights are obtained using historical rainfall data available at all the stations. The non-linear mathematical programming formulations and their minor variants developed are discussed next.

3.20.2 Model I

Minimize

$$\sqrt{\frac{1}{no} \sum_{i=1}^{no} \left(\hat{\phi}_i^m - \phi_i^m \right)^2} \tag{3.47}$$

subject to:

$$\phi_i^m = \frac{\sum_{j=1}^{ns-1} w_{mj}^k \theta_i^j}{\sum_{j=1}^{ns-1} w_{mj}^k} \quad \forall i \tag{3.48}$$

$$wl_j \leq w_{mj}^k \leq wu_j \quad \forall j \tag{3.49}$$

The variable ϕ_i^m is the estimated value of precipitation at the base station m and $\hat{\phi}_i^m$ is the actual observation in time interval i; ns is the total number of stations including the base station; θ_i^j is the observation at station j; w_{mj}^k is the weight associated in relation to station i to the station m; and k is referred to as the friction distance (Vieux, 2004), which usually ranges from 1.0 to 6.0 in distance-based weighting methods. The mostly commonly used value for k is 2. In the current context, the k value is assumed to be a variable and is optimized. The variables wl_j and wu_j represent the lower and upper bounds on the weights. The variable no is the number of time intervals.

The objective function used in this formulation is RMSE, which is generally used as a performance measure for evaluation of interpolation methods. Another performance measure, MAE, as shown in Equation 3.50, can be used as an objective function. The variation of weights and the selection of stations using these two objective functions should be evaluated.

3.20.3 Model IA

Model IA is similar to model I except that the objective function is now defined as the minimization of MAE. The objective function is given by:

Minimize

$$\frac{\sum_{i=1}^{no} \left| \hat{\phi}_i^m - \phi_i^m \right|}{no} \tag{3.50}$$

3.20.4 Model II

The formulation for model II is similar to model IA, with a change in the constraint and the inclusion of binary variables. The constraints are given by Equations 3.52 and 3.53:

$$\phi_i^m = \frac{\sum_{j=1}^{ns-1} \lambda_j (\theta_i^j w_{mj}^k)}{\sum_{j=1}^{ns-1} \lambda_j w_{mj}^k} \quad \forall i, j \tag{3.51}$$

$$\lambda_j \le w_{mj}^k \le \lambda_j w u_j \quad \forall j \tag{3.52}$$

$$w_{mj}^k - \lambda_j \le \lambda_j M \quad \forall j \tag{3.53}$$

In Equation 3.51, λ_j is a binary variable associated with station j. Inequality 3.52 is referred to as an activation and de-activation constraint. For $\lambda_j = 0$, the constraint 3.52 will translate to $0 \le w_{mj}^k \le 0$ and therefore the weight associated with the station is zero. When $\lambda_j = 1$, then the constraint $wl_j \le w_{mj}^k \le wu_j$ holds, and the weight value is now limited by lower and upper bounds. Once the binary variable is assigned a zero value the station will not be used in the weighting scheme. The binary variables will help in the objective selection of the stations in the interpolation. Inequality 3.53 can also be used for imposing the conditions of constraint 3.52, except when the lower and upper bounds of the weights are confined to 0 and 1 respectively. The variable M is a positive number that helps to assign a zero value to weight when a station is not selected for interpolation. Basically, model II and model I are identical if the weights in model I are allowed to be equal to zero. The flexibility introduced in the formulation of model II allows the selection and non-selection of stations. Strict binary variable (0, 1) formulation is not possible for model I since the weight values can take on any real values between 0 and the upper bound.

3.20.5 Model IIA

Model IIA has the same formulation as model II with an additional constraint defined by either Equation 3.54 or 3.55:

$$\sum_{j=1}^{ns-1} \lambda_j \le np \tag{3.54}$$

$$\sum_{j=1}^{ns-1} \lambda_j = np \tag{3.55}$$

In this formulation the number of stations used for estimation is constrained by an upper limit, np. The variable np is the number of stations. Two types of constraints are possible: (1) a *soft* constraint on the number of stations given by inequality 3.54; and (2) a *hard* constraint (Equation 3.55) on the number of stations, an equality constraint.

The size of the precipitation monitoring networks in space and observations available in time are seldom constant. This poses an impediment in dealing with changing networks as the weights and stations used in the interpolation are selected based on the assumption that the network is stationary in space and time. A specific set of stations in a monitoring network can always be used for interpolation, as long as the set of stations remains unchanged for those periods in which missing data need to be estimated. The constraint defined by Equation 3.56 can help in selecting a discrete set of stations in the interpolation process.

$$np^r = \sum_j \lambda_j j \in j_s \tag{3.56}$$

The variable np^r is the number of discrete sets of stations, λ_j is a binary variable associated with a specific selected station, and j_s is the pre-selected set of stations to which the station j belongs.

3.20.6 Model IIB: Stratification of data

Model IIB utilizes a formulation in which the precipitation data are classified into different pre-defined classes. A separate model is solved for each set of data belonging to a specific class defined by the magnitude of average precipitation values. The formulation is given by Equations 3.57–3.61.

Minimize

$$\sqrt{\frac{1}{no} \sum_{i=1}^{no} (\hat{\phi}_i^m - \phi_i^m)^2} \quad \forall c \tag{3.57}$$

subject to:

$$\phi_{i,c}^m = \frac{\sum_{j=1}^{ns-1} \lambda_j (\theta_i^j w_{mj}^k)}{\sum_{j=1}^{ns-1} \lambda_j w_{mj}^k} \quad \forall i, j, c \tag{3.58}$$

$$0 \le w_{mj}^k \le 1 \quad \forall m, j \tag{3.59}$$

$$\phi_{i,a} = \frac{\sum_{j=1}^{ns-1} (\theta_i^j)}{ns - 1} \quad \forall i, j \tag{3.60}$$

$$\text{if } (\delta_{l,c} \le \phi_{i,a} \le \delta_{u,c}), \text{ then } \phi_i \in \phi_{i,c} \quad \forall i, c \tag{3.61}$$

The variable $\phi_{i,a}$ is the average value of precipitation for a given time interval belonging to a specific class c; $\delta_{l,c}$ and $\delta_{u,c}$ are the lower and upper limits of a pre-set class interval. These limits can be adjusted based on considering the available data. The models developed for classes are referred to as local models. The definition of local models is confined to limiting of data sets to specific ranges of average precipitation values in the region. This partitioning of data sets based on the range of values and models associated with these data sets may not show improvement over global models that consider the entire data sets available. One reason for the failure of local models and the lack of improved performance in some instances can be attributed to the size of the window (i.e., range of precipitation values) or cluster size selection. Lloyd (2007) indicates that local models may fail if the data sets are not partitioned correctly. The local models can be referred to as conditional models, as these are conditioned on the partition of numeric data as opposed to partitioning in observational space.

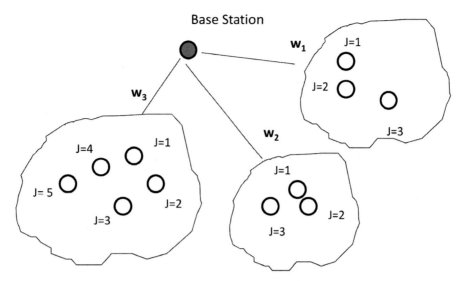

Figure 3.27 Definition of precipitation gage clusters and associated weights with base station where data are missing.

Fortin and Dale (2005) indicate that the analysis of spatial data at two levels (global and local) in space and time should be guided by knowledge about the stationarity of data. The concept of separate models for different classes of precipitation is determined by precipitation intensity in this formulation. However, season, weather type (convective/frontal precipitations), etc. can be evaluated using a similar methodology. The optimized weights derived using any mathematical programming formulation will differ and depend on these classes.

3.20.7 Model III

Model III combines Euclidean distance (d) and correlation coefficient (ρ) in a functional form that is given by Equation 3.62. Distance and spatial autocorrelation (O'Sullivan and Unwin, 2010) are respectively surrogate and real measures of strength of correlation among spatially observed data sets. The weighting factor $\left(\frac{\rho}{d_{mj}^k}\right)$ is raised by an exponent, p, and its optimal value is either fixed or solved using a mathematical programming formulation. Recent work by Teegavarapu *et al.* (2009) utilized distance and correlation in optimized functional forms for estimating missing precipitation data. The k value can be fixed equal to 2. The estimated value of the missing precipitation record is given by:

$$\phi_i^m = \frac{\sum_{j=1}^{ns-1} \left(\frac{\rho}{d_{mj}^k}\right)^p \theta_i^j}{\sum_{j=1}^{ns-1} \left(\frac{\rho}{d_{mj}^k}\right)^p} \quad \forall i \tag{3.62}$$

3.20.8 Model IIIA

Model IIIA is minor variant of model III in which the distance and correlation coefficient are used in a different functional form

given by Equation 3.63. In this model both friction distance (k) and exponent (p) are estimated.

$$\phi_i^m = \frac{\sum_{j=1}^{ns-1} \left(\rho d_{mj}^k\right)^p \theta_i^j}{\sum_{j=1}^{ns-1} \left(\rho d_{mj}^k\right)^p} \quad \forall i \tag{3.63}$$

3.20.9 Model IV: Rain gage clusters

Spatial partitioning of regions to understand processes at multiple scales is common in ecological studies (Dungan *et al.*, 2002). The formulation described in model IV explores the utility of partitioning the observation network into clusters for improved spatial interpolation of precipitation data. These clusters are not pre-fixed and are identified by the mathematical programming formulation. This formulation is referred to as the "cluster approach," in which binary variables are used to (1) select the cluster of stations that can be used for the estimation process; (2) define the number of clusters; (3) select a discrete set of stations; and (4) define the maximum number of stations that can be used in a cluster. The concept of clusters is illustrated in the Figure 3.27. The formulation is defined by equations and inequalities 3.64–3.69.

Minimize

$$\sqrt{\frac{1}{no} \sum_{i=1}^{no} \left(\hat{\phi}_i^m - \phi_i^m\right)^2} \tag{3.64}$$

subject to:

$$\phi_i^m = \frac{\sum_{l=1}^{nc} \left[w_l \left(\sum_{j=1}^{N_l} \theta_i^j \lambda_{lj}\right) \right]}{\sum_{l=1}^{nc} w_l} \quad \forall i \tag{3.65}$$

$$\sum_{l=1}^{nc} \lambda_{lj} = 1 \quad \forall j \tag{3.66}$$

$$\sum_{j\in l} \lambda_{lj} = N_l \quad \forall l \qquad (3.67)$$

$$\sum_{l=1}^{nc} N_l = ns - 1 \qquad (3.68)$$

$$\sum_{l=1}^{nc} N_l \leq ns - 1 \qquad (3.69)$$

In the formulation, λ_{lj} is a binary variable; $ns - 1$ is the number of stations used in interpolation; nc is the cluster size; np is the number of stations to be used; w_l is the weight associated with cluster l. The constraint given by Equation 3.66 ensures that once a station is selected to belong to one cluster it will not belong to any other cluster. Equation 3.67 provides the condition that the number of stations selected in a cluster is equal to the number of total stations specified. Constraints 3.68 and 3.69 provide the limits on the total number of stations used in cluster formation. Only one constraint out of the two (i.e., Equations 3.68 and 3.69) is used in any formulation. The number of stations in each cluster can also be pre-fixed using the following constraint:

$$\sum_{l=1}^{nc} N_l = nl_1 + nl_2 + \cdots + nl_{nc} \qquad (3.70)$$

The variables nl_1, $nl_2, \ldots,$ nl_{nc} are referred to a pre-specified number of stations in clusters 1, 2, and nc respectively. The cluster-based formulation described in this section is a general case of the quadrant method, where the clusters and cluster sizes are not limited to four quadrants. McCuen (2005) suggests that delineation of quadrants does not have to be based on a north–south, east–west axis system. A non-Cartesian coordinate system can be used if a proper justification for an axis system exists due to hydrometeorological and any other considerations. Factors such as local topography, seasonal variations of rainfall, homogeneous rain areas, and orographic effects can be considered for selection and orientation of the quadrant system.

3.20.10   Model VA

A variant of the IDWM is achieved by keeping the weights as defined by inverse squared distances and obtaining the optimal number of stations to be used. The variant is similar to formulation model IIB and is given by Equation 3.71. This variant helps in relaxing the size of the neighborhood for selection of the interpolation points, which is generally achieved in traditional IDWM by a trial-and-error method of jack-knifing or cross-validation.

$$\phi_i^m = \frac{\sum_{j=1}^{ns-1} \theta_i^j d_{m,j}^{-k} \lambda_j}{\sum_{j=1}^{ns-1} d_{m,j}^{-k} \lambda_j} \quad \forall i \qquad (3.71)$$

This formulation will use the constraint expressed in Equation 3.54 and select the optimal number of stations for interpolation. The use of stations that are fewer than the total number

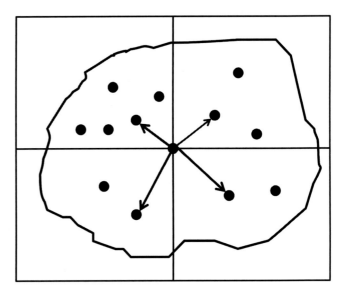

Figure 3.28 Selection of rain gage stations in each quadrant defined by the base station as a central point.

of stations, $ns - 1$, creates a local interpolation method that is not selected based on a fixed distance or pre-selected subset of $ns - 1$ stations based on any other criteria.

3.20.11   Model VB

A variant of the IDWM is defined as the quadrant method by McCuen (2005). In the quadrant method only one station closest to base station, based on Euclidean distance, is generally selected (see Figure 3.28). Modification can be made to the quadrant method and this is achieved by keeping the weights as defined by inverse squared distances and obtaining the optimal number of stations to be used in each quadrant. This modified quadrant method is similar to formulation model IIB and is given by Equations 3.72–3.74. This variant helps in relaxing the size of the neighborhood for selection of the interpolation points, which is generally achieved in traditional IDWM by trial-and-error method of jack-knifing or cross-validation.

$$\phi_i^m = \frac{\sum_{J=1}^{4} \sum_{j\in J} \theta_i^j d_{m,j}^{-k} \lambda_j}{\sum_{J=1}^{4} \sum_{j\in J} d_{m,j}^{-k} \lambda_j} \quad \forall i \qquad (3.72)$$

$$\sum_{j\in J} \lambda_{j,J} \leq np_J \quad \forall J \qquad (3.73)$$

$$\sum_{j\in J} \lambda_{j,J} = 1 \quad \forall J \qquad (3.74)$$

The variable J is the quadrant index and np_J is the total number of stations in each quadrant. This formulation uses the constraint expressed in Equation 3.72 and selects the optimal number of stations for interpolation in each quadrant J. The variable $\lambda_{j,J}$ is the binary variable used for selection of station j in quadrant J. The use of stations that are fewer than the total number of stations, $ns - 1$, creates a local interpolation method that is not selected based on a fixed distance from the base station or pre-selected subset of $ns - 1$ stations based on any other criteria. The formulation can also be extended to the normal ratio method (ASCE, 1996) to select the optimal number of stations in interpolation processes for estimating missing precipitation data.

3.20.12 Optimal gage mean and single best estimator

The optimal number of rain gages for the GME method and the best gage for SBE can be obtained using the following proposed MINLP formulation with binary variables. The estimated value of missing precipitation data is given by Equation 3.76, which involves the product of a binary variable ($\varpi_{\alpha,j}$) associated with rain gage j and the observed precipitation value at that gage. The binary variable helps in selection of stations. The variable ng is the number of stations in Equation 3.77 with upper limit value equal to $ns - 1$. When the value of ng is equated to 1 using a hard constraint (i.e., equality), the formulation becomes an optimal SBE that is not based on simple Euclidean distance or a correlation criterion.

Minimize

$$\sum_{n=1}^{no} \left| \theta'_{m,n} - \theta_{m,n} \right| \qquad (3.75)$$

subject to:

$$\theta'_{m,n} = \frac{\sum_{j=1}^{ns-1} \theta_{j,n} \varpi_{m,j}}{\sum_{j=1}^{ns-1} \varpi_{m,j}} \quad \forall n \qquad (3.76)$$

$$\sum_{j=1}^{ns-1} \varpi_{m,j} \leq ng \quad \forall n \qquad (3.77)$$

The formulation is solved using historical spatio-temporal precipitation data to minimize the objective function and obtain the optimal number of gages. An optimization solver capable of solving the MINLP formulation with binary variables is required for solution.

3.20.13 Correlation weighting approach using best estimators

The optimal CCWM (Teegavarapu and Chandramouli, 2005) can be modified to select one or more best estimators to revise the precipitation estimates. Revision of estimates was carried out using the ARM-based interpolation discussed in Section 3.21.5. The formulation uses a MINLP approach. The MINLP uses binary variables that will help in the selection of the stations as best estimators for the post-interpolation correction process.

Minimize

$$\sqrt{\frac{1}{n} \sum_{l=1}^{n} \left(\theta_m^{\prime l} - \theta_m^l \right)^2} \qquad (3.78)$$

subject to:

$$\hat{\theta}_m^l = \frac{\sum_{j=1}^{ns-1} w_{mj}^k \theta_l^j}{\sum_{j=1}^{ns-1} w_{mj}^k} \quad \forall l \qquad (3.79)$$

$$\vartheta_l = \sum_{j=1}^{ns-1} \theta_l^j \lambda_j \quad \forall l \qquad (3.80)$$

$$\sum_{j=1}^{ns-1} \lambda_j = \partial \qquad (3.81)$$

$$\text{if } (\vartheta_l = 0), \text{ then } \theta_m^{\prime l} = 0, \text{ else } \theta_m^{\prime l} = \hat{\theta}_m^l \quad \forall l \qquad (3.82)$$

The variables ϑ_l, λ_j, and ∂ are the product of the observed precipitation value at a station j and binary variable (λ_j), binary variable, and number of best estimators, respectively. When ∂ is equal to 1, then the formulation is a SBE considering zero precipitation conditions. Multiple best estimators are selected for revisions of precipitation estimates when ∂ is greater than 1. The variable $\theta_m^{\prime l}$ is the revised precipitation estimate using SBE or multiple best estimators and $\hat{\theta}_m^l$ is the initial estimate provided by the weighting method that uses Equation 3.79.

3.20.14 Issues with objective functions

The objective functions selected for mathematical programming models will have influence on the weights obtained and therefore the results obtained from the models. All possible objective functions should be experimented with before the formulations can be used for filling missing precipitation data. The length of the historical data used will also affect the performance of the models. Data used for model development and testing should be obtained from two different time periods, assuming stationarity of precipitation time series. Robust model development may need data that span over both testing and training periods. This can be achieved by creating or adopting a non-chronological data set from which training and testing data sets are selected. This ensures that the stationarity assumption is not required and improvements in predictions are possible. The optimization formulations are data intensive and sensitive. The optimal weighting methods described previously do not need any a-priori knowledge of spatial patterns of observations, as the interpolation is based on models developed from historical data. Precipitation data, like any other spatial data, exhibit spatial dependence and therefore the analyses are subject to a modifiable areal unit problem (MAUP), for which the results depend on division of space (Lloyd, 2007).

3.21 EMERGING INTERPOLATION TECHNIQUES

Underestimation of precipitation resulting from equating trace amounts of precipitation to a zero value leads to significant errors in regional water balance assessments (Dingman, 2008). Brown *et al.* (1968) reported that trace amounts of precipitation accounted for 10% of the summer precipitation on average for a region in Alaska, which might be equal to one-third of the total precipitation observed over a few years. Underestimation of precipitation in minute amounts still leads to significant errors in water balance studies or in hydrologic modeling. Dingman *et al.* (1980) and Yang *et al.* (1998) suggest that corrections have to be made to account for underestimation associated with trace amounts of precipitation. Overestimation of precipitation amounts obtained by spatial interpolation techniques is not unusual. The following sections provide details of a data mining approach for correction of estimates already provided by spatial interpolation methods.

3.21.1 Data mining

Data mining is the process of extracting interesting (non-trivial, implicit, previously unknown, and potentially useful) information or patterns from large information repositories (Chen *et al.*, 1996). It is also used to extract information and patterns derived by any knowledge discovery methods (Dunham, 2002). Knowledge discovery and data mining terms are often used interchangeably in data mining literature.

3.21.2 Association rule mining

Association rule mining (ARM) is one of the popular data mining methods mainly aimed at extracting interesting correlations, frequent patterns, associations, or causal structures among data available in databases (Zhao and Bhowmick, 2003; Zhang and Zhang, 2002). Association rule mining is regarded as an unsupervised knowledge discovery process. It has been successfully applied for deriving spatio-temporal relationships hidden in earth science data (Zhang *et al.*, 2005). Li *et al.* (2003) used data mining algorithms in conjunction with spatial interpolation to facilitate drought risk assessment using temperature and precipitation data.

Association rule mining is carried out using an a-priori algorithm developed by Agrawal *et al.* (1993). It is also referred to as a support–confidence framework for discovering association rules within a database. Association rules take the form "if antecedent, then consequent." The format is generally expressed as $X \Rightarrow Y$ suggesting that event Y is expected to occur whenever event X is observed. The events X, Y are generally referred to as items or itemsets in traditional ARM literature. In the current context, an item refers to specific event (e.g., occurrence or non-occurrence of rain at a station) and an itemset refers to a set of events (i.e., series

of stations with occurrence or non-occurrence of rain). Before the details of the algorithm are explained, two important measures for association rules need to be discussed. These measures are support and confidence, and they are discussed in relation to the current context.

3.21.3 Support

The support for an association rule $X \Rightarrow Y$ is the proportion of time intervals (D) that contain both X and Y and is expressed by:

$$\alpha = p(X \cap Y) \tag{3.83}$$

Support is defined using itemsets and indicates the proportion of the total number of time intervals that contain both X and Y. It is a measure of the significance or importance of an itemset. An itemset with a support greater than a minimum support threshold (α_m) value is called a frequent or large itemset. One important property of support is the downward closure property, which suggests that all subsets of a frequent set are also frequent. This property (i.e., no superset of an infrequent set can be frequent) is mainly used to reduce the search space in the a-priori algorithm and prune the association rules.

3.21.4 Confidence

Confidence is defined as the probability of seeing the rule's consequent under the condition that the transactions also contain the antecedent. It is important to note that confidence is directed and gives different values for the rules $X \Rightarrow Y$ and $Y \Rightarrow X$. Confidence (given by Equation 3.84) is not downward closed and was developed together with support by Agrawal *et al.* (1993).

$$\beta = p\left(\frac{Y}{X}\right) = \frac{p(X \cap Y)}{p(X)} \tag{3.84}$$

Support is initially used to find frequent (significant) itemsets exploiting its downward closure property to prune the search space. Then confidence is used in a second step to produce rules from the frequent itemsets that exceed a minimum confidence threshold. One main limitation of the confidence parameter is that it is sensitive to the frequency of the consequent (Y) in the database. Consequents with higher support will automatically produce higher confidence values even if there is no association between the items.

3.21.5 Association rule mining-based spatial interpolation

A general scheme of ARM-based spatial interpolation (Teegavarapu, 2009) is shown in Figure 3.29. A spatial interpolation technique is applied first to estimate missing data

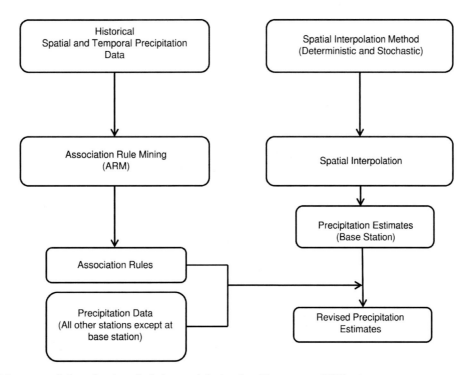

Figure 3.29 Data mining approach for estimation of missing precipitation data (Teegavarapu, 2009).

at a gaging station (i.e., base station) based on observations available at all other stations.

The spatio-temporal database of historical precipitation observations is mined separately using an a-priori algorithm and several rules are derived. The a-priori algorithm requires implementation of two major steps: (1) generation of frequent itemsets and (2) generation of association rules.

The algorithm format used by Teegavarapu (2009b) follows similar steps to that presented by Zhao and Bhowmick (2003). In the first phase all the candidate one-itemsets are identified. Using the parameters of minimum support and confidence, large one-itemsets are then selected. The process is continued to obtain two-item candidates and two-item large itemsets. Using the minimum confidence limit, association rules are derived from the large one-, large two-, and large k-itemsets. Rules are referred to as single consequent rules when only one item features in the consequent part of all the rules.

Corrections are applied to the precipitation estimates provided by the spatial interpolation methods using ARM-derived rules. The ARM-based rules can be translated to mathematical forms as described by Equation 3.85 and conditions specified by inequalities 3.86, 3.87, and 3.88:

$$\text{if } (\cap(\theta_i = 0)), \text{ then } \theta_m^o = 0, \text{ else } \theta_m^o = \theta_m \quad \forall i, i \neq m \quad (3.85)$$

$$i \leq ns - 1 \quad (3.86)$$

$$\alpha \geq \alpha_m \quad (3.87)$$

$$\beta \geq \beta_m \quad (3.88)$$

The variable i is the number of stations identified based on ARM of the database, θ_m is the estimated value using a spatial interpolation method at the base station m, θ_m^o is the revised estimate of the precipitation value, ns is the total number of stations, α, α_m, β, and β_m are the support and minimum support, and confidence and minimum confidence levels, respectively. The ARM is carried out using the WEKA (Waikato Environment for Knowledge Analysis) (WEKA, 2011) modeling environment. Historical precipitation data at all stations are converted from continuous numeric values to a categorical type of data. This process is referred to as discretization and it is needed for spatial ARM. The categorical data are specified as "no rain" and "rain" within the database of the modeling environment and they constitute only two class intervals. The ARM is used in a supervised manner in which the consequent (generally for a gage with missing data) is pre-selected. Once the association rules are extracted, they are applied to the estimates of the missing precipitation data to obtain revised estimates.

The threshold values of support and confidence factors influence the number of best rules found by the ARM process. Selection of final rules at the end of the ARM completion process can be a contentious yet crucial issue that might affect the revised estimates obtained from spatial interpolation techniques. However, the rules obtained from the historical data can be used to obtain revised estimates from interpolation methods, and performance of these methods can be evaluated before they can be applied to the test data. The rules generated based on the ARM are limited in number and are case study specific and gage network dependent.

However, the case may be different when an entirely separate rain gaging network is used. It is possible that many interesting rules are pruned or not reported as the support and confidence values are restricted to a few pre-specified limits. The number of association rules grows exponentially based on the number of stations and the categorical attribute values. For example, with n stations with binary attributes (i.e., yes or no) associated with rainfall occurrence and non-occurrence, a total of $n2^{n-1}$ association rules are possible. Also, the discretization scheme will affect the nature of the association rules. Exhaustive studies using ARM concepts need to be conducted before any recommendations can be made about the transferability of the approach to other climatic regions under different meteorological conditions.

In general, precipitation time series have a high percentage of observations with zero values and a low number of extreme values. Overestimation in all spatial interpolation models is impossible to avoid, as observations from other rain gages are used in the estimation of missing precipitation data. Overestimation is possible due to non-zero values at the other stations when the base station (i.e., the station with missing data) has registered no rain. Overestimations are not confined to only these specific events of no rain. Instances of overestimation can be reduced to a certain extent if a SBE (nearest gage by Euclidean or any other criterion) is used to obtain the rain or no rain condition and corrections of estimates using association rules are made (Teegavarapu, 2009). Underestimation at high values of precipitation is again possible as observations at several sites (i.e., rain gages) are used to obtain the interpolated values at the missing data site. Unless extreme events (high precipitation events) occur at all sites in the region, there is no way the estimates can be close to the observed values in all instances. As spatial interpolation models use weighted global or local (based on optimal number of gages) spatial interpolation, underestimation cannot be altogether avoided. Again in these instances, a local model with nearby stations may benefit the estimation. Under- and overestimated values may be reduced if (1) SBEs are used in conjunction with interpolation methods; (2) different sets of gages are selected; and (3) the optimization models are developed using transformed observations.

3.21.6   Data mining tool: WEKA

The data mining tool discussed in the previous section is public domain software referred to as the Waikato Environment for Knowledge Analysis (WEKA) (WEKA, 2011). The tool contains several modules that help in data pre-processing, classification, regression, cluster evaluation, association rule development, and visualization. The tool also provides prediction tools for continuous data and it is an ideal tool for estimation of missing precipitation data at a point. Data can be provided in the form of simple text files with values specified in the form of matrices.

Rule development using ARM concepts requires the data files to be submitted in the attribute–relation file format (ARFF). An example of the ARFF format is provided below to obtain association rules using different rainfall conditions referred to as *none*, *normal*, *low*, and *high* based on user-defined thresholds. The numeric rainfall data are transformed into symbolic values for attributes for each rain gage and this data set is then used in the WEKA environment. Different association rules can be developed for different months as "month" is also used as an attribute in the input file.

WEKA data file format for association rules:

@RELATION rain.Symbolic
@ATTRIBUTE month{January, February, March, April, May, June, July, August, September, October, November, December}
@ATTRIBUTE gage 1{none, normal, low, high}
@ATTRIBUTE gage 2{none, normal, low, high}
@ATTRIBUTE gage 3{none, normal, low, high}
@ATTRIBUTE gage 4{none, normal, low, high}
@ATTRIBUTE gage 5{none, normal, low, high}
@DATA
September,none,none,none,low,none
September,none,normal,low,low,none
September,normal,normal,low,low,none
October,none,none,none,none,normal

. . . .

WEKA data file format for prediction:

@RELATION rain.numeric
@ATTRIBUTE gage 1 real
@ATTRIBUTE gage 2 real
@ATTRIBUTE gage 3 real
@ATTRIBUTE gage 4 real
@ATTRIBUTE gage 5 real
@DATA
0,0,10,20,15
0,1,1,1,1
10,15,15,10,12

. . . .

In the format to be used for prediction, gage 5 is the base station where missing data are known to exist. Once the data are trained with historical data, prediction is possible. The WEKA environment can also be used for development of several data-driven models for estimation of missing precipitation data. Methods such as simple linear regression, additive regression, pace regression, regression by discretization, and several others are available under this environment. The ARFF files should be created accordingly to accommodate numerical values for estimation or prediction. The

Inputs

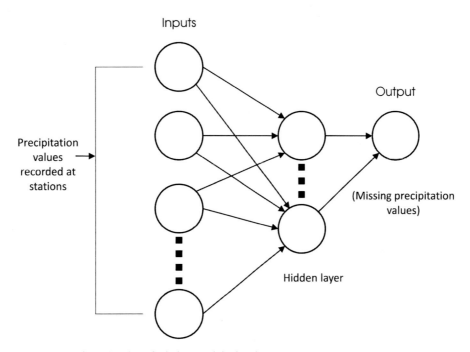

Figure 3.30 Artificial neural networks for estimation of missing precipitation data.

WEKA environment also provides universal functional approximation approaches referred to as artificial neural networks.

3.22 ARTIFICIAL NEURAL NETWORKS

Artificial neural networks (ANN) have been applied in the field of hydrology for estimation and forecasting of hydrologic variables (Govindaraju and Rao, 2000; ASCE, 2001a, b). A wealth of literature on the architecture of neural networks, training, and testing can be found elsewhere (e.g., Freeman and Skapura, 1991). ANN can be used to develop a model for estimation of missing rainfall values. A typical architecture of ANN that can be used is shown in Figure 3.30. The architecture is a feed-forward network and consists of one hidden layer with a number of hidden layer neurons. The hidden layer neurons are selected using a trial-and-error procedure. The input neurons use values from all the stations other than the station of interest and the output neuron of the ANN provides the missing value at the station of interest. The neural network training can be done using standard error, a supervised back-propagation training algorithm, or any other algorithm.

ANNs, like any other function approximators, may provide negative values of estimates when used in the estimation of missing precipitation data. As negative values are not acceptable, a post-correction scheme may be employed to convert the negative values to zero.

A few works on ANN applications relevant to the understanding and prediction of hydrologic processes are discussed

here. Applications of ANNs in hydrology are mainly related to streamflow (Govindaraju and Rao, 2000) and rainfall prediction (e.g., French *et al.*, 1992; Teegavarapu and Mujumdar, 1996), and spatial interpolation (Teegavarapu and Chandramouli, 2005). French *et al.* (1992) used a feed-forward neural network with a back-propagation training algorithm for forecasting rainfall intensity fields at a lead time of 1 hour with the current rainfall field as input. Navone and Ceccatto (1994) used an ANN model to predict summer monsoon rainfall over India.

3.23 UNIVERSAL FUNCTION APPROXIMATION-BASED KRIGING

The major limitations of choosing a pre-specified model and parameters for a semi-variogram have been discussed in the section on kriging. Authorized semi-variogram models can be developed using other bounded functional forms. An ANN model as a function approximator can be used to model the semi-variogram. This approach is referred to as universal approximation-based ordinary kriging (UOK) (Teegavarapu, 2007). The conceptual difference between traditional ordinary kriging and UOK is shown in Figure 3.31. Measures of spatial variability of observed data are generally expressed by semi-variograms in the field of geostatistics. However, many other measures are also possible, as suggested by Deutsch and Journel (1998). They discuss ten different measures of spatial variability and indicate that practitioners should spend more time on data analysis and variogram modeling than on all kriging and simulation runs combined.

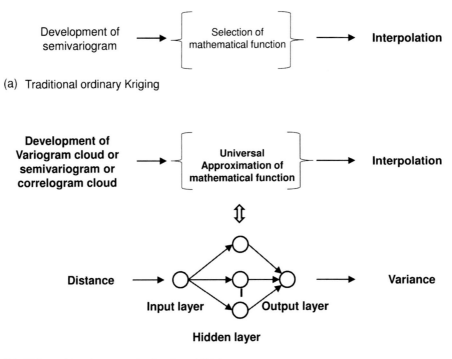

(a) Traditional ordinary Kriging

(b) Universal function approximation based Kriging

Figure 3.31 Comparison between traditional kriging and UOK (Teegavarapu, 2007).

Spatial variability can also be expressed as a function of correlation between point observations in space. Dingman (2008) suggests the use of spatial correlation coefficients to characterize relationships among observed data in space. The plot of correlation coefficients and the distance is referred to as a correlogram. The use of a correlogram in ordinary kriging and ANNs for modeling the semi-variogram is referred to as a universal function approximation-based ordinary kriging using correlogram (UOKC). In principle, estimation of missing precipitation data at one specific location in space does not require the development of an experimental variogram and a mathematical model for the semi-variogram. However, the use of a universal function approximator to fit a black-box model for the variogram has to satisfy the condition set for an authorized semi-variogram.

The functional form ultimately derived from ANN architecture may not be as simple in structure (mathematical form) as any authorized functional form (i.e., spherical or exponential). The neural network architecture (hidden layers and neurons in each layer) will define and affect the functional form that will ultimately represent the experimental variogram. The functional form in the case of ANN in general is an additive form of non-linear and linear combination of weights and sigmoidal activation functions. By using different architectures, several covariance matrices can be derived, which can be checked for conditions that can guarantee the existence of unique and stable solutions. Webster and Oliver

(2007) suggest and illustrate the use of Akaike's information criterion (AIC) to select a parsimonious semi-variogram model. A parsimonious model in the case of ANN can be obtained by limiting the number of neurons (i.e., degrees of freedom) in the hidden layers to reduce the over-fit or over-training (Maier and Dandy, 1998) of the ANN. The over-training process will reduce the generalization capabilities of an ANN. However, it should be noted that a minimum number of hidden layer neurons are required for an ANN to approximate the function.

Ordinary kriging is computationally intensive when a surface interpolation problem with several spatial data points is considered. However, the time required to develop a kriging interpolation model is generally invested in the development and assessment of semi-variogram model parameters and cross-validation to accept a specific model for an interpolation problem. In the case of UOK and UOKC, time is invested in the training and testing of the ANN model and finally checking the conditions that need to be fulfilled to obtain unique and stable weights for calculations of kriging estimates. It can be argued that selection of the number of hidden layers and the number of neurons is a trial-and-error process that has similar disadvantages to ordinary kriging. This is true in cases where the modeler has to select the number of neurons and the hidden layers by a trial-and-error approach and not by an automatic *best neural network* search method. The computational time in the case of UOK depends on the architecture of the network and the training parameters.

In the case of ordinary kriging, several semi-variograms should be experimented and checked using the cross-validation process before the best semi-variogram can be selected for a spatial interpolation process. The evaluation of the semi-variogram model also involves selection of appropriate parameters that ultimately define the nature of the function. In the case of UOK or UOKC, a split-sample method that divides the variogram cloud data into two sets (training and testing sets) is used. Therefore in UOK or UOKC, the selection of the semi-variogram model is not based on quantification of performance measures that indicate how well the interpolation performed, but how well the variogram model can explain the spatial variability of the data. Use of a variogram cloud for function approximation using ANN can be justified when the number of data points is limited. The number of training data used from a variogram cloud for the development of a variogram model using ANN affects the performance of kriging. Also, the training data selected should be representative of variances at several separation distances and not confined to a specific distance range.

Practicing hydrologists and modelers involved in interpolation of data or estimation of missing precipitation data for a specific location are faced with two options: (1) use of an authorized semi-variogram model in ordinary kriging or (2) use of a universal function approximation-based model for the semi-variogram. Adoption of any one of these options mainly depends on the data availability, location of spatial data, and an appropriate functional form to generalize the variability in the spatial data. Hydrologists and modelers should not lose focus of the underlying assumptions of kriging and conditions relevant to minimum positive variance guaranteed by it. It is easy to assume a universal function approximator such as ANN within kriging as a quick panacea.

Kriging has been successfully applied in the past to spatial interpolation of geophysical variables. The success is partly attributed to the fact that temporal variability of these variables is often assumed to be negligible. Also, kriging mainly originated from the need for characterizing and modeling the spatial variability of geological processes. In the case of hydrologic variables such as precipitation, the interpolation techniques have to account for spatial and temporal variability and also take into consideration the effect of secondary variables (e.g., elevation). Temporal variability of precipitation data in space may warrant the development of time-dependent or seasonal semi-variograms for spatial interpolation or estimation of missing data at a specific point in space.

3.24 CLASSIFICATION METHODS

Classification methods for rain or no rain conditions can be used in two ways. First, they can be used as stand-alone estimators, which are only used to estimate the occurrence or non-occurrence of rainfall in a given time interval at a specific rain gage. This

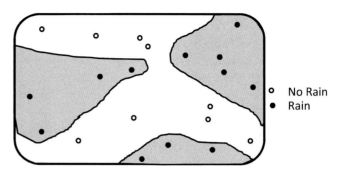

Figure 3.32 Classification of rain and no rain conditions using the SVM approach.

process produces a binary estimate identifying the state as "rain" or "no rain" or zero/one. Classification methods can also be used in conjunction with any other estimation method as a post- or pre-processor, in which the estimation method is used to produce rainfall estimates only at gages identified as having "rain." In essence, the classification method is used to draw the boundaries of the rain area(s) for the temporal frame in question. Then the estimation method is only needed to determine the expected precipitation amount for gages within areas declared as having "rain." Two classification methods can be adopted, and they are support vector machines and a single best classifier. These methods are summarized in the following sections. Radar data, if available, can be used for classification for rain or no rain conditions.

3.24.1 Classification using support vector machines

Support vector machines (SVMs) are modeling tools that can be applied to both regression and classification tools (Torgo, 2011). SVMs initially map original data sets into a new, high-dimensional space. Using the mapped data, linear models are used to separate the data into different classes using a hyper-plane. The mapping of the original data into a new space is carried out using kernel-based functions (Torgo, 2011; Shawe-Taylor and Cristianini, 2000). SVMs can be applied to identify a non-linear boundary between gages using only the data from gages with recorded rainfall in the temporal frame in question. The identified boundary maximizes the separation distance between the boundary and all available rain gages. Typical boundaries identified by the SVM method are shown in Figure 3.32.

3.24.2 Classification using single best classifier

A classifier seeks to declare a given gage as having rain or no rain in a particular time interval. In its simplest form, a classifier declares that the station does or does not have rain depending on whether its highest correlated neighbor does or does not have rain in a given time interval. The "highest correlated neighbor" refers to the neighbor having an observation in a given time frame, as selected from the rank-ordered list of the N highest correlated neighbors, where N defines the search radius for the algorithm. If N is infinite, the classifier will never fail as long as one gage has a recorded rainfall value in a given time interval of interest. If N is smaller than the number of rain gages, the classifier can fail. When this occurs, another method must be used to estimate daily precipitation at the gage of interest. It should be noted that the SVM classifier described above does not employ a search radius and, therefore, implicitly defines the search radius to be infinite.

3.24.3 Support vector machine–logistic regression-based copula method

The support vector machine–logistic regression-based copula (SVMLRC) method, which uses the concept of SVM, was developed by Landot *et al.* (2008). The method essentially involves a two-step procedure. In the first step, the probability of a rain/no rain condition is forecast at each location. During the second step, the expected rainfall amount is estimated if rain is indicated. Landot *et al.* (2008) considered several ways to estimate the rainfall amount. One is to use an SVM concept to classify days into rain/no rain. Once the classification is performed, a scheme based on bivariate density estimation using copulas, followed by a best linear unbiased combination of pairwise rainfall estimates determines the missing precipitation value. Additional details about this method can be found in Landot *et al.* (2008). Recent evaluation of this method by Aly *et al.* (2009) indicated inferior performance of this method compared to several deterministic methods discussed in this chapter.

3.25 DISTANCE METRICS AS PROXIMITY MEASURES

Moving away from the concepts of using Euclidean distances as surrogates for spatial correlations, numerical distances or distance metrics strictly based on observations can be developed. Ahrens (2006) identified one such distance, referred to as statistical distance, based on common variance of precipitation time series and reported improvements in estimation compared to geographical distance-based methods. The following statement serves as a motivation to develop new interpolation methods based on distance metrics. If a rainfall surface or field estimation is not of interest and only estimation of missing data at one single location in a region is essential, methods that rely directly on observed data can be well suited for interpolation schemes. Proximity of two observations based on real or categorical attributes can be calculated using numerical distances that are functions of these observations (Myatt and Johnson, 2009). Identification of nearest neighbors in K-NN (K-nearest neighbor) classification and

clusters of observations in space and time using K-means clustering techniques requires definition of a distance measure. This distance metric is used as a measure of proximity between any two observed data sets and in turn can be used to assign weights in the interpolation schemes to estimate missing data in the current study. Spatio-temporal precipitation data classified based on nearest neighbor schemes and clustered sets of rain gages can be used in optimal interpolation schemes for estimation of missing precipitation data. The methodologies (Teegavarapu, 2012b) using proximity metrics, K-NN, and K-means clustering for estimation of missing precipitation data are explained in the next few sections.

Distance measures based on observations at two rain gage stations, β and α (θ_β and θ_α), can be defined as real-valued functions. The functions are referred to as distance metrics if they satisfy several conditions given by the following inequalities:

$$d_{\beta,\alpha} \gg 0 \tag{3.89}$$

$$d_{\beta,\alpha} = 0 \text{ if and only if } \theta_{\beta,n} = \theta_{\alpha,n} \quad \forall n \tag{3.90}$$

$$d_{\beta,\alpha} = d_{\alpha,\beta} \tag{3.91}$$

$$d_{\beta,\omega} \ll d_{\beta,\alpha} + d_{\alpha,\omega} \tag{3.92}$$

Inequality 3.89 indicates that distance measure is always nonnegative. Equation 3.90 indicates that distance measure is equal to zero if and only if all the observations at station β are exactly equal to observations at station α. The variable n is the observation number. Distance measures between any two stations are equal, showing the property of commutativity. Inequality 3.92 indicates the property referred to as triangular inequality defined based on distances between stations β and ω, β and α, and α and ω.

3.26 DISTANCE METRICS FOR PRECIPITATION DATA

Distance metrics refer to those measures that satisfy conditions defined by inequalities and equalities 3.89–3.92. The distance metrics developed for precipitation data used in the current study are defined in the following sections. Distance metrics are calculated using pairs of observations at any two rain gages. One of the rain gages always used in the calculations is referred to as the base station at which missing precipitation data are known to exist. These metrics are calculated based on historical precipitation data available at all stations including the base station.

3.26.1 Euclidean

The Euclidean distance (Tan *et al.*, 2006; Myatt and Johnson, 2009) or L_2 norm is the simplest measure of distance between a pair of rain gage observations. This measure, represented by

Equation 3.93, is used in many data mining applications for proximity calculations and in classification and clustering schemes.

$$d_{\beta,\alpha} = \sum_{n=1}^{no} \sqrt[2]{(\theta_{\beta,n} - \theta_{\alpha,n})^2} \qquad (3.93)$$

3.26.2 Squared Euclidean

The squared Euclidean is a minor variant of the Euclidean distance. The square Euclidean (Myatt and Johnson, 2009) is the sum of the squares of the difference between the two rain gage observations, and it is given by Equation 3.94. This metric magnifies distances between observations that are further apart.

$$d_{\beta,\alpha} = \sum_{n=1}^{no} (\theta_{\beta,n} - \theta_{\alpha,n})^2 \qquad (3.94)$$

3.26.3 Manhattan

The Manhattan distance (Krause, 1987) is also referred to as the city block or taxicab distance or L_1 norm (Tan et al., 2006; Myatt and Johnson, 2009). This distance is less affected by outliers compared to Euclidean and squared Euclidean (Fielding, 2007).

$$d_{\beta,\alpha} = \sum_{n=1}^{no} |\theta_{\beta,n} - \theta_{\alpha,n}| \qquad (3.95)$$

3.26.4 Maximum

Maximum distance (Tan et al., 2006; Myatt and Johnson, 2009) is the maximum distance between two observations, the absolute difference between each variable is determined and the highest difference given by:

$$d_{\beta,\alpha} = \max |\theta_{\beta,n} - \theta_{\alpha,n}| \quad \forall n \qquad (3.96)$$

3.26.5 Minkowski

The Minkowski distance (Basilevsky, 1983) is a generalized distance measure given by Equation 3.97. The value of λ defines a specific distance. When λ is equal to 1, the measure is Manhattan and when λ is equal to 2, the measure becomes the Euclidean distance or L_2 norm.

$$d_{\beta,\alpha} = \sqrt[\lambda]{\sum_{n=1}^{no} |\theta_{\beta,n} - \theta_{\alpha,n}|^\lambda} \qquad (3.97)$$

3.26.6 Gower

The Gower distance (Gower and Legendre, 1986) is used for mixed variables (continuous and discrete) and is given by Equation 3.98. The variable ω_n is the weight, which is equal to one when both observations in a given time interval, n, are

available and zero when one of them is not available. The variable τ_n is the maximum range value based on the observations.

$$d_{\beta,\alpha} = \sqrt{\frac{\sum_{n=1}^{no} \omega_n d_n^2}{\sum_{n=1}^{no} \omega_n}} \qquad (3.98)$$

$$d_n = \frac{|\theta_{\beta,n} - \theta_{\alpha,n}|}{\tau_n} \quad \forall n \qquad (3.99)$$

3.26.7 Cosine

The cosine distance (Basilevsky, 1983; Tan et al., 2006) is based on a cosine similarity measure that measures similarity between sets of observations. The cosine similarity is a measure of the (cosine of the) angle between two vectors. The distance is calculated by:

$$d_{\beta,\alpha} = 1 - \frac{\sum_{n=1}^{no} \theta_{\beta,n} \theta_{\alpha,n}}{\sqrt{\sum_{n=1}^{no} \theta_{\beta,n}^2 \sum_{n=1}^{no} \theta_{\alpha,n}^2}} \qquad (3.100)$$

3.26.8 Canberra

The Canberra distance (Lance and Williams, 1966) given by Equation 3.101 defines the sum of the fractional differences for each variable (Myatt and Johnson, 2009).

$$d_{\beta,\alpha} = \sum_{n=1}^{no} \frac{|\theta_{\beta,n} - \theta_{\alpha,n}|}{(|\theta_{\beta,n}| + |\theta_{\alpha,n}|)} \qquad (3.101)$$

3.26.9 Correlation distance

The correlation distance is conceptually based on similarity between observations as defined by the correlation coefficient or a measure of the linear relationship between two data sets. The correlation distance is given by:

$$d_{\beta,\alpha} = 1 - \rho_{\beta,\alpha} \qquad (3.102)$$

3.26.10 Mahalanobis

The Mahalanobis distance (Myatt and Johnson, 2009) takes into account correlations within a data set between the variables. The distance is scale independent and is calculated using Equation 3.103. The calculated distance requires covariance (S) and transformed matrices using the data sets (Tan et al., 2006).

$$d_{\beta,\alpha} = \sqrt{(\boldsymbol{\theta}_\beta - \boldsymbol{\theta}_\alpha) S^{-1} (\boldsymbol{\theta}_\beta - \boldsymbol{\theta}_\alpha)^T} \qquad (3.103)$$

The variables $\boldsymbol{\theta}_\beta$ and $\boldsymbol{\theta}_\alpha$ are vectors and not individual observations (i.e., single observation at each rain gage) and S^{-1} is the inverse of the covariance matrix of the data.

3.26.11 Cosine and correlation similarity measures

Cosine and correlation distances are based on similarity measures discussed previously in Sections 3.26.7 and 3.26.9 respectively.

Table 3.15 *Example data set for calculation of correlation and cosine distances*

X	Y	ZX	ZY
1	1	− 0.1762	0.3644
2	0	0.4111	− 0.8503
0	0	− 0.7634	− 0.8503
5	2	2.1727	1.5791
3	1	0.9983	0.3644
0	0	− 0.7634	− 0.8503
0	0	− 0.7634	− 0.8503
2	1	0.4111	0.3644
0	0	− 0.7634	− 0.8503
0	2	− 0.7634	1.5791

The measures and therefore distances are the same when standardized z-scores are used to calculate cosine and actual data for correlation similarity respectively. An example using the data set provided in Table 3.15 is used to calculate the correlation and cosine distances with original data and standardized z-scores of the data.

The mean and standard deviation values for the series X and Y are 1.30 and 0.7, and 1.703 and 0.823, respectively. The correlation and cosine distances based on original data are 0.453 and 0.264 respectively. The cosine distance calculated based on the transformed values (i.e., z-scores) is the same as the correlation distance. The cosine distance can be defined in an R^n space, where each direction corresponds to an observation and, therefore, the two sets of observations are two vectors in this space. If the two sets of observations are further centered by their respective means and scaled by their respective standard deviations, then the two vectors start from the origin. When their scalar product is estimated and translated to the cosine of their angle, the numerical value obtained is same as the correlation coefficient (or correlation similarity measure). In estimation of metrics, original data are used and therefore the use of the two different distance measures will result in different estimates.

3.27 BOOLEAN DISTANCE MEASURES FOR PRECIPITATION DATA

Boolean distance measures are applied to observations that are binary variables (i.e., 0 and 1) and also to categorical variables when these variables are expressed as binary variables. Similarity measures indicate how alike two observations are to each other, with high similarity values representing situations when the two observations are alike (Myatt and Johnson, 2009). Similarity measure differs from distance measure, where low magnitudes of the latter measure indicate that observations from the two series are alike. The precipitation data sets from any two stations contain real values with zero and positive values above zero.

The real-valued precipitation time series at any station can be converted into binary form using the following expression:

$$\text{if } \left(\theta_{j,n} > \theta_{th}\right), \text{ then } \theta_{j,n}^* = 1, \text{ else } \theta_{j,n}^* = 0 \quad \forall j, n \quad (3.104)$$

The variable $\theta_{j,n}$ is the observed precipitation value at station j, in time interval n, and θ_{th} is the threshold value (lower limit) that defines the limit for the assignment of binary values. The process of transforming continuous variables into binary variables is called binarization. The similarity and distance calculations for binary variables are based on the number of common and different values in the four situations as described in Figure 3.33. The count (number) for these specific conditions is obtained from the historical data using all the rain gage stations as well as the station (i.e., base station) with the missing precipitation data. The common and different values are evaluated, and counts (δ, γ, ϕ, and ψ) required for distance metrics are calculated for each time interval, n, using conditions expressed by 3.105–3.108. The variable $\theta_{\beta,n}^*$ is the transformed observed precipitation data (i.e., binary value) at station β and $\theta_{\alpha,n}^*$ is the similar transformed value at the base station (i.e., station with missing precipitation data with index $m = \alpha$) for the time interval n.

$$\text{if } \left(\theta_{\beta,n}^* = 1, \theta_{\alpha,n}^* = 1\right) \text{ then } \delta_n = 1, \beta \in j \quad \forall \beta, n \quad (3.105)$$

$$\text{if } \left(\theta_{\beta,n}^* = 1, \theta_{\alpha,n}^* = 0\right), \text{ then } \gamma_n = 1, \beta \in j \quad \forall \beta, n \quad (3.106)$$

$$\text{if } \left(\theta_{\beta,n}^* = 0, \theta_{\alpha,n}^* = 1\right), \text{ then } \phi_n = 1, \beta \in j \quad \forall \beta, n \quad (3.107)$$

$$\text{if } \left(\theta_{\beta,n}^* = 0, \theta_{\alpha,n}^* = 0\right), \text{ then } \psi_n = 1, \beta \in j \quad \forall \beta, n \quad (3.108)$$

The aggregated values of these counts ($\delta_{\beta,\alpha}$, $\gamma_{\beta,\alpha}$, $\phi_{\beta,\alpha}$, and $\psi_{\beta,\alpha}$) are calculated using Equations 3.105–3.108 with all the available observations, no:

$$\delta_{\beta,\alpha} = \sum_{n=1}^{no} \delta_n \quad (3.109)$$

$$\gamma_{\beta,\alpha} = \sum_{n=1}^{no} \gamma_n \quad (3.110)$$

$$\phi_{\beta,\alpha} = \sum_{n=1}^{no} \phi_n \quad (3.111)$$

$$\psi_{\beta,\alpha} = \sum_{n=1}^{no} \psi_n \quad (3.112)$$

The values of counts given by Equations 3.109–3.112 are used to define binary distance metrics. These metrics are defined in the next few sections.

3.27.1 Simple matching

The simple matching metric (Myatt and Johnson, 2009) calculates the similarity measure linked to the total number of times any two

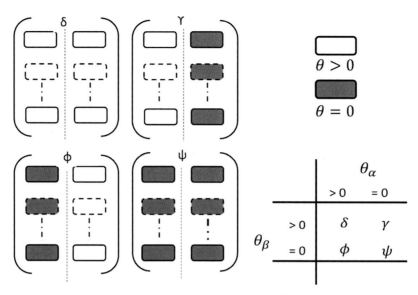

Figure 3.33 Binary transformations of precipitation data at two stations and the corresponding similarity measures and counts.

rain gage stations have conditions 3.109 and 3.110 satisfied. The distance is given by subtracting this similarity coefficient from 1:

$$d_{\beta,\alpha} = 1 - \frac{\delta_{\beta,\alpha} + \gamma_{\beta,\alpha}}{\delta_{\beta,\alpha} + \gamma_{\beta,\alpha} + \phi_{\beta,\alpha} + \psi_{\beta,\alpha}} \quad (3.113)$$

3.27.2 Jaccard

The Jaccard distance (Jaccard, 1908) calculates the total number of times any two stations have conditions 3.110 and 3.111 satisfied. The lower the number of dissimilar values (i.e., binary values), the closer are the time series observations as indicated by Equation 3.114. In this metric, joint absences are excluded from consideration and equal weight is given to matches and mismatches.

$$d_{\beta,\alpha} = \frac{\gamma_{\beta,\alpha} + \phi_{\beta,\alpha}}{\delta_{\beta,\alpha} + \gamma_{\beta,\alpha} + \phi_{\beta,\alpha}} \quad (3.114)$$

3.27.3 Rusell and Rao

The Russell and Rao distance (Russel and Rao, 1940) calculates a similarity coefficient that relates to the total number of times any two stations have condition 3.109 satisfied compared to all conditions. The similarity coefficient is then transformed to a distance measure as given by:

$$d_{\beta,\alpha} = 1 - \frac{\delta_{\beta,\alpha}}{\delta_{\beta,\alpha} + \gamma_{\beta,\alpha} + \phi_{\beta,\alpha} + \psi_{\beta,\alpha}} \quad (3.115)$$

3.27.4 Dice

The Dice distance (Dice, 1945; Sorensen, 1948) uses a similarity index that accounts for common values (δ, γ) between two observation time series. The distance, as calculated by Equation 3.116,

gives more importance to common ones. The Dice distance is similar to the Jaccard distance and was developed to measure ecological association between species. This is a measure in which joint absences are excluded from consideration and matches are doubled (Fielding, 2007). The measure is also known as the Czekanowski measure.

$$d_{\beta,\alpha} = \frac{\gamma_{\beta,\alpha} + \phi_{\beta,\alpha}}{2\delta_{\beta,\alpha} + \gamma_{\beta,\alpha} + \phi_{\beta,\alpha}} \quad (3.116)$$

3.27.5 Rogers and Tanimoto

The Rogers and Tanimoto distance (Rogers and Tanimoto, 1960) gives importance to common ones and zeros. The distance metric was mainly developed for classifying plant species. The distance measure is given by:

$$d_{\beta,\alpha} = \frac{2(\gamma_{\beta,\alpha} + \phi_{\beta,\alpha})}{\delta_{\beta,\alpha} + 2\left(\gamma_{\beta,\alpha} + \phi_{\beta,\alpha}\right) + \psi_{\beta,\alpha}} \quad (3.117)$$

3.27.6 Pearson

The Pearson distance (Ellis *et al.*, 1993), as defined by Equation 3.118, was initially used to measure the degree of similarity between objects in text retrieval systems.

$$d_{\beta,\alpha} = \frac{1}{2}$$
$$- \frac{\delta_{\beta,\alpha}\psi_{\beta,\alpha} + \phi_{\beta,\alpha}\gamma_{\beta,\alpha}}{2\sqrt{(\delta_{\beta,\alpha} + \gamma_{\beta,\alpha})(\delta_{\beta,\alpha} + \phi_{\beta,\alpha})(\gamma_{\beta,\alpha} + \psi_{\beta,\alpha})(\phi_{\beta,\alpha} + \psi_{\beta,\alpha})}}$$
(3.118)

3.27.7 Yule

The Yule distance (Yule, 1912) was initially developed as a method for identifying association between two attributes. The distance is given by:

$$d_{\beta,\alpha} = \frac{\gamma_{\beta,\alpha}\phi_{\beta,\alpha}}{\delta_{\beta,\alpha}\psi_{\beta,\alpha} + \gamma_{\beta,\alpha}\phi_{\beta,\alpha}}$$
(3.119)

3.27.8 Sokal–Michener

The Sokal–Michener distance (Sokal and Michener, 1958; Sneath and Sokal, 1962) is one of the distance metrics used in numerical taxonomy for classifying organisms and building evolutionary trees. The distance is defined by:

$$d_{\beta,\alpha} = \frac{\delta_{\beta,\alpha} + \psi_{\beta,\alpha}}{\delta_{\beta,\alpha} + \gamma_{\beta,\alpha} + \phi_{\beta,\alpha} + \psi_{\beta,\alpha}}$$
(3.120)

3.27.9 Kulzinksy

The Kulzinsky distance (Holliday *et al.*, 2002), as defined by Equation 3.121, is used in ecology for finding similarity in sites with similar species and also in calculation of intermolecular similarity and dissimilarity.

$$d_{\beta,\alpha} = \frac{2\gamma_{\beta,\alpha} + 2\phi_{\beta,\alpha} + \psi_{\beta,\alpha}}{\delta_{\beta,\alpha} + 2\gamma_{\beta,\alpha} + 2\phi_{\beta,\alpha} + \psi_{\beta,\alpha}}$$
(3.121)

3.27.10 Hamming

The Hamming distance (Hamming, 1950), as given by Equation 3.122, was originally developed to identify and correct errors in digital communication systems.

$$d_{\beta,\alpha} = \gamma_{\beta,\alpha} + \phi_{\beta,\alpha}$$
(3.122)

3.28 OPTIMAL EXPONENT WEIGHTING OF PROXIMITY MEASURES

The proximity measures are used to develop an optimal weighting method for which the objective function and the constraints are specified by Equations 3.123–3.125. This formulation helps in obtaining an optimal exponent for the reciprocal-distance measure.

Minimize

$$\sum_{n=1}^{no} \left| \theta'_{\alpha,n} - \theta_{\alpha,n} \right|$$
(3.123)

subject to:

$$\theta'_{\alpha,n} = \frac{\sum_{j=1}^{ns-1} w_{\alpha,j}^k \theta_{j,n}}{\sum_{j=1}^{ns-1} w_{\alpha,j}^k} \quad \forall nj \notin \alpha$$
(3.124)

$$w_{\alpha,j} = \frac{1}{d_{j,\alpha}} j \notin \alpha, \ j \in ns-1$$
(3.125)

The formulation is solved using historical spatio-temporal precipitation data to minimize the objective function (Equation 3.123) and obtain an optimal exponent value (k) using a non-linear optimization solver.

3.29 OPTIMAL K-NEAREST NEIGHBOR CLASSIFICATION METHOD

The optimal K-nearest neighbor (K-NN) classification method combines classification and an optimal weighting scheme. The objective is to see if classification of precipitation data into several groups to develop separate (i.e., local) optimization models yields better results compared to those from a global model. Arbitrary definitions of classes have yielded inferior results in a previous study (Teegavarapu, 2012a). Initially, the K-NN method is used as a classifier to group spatial and temporal precipitation data into several pre-defined classes. These defined classes are set as a training data set. This data set contains training tuples (Han and Kamber, 2006; Tan *et al.*, 2006) and the classification is achieved based on learning by analogy, by comparing a given test tuple with training tuples that are similar to it. Each tuple indicates a point in an n-dimensional space. In this way, all of the training tuples are stored in an n-dimensional pattern space. When given an unknown tuple, a K-NN classifier searches the pattern space for the k training tuples that are closest to the unknown tuple. These k training tuples are the k "nearest neighbors" of the unknown tuple (Fielding, 2007; Han and Kamber, 2006). Closeness is defined in terms of any distance metric (e.g., Euclidean distance). The distance between two points or tuples, say, $X = (\theta_{1,1}, \theta_{1,2}, \ldots, \theta_{1,ns-1})$ and $X_{c=1} = (\theta_{1,1}^\circ, \theta_{1,2}^\circ, \ldots, \theta_{1,ns-1}^\circ)$ is a function given by Equation 3.128.

$$X = \begin{pmatrix} \theta_{1,1} & \cdots & \theta_{1,ns-1} \\ \vdots & \ddots & \vdots \\ \theta_{no,1} & \cdots & \theta_{no,ns-1} \end{pmatrix}$$
(3.126)

$$X_c = \begin{pmatrix} \theta_{1,1}^\circ & \cdots & \theta_{1,ns-1}^\circ \\ \vdots & \ddots & \vdots \\ \theta_{c,1}^\circ & \cdots & \theta_{c,ns-1}^\circ \end{pmatrix} \qquad (3.127)$$

X represents the $n \times ns - 1$ matrix of observed precipitation values at $ns - 1$ stations for no time intervals and X_c represents $c \times ns - 1$ pre-defined classes of precipitation data. The historical observations at $ns - 1$ stations are evaluated for their proximity to each of the classes defined in X_c and each observation is designated to a specific class, c, using a distance metric, D.

$$D = f(X, X_c) \qquad (3.128)$$

Once the observations are designated to specific classes, then correlation between the observations in a specific class, c, and observations at stations at which missing data exist are obtained. The variable nc is the index for observations that belong to class c, and tnc is the total number of observations that belongs to the class c.

$\rho_{\alpha,j,c}$

$$= \frac{\sum_{nc=1}^{tnc} \theta_{nc,j}\theta_{\alpha,nc} - \sum_{nc=1}^{tnc} \theta_{nc,j} \sum_{nc=1}^{tnc} \theta_{\alpha,nc}}{\sqrt{\sum_{nc=1}^{tnc} \theta_{nc,j}^2 - \left(\sum_{nc=1}^{tnc} \theta_{nc,j}\right)^2} \sqrt{\sum_{nc=1}^{tnc} \theta_{\alpha,nc}^2 - \left(\sum_{nc=1}^{tnc} \theta_{\alpha,nc}\right)^2}} \quad \forall j, \forall c$$

$$\qquad (3.129)$$

The correlations obtained from Equation 3.128 are then used in a CCWM (Teegavarapu and Chandramouli, 2005) for estimation of missing precipitation data at station α, and given by:

$$\theta_{\alpha,n}' = \frac{\sum_{j=1}^{ns-1} \rho_{\alpha,j,c}\theta_{j,nc}}{\sum_{j=1}^{ns-1} \rho_{\alpha,j,c}} \qquad \forall nc \in n, j \notin \alpha \qquad (3.130)$$

Alternatively, optimal weights for each class can also be obtained by solving an optimization formulation given by the following objective function and constraint. The formulation is solved for each class, c.

Minimize

$$\| A.x - B \|_2^2 \qquad (3.131)$$

subject to:

$$\phi_{\alpha,j,c} \geq 0 \quad \forall j, \forall c \qquad (3.132)$$

A is the $nc \times j$ matrix of $\theta_{c,j}$ values; x is the matrix $j \times 1$ of $\phi_{\alpha,j,c}$ weight values; and B is the matrix of $nc \times 1$ values of data at the base station ($\theta_{\alpha,c}$) belonging to a specific cluster, c. The formulation minimizes the norm given by Equation 3.131 with constraint on the weights ($\phi_{\alpha,j,c}$) in inequality 3.132. This formulation provides non-negative optimal coefficients when solved. The missing precipitation data are estimated by replacing correlation coefficients ($\rho_{\alpha,j,c}$) by weights ($\phi_{\alpha,j,c}$) in Equation 3.132.

3.30 OPTIMAL K-MEANS CLUSTERING METHOD

Clustering of observations is quite common in spatial ecology (Dungan et al., 2002; Fortin and Dale, 2005), geosciences, and data mining applications (Jain and Dubes, 1988). In the current study, clustering was used to create one "virtual" station out of a number of stations (i.e., cluster) where only one weight is attached to this station (i.e., group of stations). Grouping (clustering) of stations can be achieved by K-means clustering algorithms (Johnson, 1996; Larose, 2005; Han and Kamber, 2006; Tan et al., 2006), where proximity (distance-based) metrics are used to define the clusters and the members belonging to each cluster. In a previous study by Teegavarapu (2012), clustering was achieved by optimization using binary variable–mixed integer non-linear programming formulations. The K-means clustering method can be used to identify spatial clusters of rain gage stations from the network of stations in a region. Initially, the method is used to obtain a specific number (i.e., k) of spatial clusters of precipitation stations. The steps involved in the K-means clustering method are (1) initial spatial partition of precipitation stations into k random clusters; (2) re-partition of the stations by assigning each station to the nearest center of a cluster by using a proximity measure (e.g., distance metric); and (3) re-calculation of the cluster centers as centroids. Steps two and three are repeated until a stopping criterion of no changes in centroid positions is reached. Once the clusters are identified, weights are assigned to these clusters for interpolation based on two procedures discussed in the next section.

3.30.1 Optimal weights for selected clusters

The K-means clustering method is used to identify the spatial clusters of rain gages, and weights are assigned to these clusters using an optimization formulation. Non-negative constraint requirements to obtain positive weights are enforced using the non-linear least squares constraint formulation defined by Equation 3.133.

Minimize

$$\| E.x - F \|_2^2 \qquad (3.133)$$

subject to:

$$\xi_{\alpha,cl} \geq 0 \quad \forall cl \qquad (3.134)$$

E is the $no \times cl$ matrix of $\theta_{cl,n}$ values; x is the matrix $cl \times 1$ of $\xi_{\alpha,cl}$ weight values; and F is the matrix of $no \times 1$ values of data at base station ($\theta_{\alpha,n}$). The formulation minimizes the norm given by Equation 3.133 with constraint on the weights (inequality 3.134).

This formulation provides non-negative optimal coefficients when solved. The variable m_{cl} is the number of stations in a cluster cl.

$$\theta_{cl,n} = \sum_{i=1}^{m_{cl}} \theta_{i,n} \quad \forall cl, \forall n \tag{3.135}$$

The variable $\theta_{cl,n}$ is the sum of all the observations from m_{cl} in a cluster cl. The variable $\theta_{k,n}$ is the observation at station k belonging to a specific cluster of stations with maximum correlation with the station missing data.

$$\theta_{cl,n} = \theta_{k,n} k \in m_{cl} \quad \forall cl, \forall n \tag{3.136}$$

The estimation of missing precipitation data is given by Equation 3.137 using the weights obtained by solution of the optimization formulation given by Equations 3.133–3.134. The value of $\theta_{cl,n}$ can be obtained by either Equation 3.135 or 3.136. The variable Ncl is the number of clusters.

$$\theta'_{\alpha,n} = \sum_{cl=1}^{Ncl} \theta_{cl,n} \xi_{\alpha,cl} \quad \forall n \tag{3.137}$$

3.31 PROXIMITY MEASURES: LIMITATIONS

The proximity measures calculated using distance metrics depend on the length of the historical data. As the metrics are used as weights in the imputation of missing precipitation data, the longer the length of historical data, the more robust are the metrics. The threshold value used for the assignment of binary metrics for historical data affects the performance of the methods. Conversion of real-valued precipitation data to binary values by defining a threshold value is generally recommended. The threshold value can be varied and the performance of the metrics can be tested before the methods using these metrics can be applied for estimation of missing data. A threshold value of zero is ideal for precipitation data as the value in essence suggests rain or no rain conditions. These conditions can help establish the proximity measures easily in interpolation schemes. The estimation of missing precipitation data from real-valued and binary distance metrics can be improved using data that are representative of specific storm types (frontal or convective) or a season and a fixed temporal window (i.e., month) thus building several local models as opposed to one global model. The proximity measure-based optimization methods are data intensive and sensitive.

3.32 USE OF RADAR DATA FOR INFILLING PRECIPITATION DATA

Radar data can be used for infilling of missing rainfall data. Several issues that affect the infilling methods include the historical rain gage and radar data, spatial and temporal variability of rainfall, radar–rain gage relationships, and selection of spatial extent of radar data. Before embarking on data filling using radar, one should evaluate the influence of spatial and temporal variability of rainfall processes on the performance of spatial interpolation algorithms. Seasonal variation of rainfall, rainfall areas that are delineated based on physical processes affecting the genesis and morphology of rainfall processes, and other factors may affect the performance of infilling methods.

There is a need to understand and quantify the relationship between the rain gage and precipitation estimates based on other forms of measurements (e.g., radar) and precipitation measurements at other observation points. The quantification of this relationship is essential to check the quality of rainfall data. This task is made possible by evaluation of radar and rain gage measurements independently and then against each other. The importance of such a check is confirmed by a state-of-the-art review of radar rainfall data use in urban drainage (Einfalt et al., 2004). Similarly, functional relationships can be developed relating precipitation measurements at two or more observation stations. This section describes the development of two types of functional forms (Teegavarapu et al., 2008): (1) fixed function form with pre-defined functional expressions; (2) flexible function form–binary formulations. Function-free forms (i.e., universal function approximators) were also tested by Teegavarapu et al. (2008) for linking radar and rainfall observations. It is important to note that the formulations discussed in the next section use radar as predictand and rain gage as predictor.

3.32.1 Rain gage–radar functional relationships

Mathematical programming models using linear and non-linear formulations can be developed to link rain gage and radar-based precipitation data. The formulations are aimed at obtaining optimal exponents of a power model and also obtaining optimal functional forms using binary variables. The models along with constraints are discussed next.

MODEL I
Minimize

$$\sqrt{\frac{1}{no} \sum_{i=1}^{no} \left(\hat{\phi}_i^m - \phi_i^m \right)^2} \tag{3.138}$$

$$\hat{\phi}_i^m = f\left(\theta_i^m, a, b \right) \quad \forall i \tag{3.139}$$

$$k_l \le a \le k_u \tag{3.140}$$

$$c_l \le b \le c_u \tag{3.141}$$

where, a and b are exponents, k and c refer to variable upper and lower bounds on the coefficients; $\hat{\phi}_i^m$ is the estimate of radar (NEXRAD) data values; ϕ_i^m is the NEXRAD data value for the

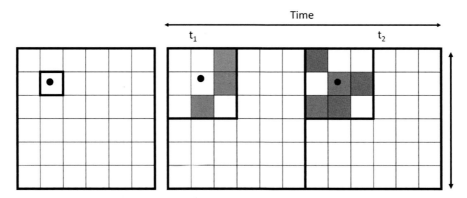

Figure 3.34 (Left) Collocation of rain gage and pixel for estimation of missing data. (Right) Variation of active weights for pixels surrounding a gage.

grid (e.g., 2 km × 2 km grid or from any resolution); θ_i^m is the rainfall value observed on a given day; and *no* is the number of observations used for obtaining optimal coefficients. The functional form (3.139) represents a power model, $\hat{\phi}_i^m = a(\theta_i^m)^b$, similar to the one used by Teegavarapu *et al.* (2008).

MODEL II

The formulation for model II is similar to model I with a change in the objective function used for optimization.

Minimize

$$\frac{\sum_{i=1}^{no} \left| \hat{\phi}_i^m - \phi_i^m \right|}{no} \qquad (3.142)$$

MODEL III

The formulation for model III uses the same objective function used in model I with modifications to constraints and the nature of function selection. The selection is possible by use of binary variables.

Minimize

$$\sqrt{\frac{1}{no} \sum_{i=1}^{no} \left(\hat{\phi}_i^m - \phi_i^m \right)^2} \qquad (3.143)$$

$$\hat{\phi}_i^m = f_1\left(\theta_i^m, a_1, b_1\right) y_1 + f_2\left(\theta_i^m, a_2, b_2\right) y_2$$
$$+ \cdots + f_k\left(\theta_i^m, a_k, b_k\right) y_k \quad \forall i \qquad (3.144)$$

$$\sum_{j=1}^{k} y_k = 1 \qquad (3.145)$$

$$k_{li}^k \le a_k \le k_{ui}^k \quad \forall i \qquad (3.146)$$

$$c_{li}^k \le b_k \le c_{ui}^k \quad \forall i \qquad (3.147)$$

where a_k and b_k are exponents or constants,; k and c refer to variable upper and lower bounds on the coefficients or constants; $\hat{\phi}_i^m$ is the estimate of radar data values; ϕ_i^m is the radar data value for a specific grid size or tessellation (e.g., 2 km × 2 km grid); θ_i^m is the rainfall value observed on a given day; *no* is the number of observations used for obtaining optimal coefficients; and $y_1, \ldots,$

y_k are the binary variables. Equation 3.145 ensures that only one functional form is selected.

3.33 GEOGRAPHICALLY WEIGHTED OPTIMIZATION

Geographically weighted optimization (GWO) (Teegavarapu *et al.*, 2010) derives motivation from the concept of geographically weighted regression (Fotheringam *et al.*, 2002). In the case of geographically weighted optimization, a moving window (fixed number of pixels or grids) is adopted for obtaining optimal weights in a scheme to interpolate missing values at a gage centered in the pixel that is the center of the window. The radar-based (e.g., NEXRAD in the USA) rainfall estimates that are available in a grid format surrounding the grid (cell) in which the rain gage is located are used in the model. The model will identify the cluster of radar data values that can be used for infilling the rain gage records. Spatial and temporal variability of weights for different clusters needs to be investigated. Figure 3.34(left) shows the collocation of rain gage and the radar pixel and Figure 3.34(right) shows the pixel grids that are active (i.e., having positive or non-zero weights that contribute to the weighting scheme) in two different time intervals. The weights can be obtained using a linear or non-linear optimization formulation. Euclidean distances in GWO, if utilized, are based on distances calculated between the centroids of the radar grids and the rain gages in the optimization schemes.

The functional relationships between rain gage and radar-based precipitation data along with cluster-based radar grids are shown in Figure 3.35.

Teegavarapu *et al.* (2010) document several variants of weighted interpolation schemes to estimate missing values at a rain gage. The variants are (1) simple inverse distance-based scheme with exponents optimized; (2) correlation-based weighting scheme using historical correlation between the rain gage and radar observations; (3) optimal weighting scheme that uses

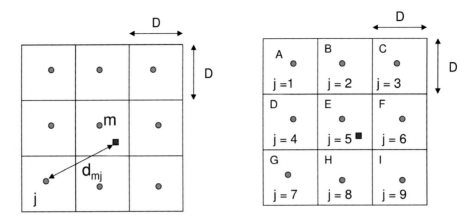

○ Centroid of Radar Grid pixel

■ Rain gage

Figure 3.35 Location of rain gage and centroids of radar grid pixels for use in interpolation of missing precipitation data.

radar data from surrounding grids and rain gage data from other rain gages in the region; and (4) interpolation schemes in which optimum number of rain gages and grid-based observations are selected. All these variants were based on a 3 × 3 grid (a total of 9 grids or cells) of radar data with a rain gage located in the center cell.

3.34 SINGLE AND MULTIPLE IMPUTATIONS OF MISSING DATA

The main disadvantage of deterministic interpolation techniques is the lack of uncertainty assessment of estimates. The use of multiple imputations of missing data based on one or more methods provides the deterministic methods with the required capabilities to provide uncertainty assessments. Imputation involves replacing an incomplete observation with complete information based on an estimate of the true value of the unobserved variable. Single imputation refers to filling of missing data using one model. Multiple imputations are possible when optimal weights in weighting methods vary based on data length, representative historical data used in the estimation of weights, and number of stations used in the interpolation. The process of filling missing data with one value for each value of missing data is generally referred to as single imputation (Junninen et al., 2004). The single imputation process has one disadvantage in that uncertainty associated with the estimate cannot be assessed and the process might provide biased estimates with either underestimation or overestimation of variance. Multiple imputations of missing data rely on development of multiple estimates for each of the missing values to mainly reflect the uncertainty associated with the missing values. Many researchers agree that this is a statistically sound approach

for imputation of missing values. Multiple imputations can be done using a single model or multiple models. The imputed data sets are referred to as "multiply imputed data" and the filled values are referred to as "imputes."

Multiple imputation inference involves three distinct phases:

- The missing data are filled m (a specific number greater than 1) times to generate m complete data sets.
- The m complete data sets are analyzed using standard statistical analyses.
- The results from the m complete data sets are combined to produce inferential results.

The multiple imputation procedure creates multiply imputed data sets for incomplete multivariate data. It uses methods that incorporate appropriate variability across the m imputations. The method of choice depends on the missing data mechanisms. Three common methods that are used for multiple imputations in social sciences are (1) mean imputation (unconditional), which uses observed data with no separation of data into classes; (2) mean imputation (conditional), which uses observed variables when the data are separated into classes (regression); and (3) stochastic regression. Stochastic regression involves replacing missing values by predicted values through a regression plus a residual (drawn to reflect uncertainty in the predicated value) (Little and Rubin, 1987). A normally distributed residual term to each predicted value restores variability to the imputed data. Other methods in social sciences include expectation maximization, maximum likelihood, hot-deck imputation, and list-wise deletion. Data sets that have no gaps or gaps filled by any interpolation method are referred to as complete data sets. Complete data sets can be evaluated by imputation, pooling, and assessment (IAP) concepts

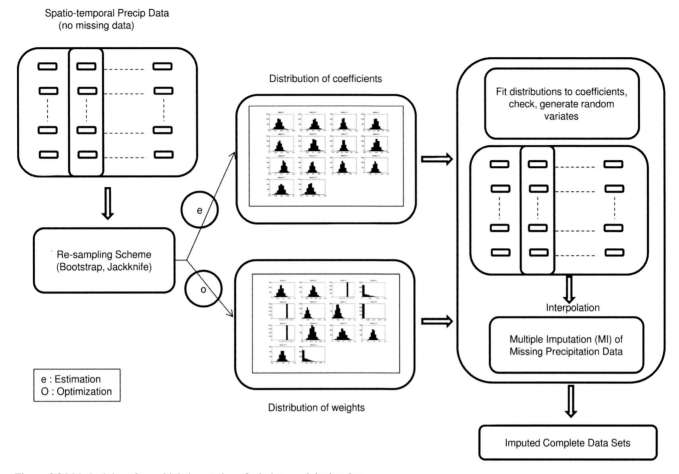

Figure 3.36 Methodology for multiple imputation of missing precipitation data.

(Enders, 2010). Total variance from the completed data sets can be obtained by within-imputation variance and between-imputation variance (Little and Rubin, 1987).

3.34.1 Methodology for multiple imputations

The methodology for multiple imputation can be described in a few steps and they are (1) use re-sampling techniques such as bootstrap and jack-knife to obtain distributions of spatial correlations between the station missing the data and each of the stations in the monitoring network using historical data; (2) evaluate the probability distributions of the correlations and confirm their fit to standard statistical distributions via goodness-of-fit tests; (3) use sampling distributions or draw random variates from these distributions to generate multiple correlation coefficient values to be used in the CCWM and as starting weights for the optimal correlation-based spatial interpolation method; (4) generate multiple imputations of missing precipitation data; and (5) evaluate the uncertainty associated with each estimate. The multiple imputations of missing data can be generated using the bootstrap

and jack-knife sampling statistics (e.g., correlation coefficients) or random variates generated from validated distributions of sampling statistics. The methodology described by the four steps is illustrated in Figure 3.36.

Step 1 described earlier can be replaced with an optimal weighting method or non-linear least squares method to obtain weights for each bootstrap sample.

3.34.2 Bootstrap re-sampling

The bootstrap method of re-sampling was proposed by Efron (1979). The samples in the bootstrap re-sampling process (Efron and Gong, 1983; Efron and Tibshirani, 1993; Chernick, 1999) are typically generated by sampling (with replacement) from the original data. In general it is possible to generate as many bootstrap samples as desired. However, when the original data set size is small, the entire universe of samples can be explored. As the original sample size increases the robustness of bootstrap sampling increases. The bootstrap distribution approximates the sampling distribution of the statistic.

3.34.3   Jack-knife re-sampling

Jack-knife re-sampling was a technique introduced by Quenouille (1949) and improved by Tukey (1958) to assess the performance of models influenced by subsets of observations when outliers are present. The jack-knife technique is similar to bootstrap, where systematic re-sampling of observations is carried out leaving one out from the sample set every time. The technique is also referred to as *leave-one-out* re-sampling. The technique helps in assessing the bias and standard error in a statistic that is critical for statistical inference. Jack-knife differs from bootstrapping in that data points are not replaced prior to each re-sampling (Emery and Thompson, 1997). The number of data sets created using this re-sampling scheme is equal to N_{P1} and the total number of samples generated is equal to N.

3.34.4   Sampling statistics using bootstrap and jack-knife schemes

The sample statistics (e.g., mean and standard deviation) can be estimated based on the pth bootstrap sample as given by following equations:

$$\hat{\phi}_{m,j,p} = \beta_j\left(\left[\theta_{m,1}...\theta_{m,i-1},\theta_{m,i},\theta_{m,i+1}...\theta_{m,p}\right],\left[\theta_{j,1}...\theta_{j,i-1},\theta_{j,i},\theta_{j,i+1}...\theta_{j,p}\right]\right) \quad \forall j,i,p \tag{3.148}$$

$$\beta\,() = \rho_{mj} \quad \forall j \tag{3.149}$$

$$\mu_{B,p,j} = f\left(\hat{\phi}_{B,p}\right) = \frac{1}{no}\sum_{i=1}^{no}\hat{\phi}_{m,j,p} \quad \forall j, \forall p \tag{3.150}$$

$$\sigma_{B,p,j} = f\left(\hat{\phi}_{B,p}\right) = \sqrt{\frac{\sum_{i=1}^{no}\left(\hat{\phi}_{m,j,p} - \mu_{B,p}\right)^2}{no - 1}} \quad \forall j \tag{3.151}$$

Random variates are generated based on the distribution of the bootstrap statistic.

$$\rho_{B,p,j}^{y} = \mu_{B,p,j}R_N + \sigma_{B,p,j} \quad \forall j, \forall y \tag{3.152}$$

$$\mu_{j,y} = \frac{1}{nr}\sum_{y=1}^{nr}\rho_{B,p,j}^{y} \quad \forall j, \forall y \tag{3.153}$$

$$\sigma_{j,y} = \sqrt{\frac{\sum_{y=1}^{nr}\left(\rho_{B,p,j}^{y} - \mu_{j,y}\right)^2}{nr - 1}} \quad \forall j \tag{3.154}$$

$$\hat{\phi}_{m,j} = \beta\left(\left[\theta_{m,1}...\theta_{m,i-1},\theta_{m,i},\theta_{m,i+1}...\theta_{m,no}\right],\left[\theta_{j,1}...\theta_{j,i-1},\theta_{j,i},\theta_{j,i+1}...\theta_{j,no}\right]\right) \quad \forall j \tag{3.155}$$

$$\hat{\phi}_{m,j,k} = \beta\left(\left[\theta_{m,1}...\theta_{m,i-1},\theta_{m,i},\theta_{m,i+1}...\theta_{m,no}\right],\left[\theta_{j,1}...\theta_{j,i-1},\theta_{j,i},\theta_{j,i+1}...\theta_{j,no}\right]\right) \quad \forall j \tag{3.156}$$

$$\rho_{B,p,j}^{y} = \mu_{B,p,j}R_u + \sigma_{B,p,j} \quad \forall j, \forall y \tag{3.157}$$

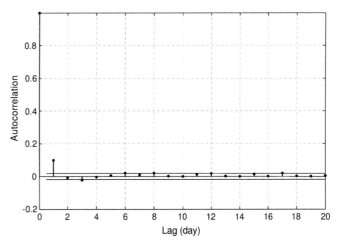

Figure 3.37 Autocorrelation function of daily rainfall data at a site in Kentucky, USA.

3.34.5   Multiple imputations: variability of parameters

This section is used to illustrate the variability of parameters for a CCWM when several bootstrap samples are used from a long-term spatio-temporal precipitation data set at 15 rain gages in the state of Kentucky, USA. Details of the location of the rain gages and the data are available from a recent study by Teegavarapu (2012a). Thirty years of historical precipitation data are used to estimate correlation between a specific rain gage (numbered 1–14) and the base station. The 30-year data in chronological order are divided into approximately 17 equal data sets and the correlations are calculated. Table 3.16 provides the details of the correlations for different data sets. High variability is seen in the correlations. Once these correlations are characterized by appropriate statistical distributions and random variates are generated, multiple imputations are possible using CCWM.

When multiple imputations are required according to the methodology introduced in Section 3.34.1, bootstrap sampling of the entire data set is carried out to obtain a vector of correlation values to use in CCWM.

3.35   TEMPORAL INTERPOLATION OF MISSING DATA

Time series with missing precipitation data can be imputed using mean imputation or cubic spline interpolation or any other method using observations in a series. This type of interpolation is referred to as time interpolation. In many instances, infilling by these methods is not advisable due to weak autocorrelation in precipitation time series especially at smaller time intervals (e.g., day). An example of daily data from a rain gage from Lexington, Kentucky, USA, is used to illustrate this point. The autocorrelation function based on the data is shown in Figure 3.37. The lag-one

Table 3.16 *Correlation coefficients estimated from 17 equally sized data sets from historical precipitation data*

	Correlation coefficients													
	Rain gages													
Data set	1	2	3	4	5	6	7	8	9	10	11	12	13	14
1	0.524	0.635	0.090	0.498	0.191	0.119	0.492	0.438	0.100	0.604	0.614	0.706	0.546	0.034
2	0.621	0.558	0.164	0.624	0.318	0.208	0.644	0.571	0.199	0.488	0.546	0.566	0.622	0.157
3	0.525	0.696	0.261	0.585	0.215	0.269	0.548	0.477	0.181	0.689	0.623	0.710	0.610	0.651
4	0.686	0.676	0.090	0.566	0.287	0.092	0.484	0.396	0.166	0.581	0.546	0.707	0.667	0.691
5	0.815	0.639	0.201	0.564	0.435	0.219	0.543	0.532	0.192	0.585	0.667	0.726	0.531	0.601
6	0.538	0.599	0.140	0.600	0.206	0.182	0.551	0.547	0.204	0.652	0.672	0.761	0.623	0.664
7	0.528	0.600	0.079	0.622	0.337	0.250	0.656	0.566	0.155	0.503	0.568	0.790	0.646	0.548
8	0.276	0.439	0.080	0.327	0.235	0.487	0.496	0.288	0.086	0.457	0.470	0.588	0.597	0.432
9	0.507	0.652	0.025	0.641	0.242	0.508	0.523	0.448	0.195	0.645	0.572	0.735	0.583	0.576
10	0.661	0.667	0.193	0.641	0.336	0.610	0.574	0.494	0.272	0.562	0.528	0.668	0.563	0.599
11	0.377	0.365	0.091	0.121	0.155	0.375	0.375	0.251	0.176	0.455	0.437	0.530	0.470	0.521
12	0.520	0.537	0.131	0.523	0.381	0.627	0.613	0.340	0.111	0.574	0.650	0.783	0.636	0.647
13	0.533	0.681	0.095	0.616	0.343	0.634	0.443	0.441	0.223	0.661	0.632	0.752	0.619	0.675
14	0.455	0.665	0.135	0.372	0.306	0.598	0.463	0.424	0.175	0.578	0.592	0.627	0.543	0.543
15	0.577	0.677	0.141	0.620	0.335	0.656	0.536	0.619	0.150	0.714	0.613	0.778	0.679	0.609
16	0.421	0.659	0.149	0.565	0.264	0.539	0.474	0.439	0.023	0.672	0.610	0.659	0.452	0.453
17	0.537	0.615	0.220	0.671	0.333	0.682	0.513	0.452	0.279	0.589	0.577	0.698	0.745	0.514
All data	0.530	0.614	0.141	0.520	0.283	0.407	0.516	0.452	0.170	0.592	0.584	0.690	0.597	0.527

autocorrelation value is 0.1003. This value is so low that any missing data estimation based on past values becomes difficult. At coarser temporal resolution higher than day (e.g., week, month), the autocorrelations might improve and provide ways to impute missing precipitation data. Teegavarapu and Mujumdar (1996) demonstrated the utility of ANNs for short-term rainfall forecasting with a lead time of 10 days. The temporal rainfall variability is often high and with weak autocorrelation at short time intervals (durations less than a week) it is advisable to rely on spatial interpolation as opposed to temporal interpolation for estimation of missing data.

3.36 DATA SET SELECTION FOR MODEL DEVELOPMENT AND VALIDATION

Several methods described in this chapter are data driven, data sensitive, and data intensive. Therefore, it is imperative that the model development and testing should involve careful selection of data sets that are accurate, without any gaps, and are time consistent. Time consistent data require that all rain gages have precipitation data for a specific time interval. Figure 3.38 shows how the calibration and test data are used for parameter estimation and model performance evaluation respectively. The process of dividing the data into two parts for evaluation of spatial interpolation models is referred to as split validation.

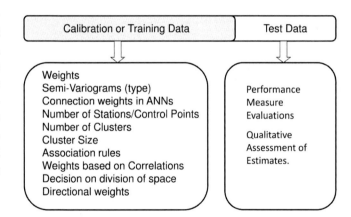

Figure 3.38 Use of calibration and test data for parameter estimation and model performance evaluation.

Cross-validation is simply the assessment of interpolation methods by leave-one-out analysis. In a spatial interpolation exercise, missing values at each of the observation points are estimated by using all the observations at the rest of the points.

Robust evaluation of models for conceptual accuracy, universal applicability, and validity in all situations requires division of available data into two (calibration and test) or three (calibration, validation, and test) clusters. Continuous chronological and non-chronological data sets may be required for testing. Figure 3.39 shows a schematic of data division for evaluation of models. When

Chronological/randomized time series

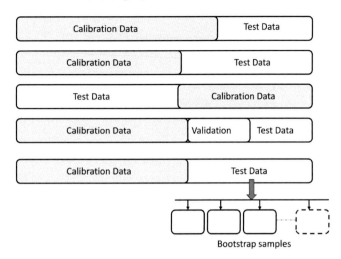

Figure 3.39 Division of data sets for robust testing and performance evaluation of models.

competing models are available for estimation of precipitation, then selection of reliable and robust performance measures is as important as the data sets used for model development and testing.

Performance evaluations using a subset of a test data set based on bootstrap samples can be conducted. The length of the bootstrap samples (random samples with replacement) can be equal to the length of the test data set or less.

Counterintuitive results suggesting improved performance of models in the test phase compared to those from the calibration phase are also possible in many instances. The notion that models always perform better in the calibration phase than in the test phase may not be valid always as the point estimation of precipitation depends on the spatial autocorrelation among observations in a given time interval or a group of time intervals.

3.37 PERFORMANCE MEASURES

Several quantitative and qualitative measures are generally used to evaluate the interpolation methods. This section describes the quantitative measures (Aly *et al.*, 2009) used in assessment of methods.

Root mean square error (RMSE)

$$RMSE = \sqrt{\frac{\sum_{m,j} \left(\theta_{m,i} - \alpha_{m,i}\right)^2}{n}} \qquad (3.158)$$

where $\theta_{m,i}$ is the observed rainfall at gage m in a time interval i, $\alpha_{m,i}$ is estimated rainfall value at the same gage, and n is the number of missing rainfall values across all gages in the evaluation data set. The RMSE is expressed in the same units as the recorded rainfall values.

Mean absolute error (MAE)

$$MAE = \frac{\sum_{m,i} |\theta_{m,i} - \alpha_{m,i}|}{n} \qquad (3.159)$$

where $\theta_{m,i}$, $\alpha_{m,i}$, and n are explained above. The MAE has the same units as the RMSE. However, the MAE is less sensitive to large error values that can dominate the calculation of the RMSE.

Mean error (ME)

$$ME = \frac{\sum_{m,i} \left(\theta_{m,i} - \alpha_{m,i}\right)}{n} \qquad (3.160)$$

The ME is also expressed in depth units, where a positive ME value will indicate underestimation, i.e., interpolated values are less than the observed. For water balance applications, a small ME value is highly desirable since small deviations from observed rainfall in small time intervals can result in large cumulative mass balance errors over long time periods.

Correlation coefficient (CC)

$$CC_{\alpha,\theta} = \frac{\text{cov}(\alpha, \theta)}{\sqrt{\text{var}(\theta)\ \text{var}(\alpha)}}$$

$$= \frac{n \sum_{m,i} \theta_{m,i} \alpha_{m,i} - \sum_{m,i} \theta_{m,i} \sum_{m,i} \alpha_{m,i}}{\sqrt{n \sum_{m,i} \theta_{m,i}^2 - \left(\sum_{m,i} \theta_{m,i}\right)^2}\sqrt{n \sum_{m,i} \alpha_{m,i}^2 - \left(\sum_{m,i} \alpha_{m,i}\right)^2}} \qquad (3.161)$$

The dimensionless correlation coefficient ranges between -1 and 1, with 1 indicating perfect correlation between the observed and predicted rainfall values. Correlation coefficients can be evaluated for different time frames (e.g., seasons or months) as well as for the entire data set.

Percentage of correct state for wet time intervals (P_w)

$$P_w = 100 \left(\frac{\sum_{m,i} I_{1,1}^m}{\sum_{m,i} I_1^m}\right) \qquad (3.162)$$

where $I_{1,1}^m$ is an indicator variable taking a value of 1 if $\theta_{m,i} > 0$ and $\alpha_{m,i} > 0$ (and zero otherwise) and I_1^m is an indicator variable having a value of 1 if and only if $\theta_{m,i} > 0$. Thus, P_w indicates the fraction of time intervals (e.g., days) on which non-zero precipitation (defined to be a specific depth) was both observed and calculated.

Percentage of correct state for dry time intervals (P_d)

$$P_d = 100 \left(\frac{\sum_{m,i} I_{0,0}^m}{\sum_{m,i} I_0^m}\right) \qquad (3.163)$$

where $I_{0,0}^m$ is an indicator having the value of 1 if $\theta_{m,i} = 0$ and $\alpha_{m,i} = 0$ (and zero otherwise) and I_0^m is an indicator with the value of 1 if $\theta_{m,i} = 0$. Similarly to P_w, P_d shows the fraction of time intervals in which no precipitation (defined to be less than a specific threshold) was both observed and estimated.

Percentage of correct state for all time intervals (P_A)

$$P_A = 100 \left(\frac{\sum_{m,i} I_{1,1}^m + I_{0,0}^m}{n} \right) \qquad (3.164)$$

where $I_{1,1}^m$ and $I_{0,0}^m$ are the indicator variables described above and n is the number of data points in the entire evaluation data set. Hypothesis testing can also be employed sometimes to test the significance of the calculated ME values. The measures used for the correct states for dry and wet conditions are similar to general contingency measures discussed later in the chapter.

Several other performance measures discussed by Li and Heap (2011), such as mean square reduced error (MSRE), mean standardized error (MSE2), root mean square standardized error (RMSSE), averaged standard error (ASE), Willmott's D, ratio of the variance of estimated values to the variance of the observed values (RVAR), and model efficiency can be used for exhaustive evaluation of spatial interpolation methods. These error measures are described by the following equations:

$$MSRE = \frac{1}{n} \sum_{m,i}^{n} (\theta_{m,i} - \alpha_{m,i})^2 / s^2 \qquad (3.165)$$

$$MSE2 = \frac{1}{n} \sum_{m,i}^{n} (\alpha_{m,i}^s - \theta_{m,i}^s) \qquad (3.166)$$

$$RMSSE = \sqrt{\frac{1}{n} \sum_{m,i}^{n} (\theta_{m,i}^s - \alpha_{m,i}^s)^2} \qquad (3.167)$$

$$ASE = \sqrt{\frac{1}{n} \sum_{m,i}^{n} \left(\alpha_{m,i}^s - \left(\frac{\sum_{i}^{n} \alpha_{m,i}^s}{n} \right) \right)^2} \qquad (3.168)$$

$$Willmott's\ D = 1 - \frac{\sum_{m,i}^{n} (\theta_{m,i} - \alpha_{m,i})^2}{\sum_{m,i}^{n} (|\alpha_{m,i}'| + |\theta_{m,i}'|)^2} \qquad (3.169)$$

$$RVAR = \frac{var(\alpha)}{var(\theta)} \qquad (3.170)$$

$$Efficiency = 1 - \frac{\sum_{m,i}^{n} (\theta_{m,i} - \alpha_{m,i})^2}{\sum_{m,i}^{n} (\alpha_{m,i} + \alpha_{m,i})^2} \qquad (3.171)$$

The variable s is the standard deviation calculated based on error values; $\alpha_{m,i}^s$ is the standardized estimated precipitation value at station m for time interval i; $\theta_{m,i}^s$ is the standardized observed value; $\alpha_{m,i}''$ is the average of observed values; $\alpha_{m,i}'$ is equal to $(\alpha_{m,i} - \alpha_{m,i}'')$; $\theta_{m,i}^i$ is equal to $(\theta_{m,i} - \theta_{m,i}'')$; and $\theta_{m,i}''$ is the average of estimated values.

Li and Heap (2011) and Hu et al. (2004) provide an extensive discussion about these measures and their utility in selection of models or methods. Criteria for judging the performance measures in selecting the best model are also discussed. A method having the performance of MSE2 close to zero and with RMSE being lowest points to the better model among those under evaluation. According to Li and Heap (2011), if ASE is greater than RMSE, then the method overestimates the MSRE, also called the studentized residuals for regression diagnostics in statistics (Venables

and Ripley, 2002) or standardized MSE (Martínez-Cob, 1996), which should approach 1. The ratio of variances given by RVAR should be closer to 1, indicating that the observed estimated variances are equal. Correlation coefficient is the most commonly used measure in assessment of interpolation methods. However, this measure is sensitive to outliers and may not fully explain the model/method capabilities. Willmott's D (Willmott, 1982) is also referred to as the index of agreement, and has the advantage of scaling the magnitude of the variables, retaining the mean information and not amplifying the outliers (Willmott, 1982; Li and Heap, 2011). The closer Willmott's D value is to 1 the better the model. The efficiency measure, if closer to zero, indicates that the mean value of the observations is more reliable than the estimated values (Vicente-Serrano et al., 2003).

Performance evaluation for spatial interpolation for estimation at multiple locations can be carried out using different cross-validation methods (Bivand et al., 2008; Hengl, 2009). These methods are k-fold, leave-one-out, and jack-knifing cross-validation methods. In the case of the k-fold method, the sample is split into k (a specific number) equal parts and then each part is used for validation. In the leave-one-out method, as the name suggests, one sampling point is eliminated in an iterative way to estimate values at all points. The eliminated point is not originally used in the model development. Jack-knifing is primarily aimed at estimating the bias and standard error (variance) of the statistic.

Precipitation depth at any temporal resolution less than or equal to one day can be regarded as a special variable with the possibility of several zero values. In the estimation process, no distinction is generally made between rain and no rain events unless a classifier is used to test these conditions. Also, in the evaluation process there is way to separate out the no rain events from the rest of the events. Inclusion of no rain events in the evaluation leads to a heavily skewed (i.e., positively skewed) distribution of values. Therefore, in many instances evaluation based on a specific set of thresholds can be used. Also, a 5-mm estimation error on a 5-mm observed rainfall is not the same as a 5-mm estimation error on a 50-mm observed rainfall. Therefore, these two errors should not be treated with the same weight or importance in a RMSE or an AE criterion in the performance evaluation phase. In these situations, weighted performance measures might be applicable.

The quantitative error measures discussed in this section may not provide a complete assessment of interpolation methods as they are average measures calculated for a specific period of time. The temporal distribution and magnitude of the precipitation affects the outputs of rainfall–runoff models, which require accurate precipitation values to simulate hydrologic processes at different spatial and temporal scales. Therefore, precipitation data series with gaps filled by interpolation methods and data series with no gaps should be used in hydrologic simulation

Table 3.17 *Contingency table for observed and estimated precipitation values*

		Estimated precipitation	
Observed	$P_o > 0$	$P_e > 0 \, (C_{11})$	$P_e = 0 \, (C_{10})$
precipitation	$P_o = 0$	$P_e > 0 \, (C_{01})$	$P_e = 0 \, (C_{00})$

models as part of a comprehensive assessment of interpolation methods.

3.37.1 Contingency tables

Four other measures used for classification abilities of any interpolation method can be developed using a contingency table. The outcomes and actual observed values are used to develop the counts for the measures. The four measures (Myatt and Johnson, 2009), concordance, error rate, sensitivity, and specificity, can be obtained from Table 3.17 using the counts for conditions. The counts, C_{11}, C_{10}, C_{01}, and C_{00}, are calculated based on conditions described in Table 3.17. The variables P_o and P_e refer to observed and estimated precipitation values.

$$Concordance = \frac{C_{11} + C_{00}}{C_{11} + C_{10} + C_{01} + C_{00}} \qquad (3.172)$$

$$Error\ rate = \frac{C_{01} + C_{10}}{C_{11} + C_{10} + C_{01} + C_{00}} \qquad (3.173)$$

$$Sensitivity = \frac{C_{11}}{C_{11} + C_{10}} \qquad (3.174)$$

$$Specificity = \frac{C_{00}}{C_{01} + C_{00}} \qquad (3.175)$$

Concordance refers to the accuracy of the model, error rate relates to prediction errors, and sensitivity and specificity measures calculate the ability of the model to estimate the correct states (i.e., rain or no rain). Accuracy is also referred to as fraction correct and error rate to fraction incorrect.

Bootstrap samples can be generated from the test data to evaluate the performance of different methods. The samples are observed and estimated data pairs are drawn randomly with replacement. The performance measures are calculated for each bootstrap sample for evaluation of multiple methods. Confidence intervals can be developed for the error performance measures. The intervals can be developed using one of four methods: (1) basic percentile, (2) bias corrected, (3) bias corrected and accelerated, and (4) studentized. The development of confidence intervals will help in assessing the differences between model performances statistically.

Tests for residuals that are generally done for regression models can be extended for the estimates obtained from different interpolation models. The tests should confirm the homoscedasticity and normality of residuals with the mean of residual values close to

zero. It is important to note that the error measures are generally calculated as average values for the calibration and test periods. A one percent difference in RMSE value suggests, on average, that one model is either over- or underestimating precipitation values by one percent. Small variations in rainfall intensity can introduce significant changes in the runoff values generated from distributed rainfall–runoff models (Vieux, 2004). Any improvement, however minute, in the rainfall estimation can therefore be considered significant, as rainfall is a crucial input that governs the response of hydrologic systems and the results of continuous simulation models.

3.37.2 Evaluation for categorical values

Categorical value evaluation is similar to assessment using contingency measures. However, this evaluation allows assessment of interpolation or estimation methods for specific categorical values of the variables of interest. In the case of rainfall estimation, classification of rain or no rain conditions by the interpolation method needs to be checked for accuracy. The Kappa statistic (Congalton and Green, 1999; Hengl, 2009) can be used to evaluate classification accuracy. The statistic can be used to measure the difference in the actual agreement between predictions and ground truth and the agreement that could be expected by chance (Hengl, 2009). A simpler Kappa statistic is a measure to evaluate the percentage of correctly classified conditions in space and time:

$$K_c = \frac{1}{n} \sum_{i=1}^{n} p_c; \ p_c = 1 \text{ if } C(e) = C(g); \ p_c = 0 \text{ if } C(e) \neq C(g)$$

$$(3.176)$$

The variable $C(e)$ is the estimated value belonging to a class C; $C(g)$ is the ground truth (i.e., observed value); and p_c is the percentage correct. The Kappa statistic discussed here is a simplified version of the original statistic discussed by Congalton and Green (1999).

3.38 QUALITATIVE EVALUATION

Qualitative evaluation of estimation models focuses on the use of visual exploratory data analysis techniques to assess the performance of the interpolation methods. Examples of these visual assessment methods include (1) scatter plots of estimated and measured precipitation; (2) distribution of the model residuals calculated from estimates and observed values; and (3) accumulated estimated and measured precipitation values and comparison of surface generation plots based on different interpolation algorithms. Visual observation of surfaces helps in the assessment of any tent-pole effect (or bull's-eye effect) evident and any artefacts in the interpolation schemes. It is recommended that histograms

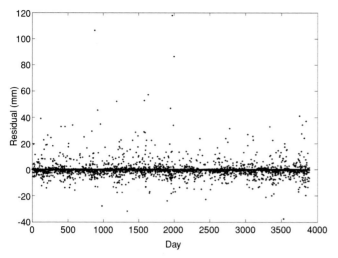

Figure 3.40 Residuals over time to assess time dependence.

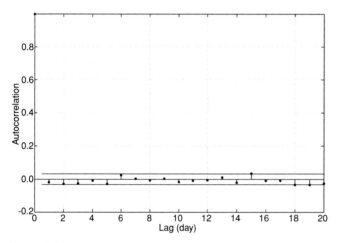

Figure 3.41 Autocorrelation function of residuals.

of residual errors be evaluated to assess the bias and precision of the estimates. Histograms of residuals will provide insights into model performance in terms of bias and variance. A plot of residuals based on estimated and observed values at a rain gage in Kentucky, USA, is shown in Figure 3.40. The residuals with respect to time show no trend or structure. An autocorrelation plot of the residuals can also be used to evaluate the time dependency of residuals as shown in Figure 3.41. The autocorrelation values are negligible, suggesting no time dependency or persistence. Kanevski and Maignan (2004) suggest that bias indicates accuracy of prediction (or estimate) and variance reflects precision of the estimate. A model with zero bias (based on distribution of residuals) and less variance is the best model. Models with negative, zero, and positive biases with small or large variances reflected in histograms of residuals are not preferable.

3.39 MODEL SELECTION AND MULTI-MODEL COMPARISON

Error statistics, namely RMSE, AE, and CC (ρ), account for most of the widely used measures to assess the performance of the methods used in estimation of missing precipitation data. Factors such as intuitive reasonableness of method or approach, and conceptual accuracy and simplicity of the method should also be considered for objective assessment and, finally, selection of the methods. Particular attention is also paid to visual observation of residuals to assess biases in estimation. When multiple performance measures are used for selection of any method from a set of equally competitive methods, identification of that best method is often difficult. Linear or non-linear weighting functions can be developed to transform the performance measures (i.e., RMSE, AE, MAE, ρ) into an accumulated dimensionless total performance measure. This total measure can then be used for objective comparison and selection of the best method from a set of competing methods. The performance measures can be added to transform them to a score (or weight) between zero and 1. The total weighted performance measure (W_{total}) is given by Equation 3.181.

$$w_\rho = 1 - \left[\frac{\Omega_\rho}{\rho'}(\rho) \right] \tag{3.177}$$

$$w_{AE} = \frac{\Omega_{AE}}{AE'}(AE) \tag{3.178}$$

$$w_{RMSE} = \frac{\Omega_{RMSE}}{RMSE'}(RMSE) \tag{3.179}$$

$$w_{MAE} = \frac{\Omega_{MAE}}{MAE'}(MAE) \tag{3.180}$$

$$W_{total} = w_\rho + w_{AE} + w_{RMSE} + w_{MAE} \tag{3.181}$$

The variables Ω_ρ, Ω_{AE}, Ω_{RMSE}, Ω_{MAE} are weighting factors, and ρ', AE', $RMSE'$ and MAE' are the maximum values performance measures from all the methods. Equations 3.177–3.180 are referred to as weighting functions, which transform performance measures to unitless constant values. These weighting functions exactly resemble the fuzzy membership functions developed by Teegavarapu and Elshorbagy (2005) for hydrologic model evaluation. The overall performance measure (W_{total}) is designed in such a way that the method with the lowest value of the total weight is the best among all competing methods. The number of performance measures in the equation can be increased or decreased and additional weighting factors or non-linear functions can be assigned to each performance measure. Multi-model assessment may also be carried out by adding normal noise of 1%, 5%, and 10% to the daily records and checking the variations of the performance measures.

A Taylor diagram (Taylor, 2001) provides a concise statistical summary of how well patterns match each other in terms of

Table 3.18 *Hypothetical multi-model estimations and observed values*

	Estimated and observed values								
Observed	2	3	5	6	7	8	10	11	9
Model A	2	2.5	4.5	6	6.7	8	11	10	9
Model B	2	3	5	5	5	6	9	11	4
Model C	2	5	5	4	7	8	9	11	4
Model D	1	2	7	7	9	4	11	9	5

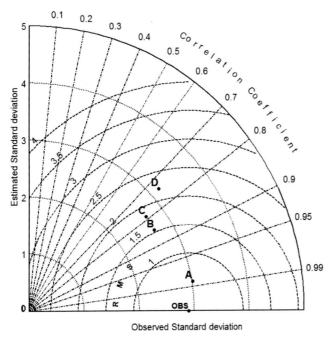

Figure 3.42 Taylor diagram based on observed and estimated data sets provided in Table 3.18.

Table 3.19 *Suitability of spatial interpolation methods for point estimation and surface generation*

Method	Point estimation	Surface generation
Inverse distance weighting method	+	+
Gage mean estimator	+	−
Single best estimator	+	−
Kriging	+	+
Correlation coefficient weighting method	+	−
Artificial neural networks	+	−
Regression models	+	+
Radial basis functions	+	+
Kernel density functions	+	+
Thin splines	+	+
Nearest neighbor techniques	+	+
Optimal functional forms	+	−
Cluster-based optimal functional forms	+	−
Thiessen polygons	+	+
Local and global polynomial functions	+	+
Universal function approximation kriging	+	+
K-means clustering approach	+	−
K-NN classification	+	−
K-NN proximity-based imputation	+	−

Possible (+) Not possible (−)

their correlation, their RMS difference, and the ratio of their variances. A Taylor diagram combines statistical measures such as standard deviations, centered RMS deviations, and correlations based on multi-model estimations in one graph. The construction of the diagram (with the correlation given by the cosine of the azimuthal angle) is based on the similarity of the equation summarizing the RMS, the correlation, the standard deviation, and the law of cosines (Taylor, 2001). The diagram is also used by the IPCC for multi-model comparisons. For illustration of the Taylor diagram concept and its utility, hypothetical data sets of monthly precipitation data (inches per month) are used in an example. The data sets are given in Table 3.18.

Based on the nine estimated values and observed values given in Table 3.18, a Taylor diagram is constructed and is shown in Figure 3.42. The observed values are provided in the first row of the table. The diagram provides the isolines of correlation for standard deviation and centered RMS deviations. They can be

omitted for clarity. As seen from the model estimations, model A is the best. Model A provides estimates with the highest correlation with observed values, and model D provides the lowest. Therefore, point A is located close to point OBS (i.e., observed) on the diagram.

3.40 SURFACE GENERATION

Surface generation is the process of using observations at specific control points in space to create a surface that is applicable to a specific grid or tessellation. Gridded precipitation data are required for distributed hydrologic modeling. Some of the spatial interpolation schemes discussed in this chapter can be used for surface generation. One critical requirement of the surface generation method is the ability to model the spatial variation of the data. The deterministic and stochastic methods of surface generation have two elements in common and they are the distance and the number of neighborhood control points. It is important to note that some methods discussed in this chapter are aimed at point estimation only, especially estimation of missing precipitation data. Table 3.19 provides the suitability of methods used for point estimation and surface generation.

Figure 3.43 Inverse distance interpolation with five nearest neighbors.

Surface generation using the IDWM, and local and global polynomial interpolations is provided in Figures 3.43–3.47. Figures 3.43 and 3.44 show interpolated surfaces with 5 and 15 neighbors selected in inverse squared-distance interpolation using ArcGIS software. Inverse distance interpolation with 15 neighbors using optimal exponents is shown in Figure 3.45. Interpolation surfaces using first-order local and global polynomial functions are shown in Figure 3.46 and Figure 3.47 respectively.

It can be noted that as the number of neighbors used in the interpolation process increases, the smoothness of the surface increases. Surface generation is an iterative process and requires a comprehensive assessment of the authenticity of the surface variation and closeness to the actual variation of the process of the physical variable that is being interpolated. The presence of outliers can influence the surface generated using any interpolation method. Outliers can be extreme values that may be part or not part of the actual observed values. Histograms of the data and variogram clouds can be used to easily identify the outliers. Any value with very low frequency lying in the tails of the data distribution requires further scrutiny.

3.41 GEO-SPATIAL GRID-BASED TRANSFORMATIONS OF PRECIPITATION DATA

Spatial interpolation using point precipitation data to generate precipitation fields or surfaces is an essential task in distributed hydrologic modeling. Also, the generation of precipitation data sets confined to a fixed tessellation (grid of specific size) is generally the final product of any processed radar data. Radar data available as gridded data can be used for estimation of missing precipitation data at a rain gage using local filters (Lloyd, 2007) with appropriate focal functions to derive the value of a cell using values from a nearest group of cells. Moving window approaches using focal operators (Lloyd, 2007, 2010) are common in image processing problems. A moving window approach was adopted in a recent study (Teegavarapu and Pathak, 2008) in which pixels (grids) with radar-based precipitation estimates surrounding a rain gage are used to estimate missing data at that gage. Spatial interpolation involving grid operators such as local, focal, zonal, and global functions (Chou, 1997) is generally used to process grids for spatial segmentation and classification.

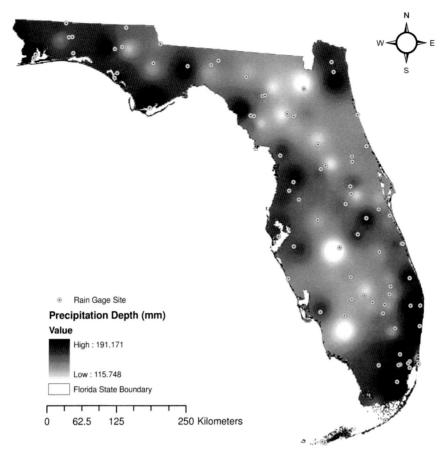

Figure 3.44 Inverse distance interpolation with 15 nearest neighbors.

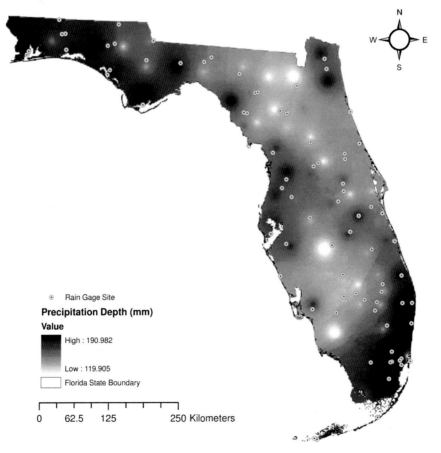

Figure 3.45 Inverse distance interpolation with 15 nearest neighbors and optimal exponent.

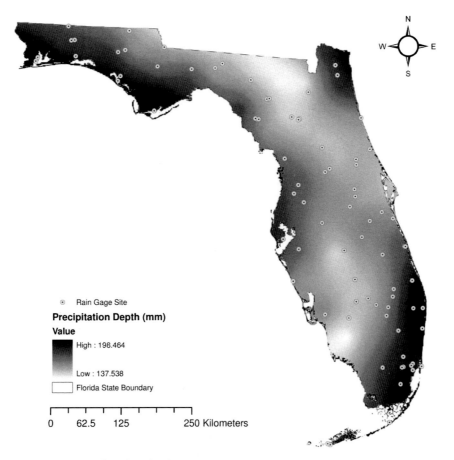

Figure 3.46 Local polynomial interpolation of precipitation data.

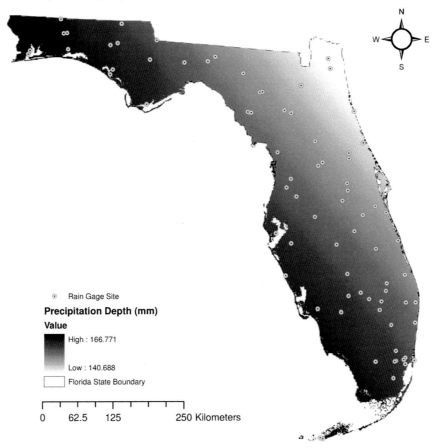

Figure 3.47 Global polynomial interpolation of precipitation data.

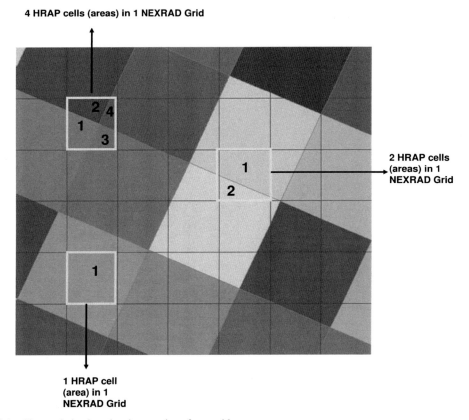

Figure 3.48 Area-weighted interpolation based on intersection of two grids.

Spatial domain filters (Mather, 2004) are used in remotely sensed image processing studies that use moving average windows to reduce the variability of images. These spatial filters adopt different forms of focal operators (Smith *et al.*, 2007) where the value of any given cell (grid or pixel) is a function of values from the surrounding pixels. However, no optimization formulations are generally incorporated into these filters. Spatial interpolation methods for grid-based transformations of precipitation data are possible by extending the principles of geo-spatial analysis. In this section, an example of transformation of precipitation estimation from one grid (HRAP 4 km × 4 km diagonal) to another (Cartesian 2 km × 2 km grid) adopted from a study by Teegavarapu *et al.* (2011b) is presented. Figure 3.48 shows the overlay of two grids of precipitation estimates obtained from multiple sources (sensors) and also shows a schematic of Cartesian grid areas intersecting the diagonal HRAP grids.

The spatial weighting methods used for this example can be referred to as geographically moving window interpolation methods. Studies involving application of grid-to-point interpolation methods were reported recently for establishing rain gage–radar relationships (Teegavarapu *et al.*, 2008) and infilling of missing precipitation data at a gage using radar data (Teegavarapu and Pathak, 2008). Re-sampling methods are ideal for problems involving geometric transformation of digital images (Chang, 2010; Mather, 2004; Lloyd, 2010). Re-sampling methods such as nearest neighbor, bilinear, and cubic convolution use

information from surrounding cells neighboring a point in weighted or non-weighted schemes for interpolation. The nearest neighbor scheme uses one control point closest to the point where the interpolated value is required and bilinear interpolation uses four nearest neighbors. Two methods used in this example are similar to nearest neighbor and bilinear interpolation schemes. The similarity is in the use of one or four nearest neighbors and not in the weighting scheme.

3.41.1 Area-weighting method

In this method, the areal extent of overlay is used as a weight using:

$$\theta_m = \frac{\sum_{j=1}^{n} A_j \phi_j}{\sum_{j=1}^{n} A_j} \quad \forall m \qquad (3.182)$$

where m is the index for the NEXRAD (e.g., 2 km × 2 km grid) Cartesian grid; j is the index for the 4 km × 4km HRAP grid; A_j is the area of the 4 km × 4 km HRAP grid j, within a 2 km × 2 km NEXRAD grid j; and n is number of distinct areas of the HRAP grid within a 2 km × 2 km NEXRAD grid j. This method is similar to the natural neighbor method or area-stealing interpolation (Sibson, 1981). In the natural neighbor method, the neighbors of any point are those associated with neighboring Thiessen polygons. Polygons constructed with and without the point of interest using all the observation points in space provide the proportions

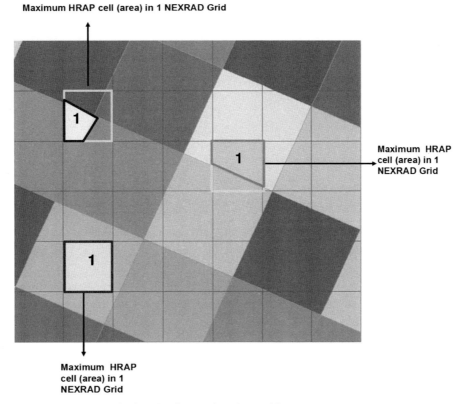

Figure 3.49 Maximum area-weighted interpolation based on intersection of two grids.

of areal overlaps. These proportions are used as the weights. The area-weighting method used does not require construction of Thiessen polygons.

3.41.2 Maximum area method

In this method, the maximum areal extent of overlay is identified (shown in Figure 3.49) and the value associated with the maximum area is assigned to the estimate required for the Cartesian grid.

$$\theta_m = \phi_j \quad \forall m \tag{3.183}$$

$$A_j = \max(A_k) \quad \forall k \tag{3.184}$$

$$j = k \tag{3.185}$$

where A_j is the maximum area occupied by the HRAP grid within a 2 km × 2 km grid. This method is similar to the nearest neighbor method (Mather, 2004). However, the nearest neighbors are not selected based on Euclidean distance, but based on the maximum areal extent of the HRAP grid. This method can be referred to as a single best estimator (SBE) as only information from one HRAP grid is always used.

3.41.3 Inverse distance weighting

Inverse distance weighting is based on the distance calculated from the centroids of the HRAP grids to the centroids of the Cartesian grids. Inverse distance weighting can be constrained to

a specific number of neighbors (in this case, four) and can be constrained by a neighborhood region, say a closed polygon (in this case, a circle of a pre-defined radius). Figure 3.50 shows the variations in the IDWM.

The method is similar to bilinear interpolation in some aspects. However, the bilinear method uses four nearest neighbors and requires three linear interpolations (Chang, 2010). In the case of IDWM, the number of neighbors need not be fixed and the reciprocal distances are used as weights. The bilinear interpolation requires more computational time compared to the IDWM.

3.41.4 Equal weights (average) method

In this method, equal weights are assigned to all the HRAP cells that intersect the NEXRAD cells. This is the simplest of all the methods. The estimate from this method is given by:

$$\theta_m = \frac{\sum_{j=1}^{n} \phi_j w_j}{\sum_{j=1}^{n} w_j} \quad \forall m \tag{3.186}$$

This method is similar to a moving average filter used in image processing and is also classified as a spatial domain filter (Mather, 2004). In the case of a spatial domain filter, the cells surrounding the central cell are used to obtain a weighted value for the central cell. In the equal weights method, only those cells (HRAP) that intersect with the central cell (NEXRAD) grid are adopted for estimation.

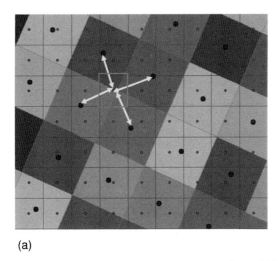
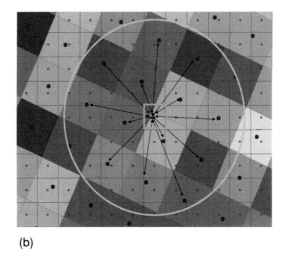

(a)   (b)

Figure 3.50 (a) Inverse distance-based interpolation (nearest four neighbors) and (b) inverse distance-based (radius limited) interpolation.

The methods discussed for grid-based interpolation are radius limited (i.e., fixed distance) as well as nearest neighbor local interpolators. Spatial analysis procedures are used to establish the nearest neighbors. The limits on Euclidean distances need not be chosen arbitrarily, but through several multiple executions of interpolation methods using different search neighborhoods. Variants of the methods developed will produce different reconstructions of precipitation fields. These variants need to be thoroughly evaluated as there is no universally best interpolation scheme (O'Sullivan and Unwin, 2010). Stochastic interpolation methods based on the concepts of geostatistics can be used for the transformation. Ordinary kriging is one such geostatistical interpolation method, which requires development of semi-variograms for each grid based on nearest HRAP grid data. While the development of semi-variograms is not difficult, the selection of an appropriate semi-variogram suitable for the spatial data structure from a set of authorized semi-variograms (i.e., exponential, Gaussian, etc.) is time and computationally intensive especially when data need to be transformed for several thousand cells.

3.42   STATISTICS PRESERVING SPATIAL INTERPOLATION

Spatial interpolation methods discussed in this chapter are valid only for estimation of missing precipitation data at a single site. The parameters (weights, binary variables, exponents) obtained through optimization and other procedures using training data are specific to an observation site and will be different if another site is used. Infilling of missing data for longer lengths of data at several stations requires the condition of maintaining spatial patterns/variability to be met. Infilling of data may introduce inhomogeneities. In general, for many precipitation data series, assessment of inhomogeneities is difficult as metadata (i.e., information about changes in measurements or observation site) are

not always available. The objective functions used in the optimization formulations and discussed in this chapter can be modified to ensure historical statistical properties at a station (or site) and station-to-station (site-to-site) relationships or spatial patterns are preserved. The spatial interpolation methods should preserve (1) a set of statistics of data at a site; (2) site-to-site relationships (e.g., cross correlations); and (3) statistical distribution characteristics. The statistics preserving properties of any method can be tested using two options: (1) compare historical site-specific and regional statistics with those of infilled data alone; and (2) compare historical site-specific and regional statistics without filled data with those from a data set that contains historical data and infilled data. El Sharif and Teegavarapu (2011, 2012) provide several optimization formulations to address the issue of preserving site- and region-specific statistics. They recommend the use of visual and statistical tests to evaluate the distributions of data using the two options discussed previously. Visual evaluation will consist of use of cumulative density function (CDF) plots of data sets, and the statistical tests are based on two-sample Kolmogorov and Smirnov (KS) tests. If missing precipitation data existing in only one specific time period of a time series are filled, then the efficacy of the interpolation method used for filling can be evaluated by tests of homogeneity for two data sets (i.e., data with no gaps and data with gaps filled). Pettitt's test (Pettitt, 1979), Buishand's test (Buishand, 1982), Alexandersson's standard normal homogeneity test (Alexandersson, 1986), and Von Neumann's ratio test (Von Neumann, 1941) can be used for assessment of homogeneity. Interestingly, Buishand and Alexandersson have used the tests that were named after them for evaluating the homogeneity of rainfall records.

3.42.1   Optimization formulations

Traditional optimization formulations discussed previously in this chapter can be modified to include multi-objectives to

Table 3.20 *Variables used in the multi-objective formulation*

Variable	Explanation
$\theta_{m,n}$	Observed values at station m
$\theta'_{m,n}$	Estimated value of precipitation at base station m
μ_m	Mean of series based on observed values
μ'_m	Mean of combine data series (observed and estimated values)
σ_m	Standard deviation of series based on observed values
σ'_m	Standard deviation of combined data series based on observed and estimated values
σ_i	Standard deviation of precipitation data series at station i
μ_i	Mean of precipitation data series at station i
$\rho'_{i,m}$	Correlation coefficient based on data series at station i and station m (observed and estimated values)
$\rho_{i,m}$	Correlation coefficient based on data series at station i and station m based on observed values
$W_{RMSE}, W_\mu, W_\sigma, W_{\sigma*}, W_{\mu*}, W_{\rho*}$	User-specified weights for each objective

preserve site-specific statistics and site-to-site statistics. El Sharif and Teegavarapu (2011, 2012) report some conceptually simple formulations that use summary statistics, such as mean, standard deviation, and correlation between stations, as properties to be preserved. The formulation is given by Equations 3.187–3.189.

Minimize

$$
W_{RMSE}\sqrt{\frac{\sum_{n=1}^{no}\left(\theta'_{m,n}-\theta_{m,n}\right)^2}{no}} + W_\mu|\mu_m - \mu'_m| + W_\sigma|\sigma_m - \alpha'_m|
$$
$$
+ W_{\sigma*}\sum_{i=1}^{ns-1}\left(|\sigma_i - \alpha'_m| - |\sigma_i - \sigma_m|\right)
$$
$$
+ W_{\mu*}\sum_{i=1}^{ns-1}\left(|\mu_i - \mu'_m| - |\mu_i - \mu_m|\right)
$$
$$
+ W_{\rho*}\sum_{i=1}^{ns-1}\left(|\rho'_{i,m}| - |\rho_{i,m}|\right) \tag{3.187}
$$

subject to:

$$
\theta'_{m,n} = \frac{\sum_{j=1}^{ns-1} w_{mj}^k \theta_n^j}{\sum_{j=1}^{ns-1} w_{mj}^k} \quad \forall n \tag{3.188}
$$

$$
w_{lj} \leq w_{mj}^k \leq w_{uj} \quad \forall j \tag{3.189}
$$

Table 3.20 provides details of the variables used for multi-objective optimization formulation.

The formulation (Equations 3.187–3.189) attempts to minimize the weighted absolute differences of:

1. Mean values of precipitation time series based on historical data and data sets with historical data and estimated data.
2. Standard deviation values of precipitation time series based on historical data and data sets with historical data and estimated data.
3. Correlations based on precipitation time series at station with missing data and observations at all other stations using historical data and data sets with historical data and estimated data.

In a multi-objective framework, the user can define the weights associated with each objective function. It should be evident that the objective function defined in Equation 3.187 also includes the traditional objective of RMSE.

3.43 DATA FOR MODEL DEVELOPMENT

Historical data, referred to as training data, are required for estimation of model parameters for several models described in this chapter. The data are primarily used for estimation of correlations; obtaining optimal number of neighbors; development and evaluation of semi-variograms; and estimation of weights in optimal weighting methods and cluster and classification methods. Some of the methods do not require historical data for estimation of missing precipitation data. The SBE, GME, trend surface, and thin-plate splines with and without tension methods do not require training data as there are no parameters that need to be estimated from historical data. These methods use only test data for estimating missing precipitation data at a base station. However, the optimal GME requires historical data (i.e., training data) to obtain the optimal number of rain gages. The normal ratio method requires average annual rainfall observed at stations based on historical data. The nearest neighbor does not require historical data. However, the method requires construction of Thiessen polygons with and without the base station in the monitoring network and the calculation of intersecting common areas. Any spatial analysis environment can be used for development of these polygons.

3.44 OPTIMIZATION ISSUES: SOLVERS AND SOLUTION METHODS

Several optimization models or formulations discussed in this chapter require an optimization solver for solutions. Any commercially available solver can be used for this purpose. Some solvers have their own language and syntax and a few of them can provide a flexible interface to solve optimization problems that resemble spreadsheet platforms. For example, a solver that uses a GRG-based algorithm for solving linear and non-linear programming problems is available in the Microsoft Excel environment. The Nelder–Mead simplex method (Lagarias *et al.*, 1988) can be used for many of the non-linear optimization formulations discussed in this chapter. Optimization solvers such as DICOPT (DIscrete Continuous OPTimizer) under GAMS (general algebraic modeling system) (Brooke *et al.*, 1996) or a GRG algorithm with branch and bound method available in the Microsoft Excel solver environment can be used for solution of formulations with integer and binary variables. In the GAMS environment, the constraints and objective function can be specified in algebraic form and the formulation can be solved with non-linear and mixed integer linear solvers. Genetic algorithms coded or developed through a programming language or available as add-ins in spreadsheet environments can be used for solution of problem formulations that require solutions for mixed integer variables.

Optimization problems with binary variables are computationally intensive. The number of enumerations required for a solution using a branch and bound method with non-linear programming formulation for no restriction on the number of binary variables (n) has an upper limit of 2^n. The solutions obtained for the optimization formulations for linear mathematical programming formulation problems are global optima and for non-linear problems are local or global optima. In the latter case, global optimum solutions are not always guaranteed. Mathematical and conceptual equivalency between optimization models with binary variables and the NLS constraint problem formulations can be easily established. However, the flexibility of defining conditional constraints in MINLP formulations is not available in NLS constraint problem formulations. As long as the deterministic optimization method with constant initial guess values is used and no stochastic search techniques are employed, the results from NLS models are reproducible. Computationally, the NLS constraint problem is easier to solve compared to the MINLP formulation. However, the MINLP formulations with binary variables may result in parsimonious models, as controls points that are least helpful in interpolation are not included in the interpolation.

3.45 SPATIAL ANALYSIS ENVIRONMENTS AND INTERPOLATION

Commercially available spatial analysis environments such as ArcGIS (ESRI, 2001) provide several tools for spatial interpolation. The interpolation is mainly implemented for one set of spatially distributed values to develop surfaces at a fixed tessellation or grid scale. Temporal interpolation or estimation of missing precipitation data for several time intervals is not easily possible using this environment. The interpolation methods available under the ArcGIS environment include (1) inverse distance weighting with flexibility to choose the number of neighbors and optimal exponent; (2) ordinary kriging with facility to remove a spatial trend before modeling the semi-variogram with several authorized semi-variogram models; (3) local and global polynomial interpolation; (4) co-kriging; and finally (5) radial basis functions. The geostatistical wizard extension under the ArcGIS (ESRI, 2001) spatial analysis environment provides all the tools required for interpolation. The extension also provides cross-validation tools to evaluate the interpolation methods using several performance measures. The geostatistical wizard has been substantially improved in the most recent versions of the ArcGIS software.

Li and Heap (2008) document public domain and proprietary software available for interpolation and point pattern analysis, including S-PLUS, Geo-EAS, GeoSTAT, GSLIb, and several scripts available under the *R* statistical package. However, much of the software does not provide the flexibility to estimate missing precipitation values in a time series.

3.46 DATA FILLER APPROACHES: APPLICATION IN REAL TIME

The interpolation approaches discussed in this chapter assume that time-consistent observations are available at all the observation sites used for estimation of missing data. In many practical situations in any given time interval, observations at all the stations may not be available in all time intervals. As the parameters for the interpolation methods or models are based on complete data (i.e., data without gaps) at all other stations excepting the base station, filler approaches are needed to fill the data at sites with missing data. Filler approaches may adopt one of the useful schemes suggested below.

- Single best estimator based on Euclidean distance or correlation criterion. Correlation criterion is generally recommended.
- Fill the gaps in a data set using any approach such as inverse distance or correlation coefficient weighting.
- Local interpolation methods using two or more nearest neighbors can be used for filling the data at a site.

3.47 LOCAL AND GLOBAL INTERPOLATION: ISSUES

The spatial interpolation methods discussed in this chapter may be implemented under local or global environments with respect to selection of neighbors. In local interpolation, neighbors can be selected by different criteria: (1) Euclidean distance, (2) correlation, (3) stations within a specific region and (4) geographic orientation, and (5) stations with reliable data. Neighbors can be ranked by proximity based on distance and correlation and other criteria. Neighbors can also be selected by using MINLP formulations with binary variables. All interpolation methods should be tested for their performance in both local and global variants.

3.48 UNDER- AND OVERESTIMATION

Under- and overestimation of precipitation values is often an unavoidable problem using spatial interpolation methods. Local models developed using a subset of data (i.e., classified based on data magnitude or temporal scale: month or season) might help in reducing the bias, especially when occurrences of precipitation depths above a specific threshold are not uniform in space or confined to a specific site. Corrections to precipitation estimates from interpolations methods using SBEs may help in reducing the under- and overestimations (Teegavarapu, 2009). In general, no rain conditions (zero precipitation values) form the majority of precipitation data when daily or finer temporal level observations are considered. Overestimation is impossible to avoid as observations from other rain gages are used in the estimation of missing precipitation data. Overestimation is due to possible non-zero values at the other stations when the base station (i.e., station with missing data) has registered no rain. Overestimations are not confined to only these specific events of no rain. Underestimation at high values of precipitation is again possible as observations at several sites (i.e., rain gages) are used to obtain the interpolated values at the site where the missing data exist. Unless the extreme events (high magnitude precipitation events) occur at all sites in the region, there is no way the estimates can be close to the observed values in all instances. As many of the optimization models discussed in this chapter use weighted local (based on optimal number of gages) spatial interpolation, underestimation cannot be altogether avoided. Again, in these instances a local model with nearby stations may benefit the estimation. Under- and overestimated values may be reduced if (1) SBEs are used in conjunction with interpolation methods; (2) different sets of gages are selected; and (3) the optimization models are developed using transformed observations.

3.49 MAIN ISSUES AND COMPLEXITIES OF SPATIAL ANALYSIS OF PRECIPITATION DATA

• Precipitation is a complex process that varies spatially and temporally and is region specific. Topography influences precipitation patterns and the interpolation methods should consider the spatial variation of precipitation with terrain.

• Determination of time-consistent observations is important for infilling or point estimation methods.

• Spatial interpolation methods may lead to over- and underestimation for low and high precipitation magnitudes. This is unavoidable due to the fact that each estimate is dependent on other observations in space.

• It is essential that the monitoring network remains stationary and the observations at all control points are available for estimation of missing data.

• Homogeneity of data needs to be checked before spatial interpolation methods can be used for estimation of missing precipitation data.

• Data-driven approaches with less reliance on surrogate parameters of spatial autocorrelation are better than methods that rely on these parameters.

• Missing precipitation data are generally considered to be missing completely at random (MCAR). This assumption needs to be validated before the estimation process is started.

• Stochastic methods that provide uncertainty assessments of the estimates should be preferred over deterministic methods.

• Methods that provide uncertainty assessments of estimates are essential. Stochastic interpolation methods provide uncertainty assessments. However, deterministic methods used in multiple imputation modes can help address this issue. Single-model and multi-model imputations should be carried out to obtain uncertainty assessments.

3.50 SPATIAL INTERPOLATION FOR GLOBAL GRIDDED PRECIPITATION DATA SETS

Spatial interpolation is also essential for conversion of point measurements to areal gridded estimates. Global gridded precipitation data sets that are now available are created using spatial interpolation methods. Detailed discussion about global precipitation data sets was provided in Chapter 2 and others of this book. In many global gridded data sets (Hulme, 1992, 1994), station data within each grid box are areally weighted in order not to give neighboring stations as much weight as isolated ones, after missing data have been estimated using an inverse-distance, angular-weighted anomaly interpolation scheme (Legates and Willmott, 1990). Many gridded data sets are produced based on precipitation anomalies relative to a standard normal period and are not based on actual precipitation observations. It is generally believed that topographic weighting schemes are not required, as the dependence of precipitation anomalies on elevation is much smaller (Hulme and New, 1997) and excluding the effects of elevation is reasonable.

Spatial interpolation for development of gridded data sets using rain gage data requires (1) critical evaluation of available rain gage data; (2) filling of missed rain gage observations; (3) assessment of precipitation anomalies as opposed to actual observations for interpolation; (4) development of methods for dealing with sparse rain gage observations in space; (5) understanding of regional climatology and precipitation patterns; and (6) comprehensive testing of available data interpolation methods.

3.51 SPATIAL INTERPOLATION OF EXTREME PRECIPITATION DATA

Extreme precipitation data collected at rain gages are often analyzed for the frequency analysis that forms the basis for hydrologic design storms. Interpolation of point measurements to areal data sets is difficult due to the limitations of interpolation methods. The tent-pole effect is quite common when iso-pluvial contours are developed for hydrologic design. Development of regional extreme value analysis and careful selection of interpolation method are two solutions to avoid unwanted artefacts associated with interpolated surfaces.

3.52 APPLICABILITY OF METHODS

Several interpolation models discussed in this chapter were evaluated in past studies by Teegavarapu (2007, 2008, 2009, 2012a) in temperate and tropical climate zones. The validity of all the optimization models discussed in this chapter for other climatic zones needs to be thoroughly assessed. However, suitability and applicability of these approaches to other climate regimes can be briefly discussed. Transferability of methods described in this chapter that depend on historical data to any other climatic region is straightforward. The historical data, when used for estimation of parameters (e.g., weights, number of neighbors, correlations, proximity metrics), provide the advantage of developing site-specific methods for estimation of missing precipitation data. Many of the methods discussed in the chapter directly deal with observed data in space rather than geographical distance (Euclidean) generally used in traditional distance-weighted methods, or spatial coordinates in trend surface and thin-plate spline methods. A few methods discussed in this chapter avoid the subjectivity of model parameter estimation and selection by directly dealing with data.

3.53 RAIN: RAINFALL ANALYSIS AND INTERPOLATION SOFTWARE

The RAIN software is a suite of deterministic and stochastic interpolation schemes based on inverse distance, SBEs, correlation weighting, and kriging methods provided with this book. The

Table 3.21 *List of the interpolation methods available in RAIN software*

Method	Variants
Gage mean estimator	Nearest neighbors (correlation)
	Nearest neighbors (distance)
	Local and global
Singe best estimator	Selection by distance
	Selection by correlation
Natural neighbor	
Quadrant	Neighbors and exponents
Inverse exponential	Nearest neighbors
	Radius limited
Inverse distance	Optimal exponent
	Nearest neighbors
	Radius limited
Correlation weighting	Neighbors
	Optimal exponent
Optimal weighting	Distance-based
	Correlation-based
Multiple linear regression	Ordinary regression (distance, correlation)
	Step-wise (distance, correlation)
	Robust-fit (distance, correlation)
Trend surface	Local (linear, quadratic, cubic)
	Global (linear, quadratic, cubic)
Thin-plate splines	Nearest neighbors
Thin-plate splines with tension	Nearest neighbors
Artificial neural networks	ANN parameters
K-NN imputation	Proximity metrics
Ordinary kriging	Variogram cloud
	Spherical, exponential, Gaussian, circular, linear

software is developed by the author of this book. Data sets useful for simple experiments are also provided on the website. The software is generic and easy to work with. A help file is provided with the software for users to apply the techniques discussed in this chapter.

The RAIN software provides several options to analyze missing precipitation data. These include (1) visualization of spatial orientation of observation sites; (2) histograms of observed data at different sites; (3) selection of a specific station from a set of stations for analysis of missing precipitation data; (4) development and evaluation of Thiessen polygons; (5) correlation among observation sites shown by scatter plots; and (6) a suite of deterministic and stochastic interpolation methods for estimation of missing precipitation data at a site. A list of interpolation methods is provided in Table 3.21.

The RAIN software also provides details of four quantitative performance measures: (1) correlation coefficient based

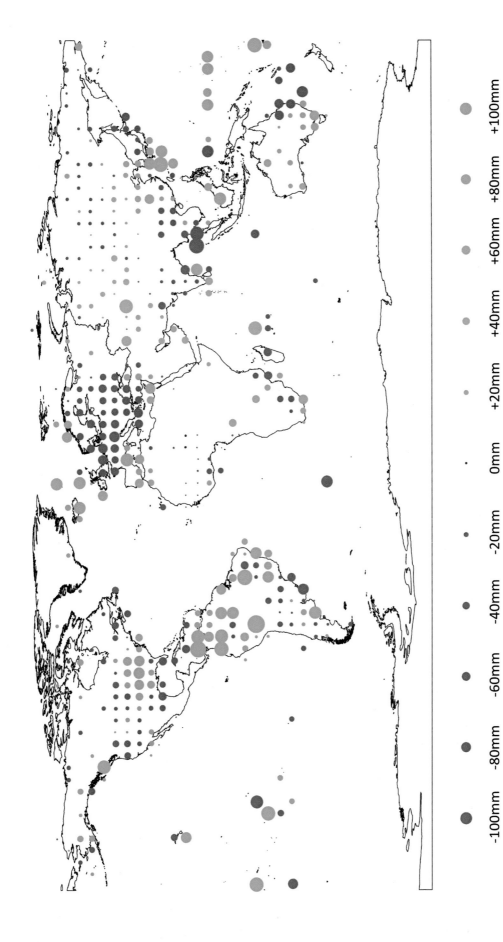

Figure 2.30 Precipitation anomalies for the month of October 2010, based on the GHCN data set using base period 1961–90.

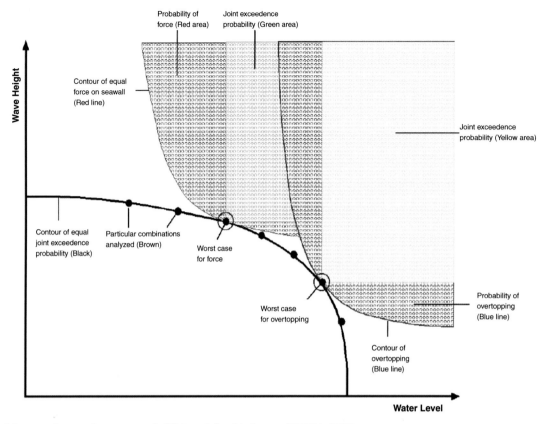

Figure 4.11 Joint exceedence and response probabilities relationship (source: DEFRA, 2005).

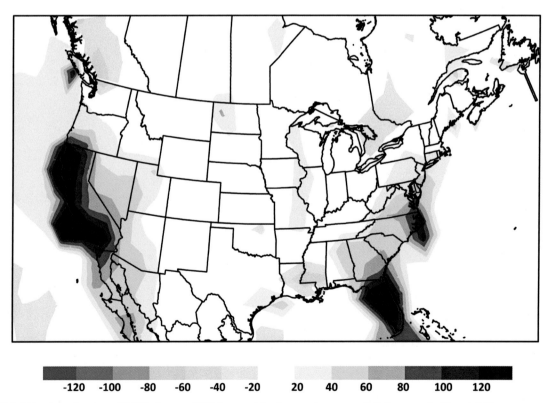

Figure 6.8 Rainfall patterns for strong El Niño for year 1998 (source: http://www.srh.noaa.gov/mlb/?n=enso_florida_rainfall).

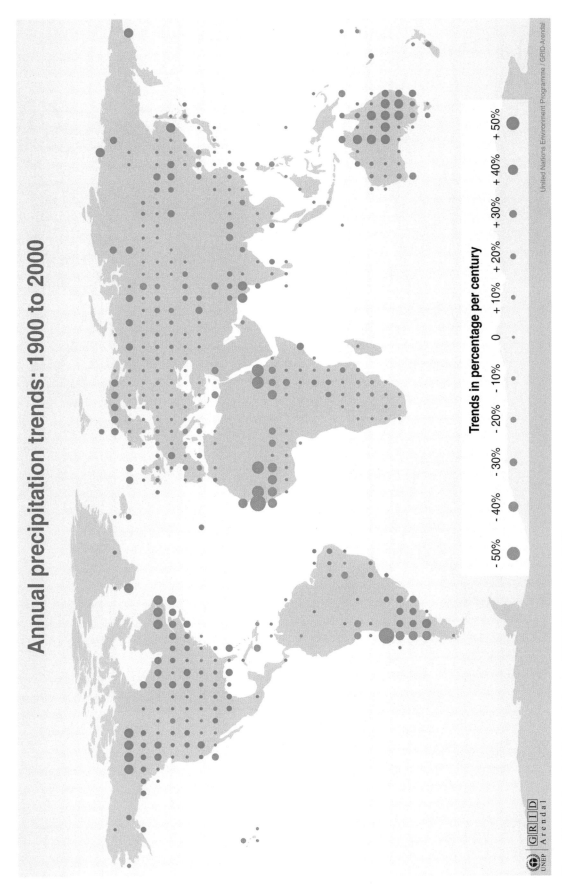

Figure 7.2 Global precipitation trends (source: UNEP/GRID–Arendal, 2005).

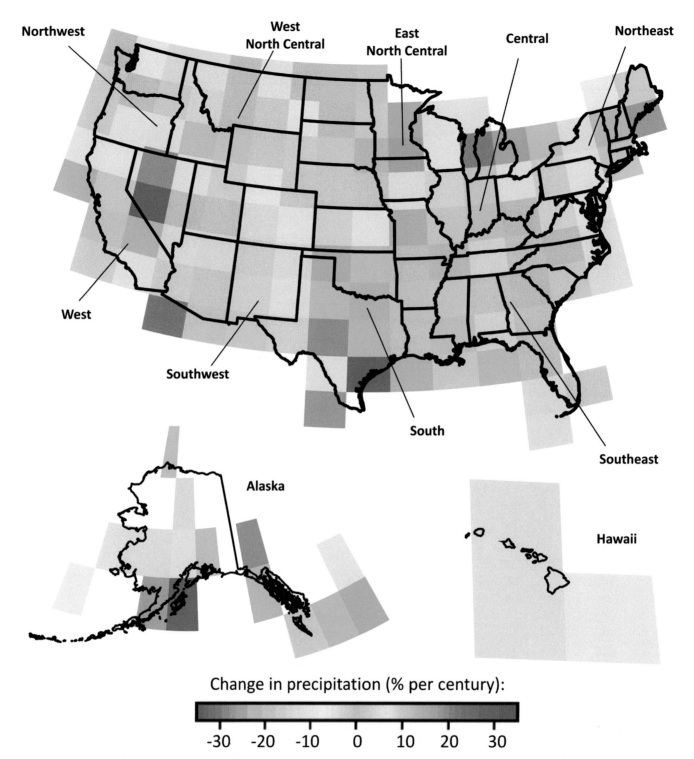

Figure 7.3 Annual precipitation trends 1901–2005 (source: EPA; data courtesy NOAA=NCDC).

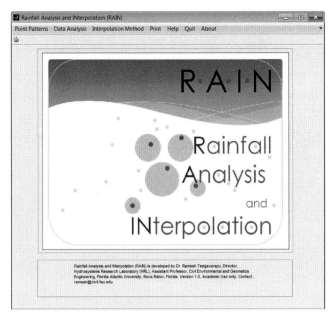

Figure 3.51 Initial screen of RAIN software.

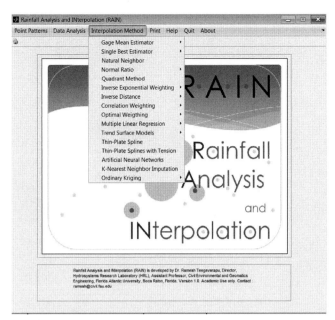

Figure 3.52 Selection of interpolation methods from RAIN software menu.

on observed and estimated precipitation values; (2) absolute error (AE); (3) mean absolute error (MAE); and (4) root mean squared error (RMSE). The software also provides three plots for each method: (1) scatter plot of estimated and observed precipitation values; (2) residual plot; and (3) histogram of residuals.

3.54 USE AND APPLICATION OF RAIN SOFTWARE

The application of RAIN software for estimation of missing precipitation data requires data sets to be prepared in a specific format. Three separate text files are required: (1) location data, (2) calibration data, and (3) validation data. Figure 3.51 shows the initial screen of the RAIN software.

The point patterns item in the menu of the software provides spatial locations of observation stations and Thiessen polygons. The polygons will aid in estimation of missing data using the natural neighbor technique. The data analysis option provides correlation structure of the data. Different interpolation methods can be selected from the interpolation methods menu item, as shown in Figure 3.52.

Screenshots of the RAIN software providing visual and quantitative performance measures and results for estimation using different neighbors are provided in Figure 3.53 and Figure 3.54 respectively.

A user's manual and sample data sets are provided on the website. The RAIN software can be used for any data set as long as the specific format for inputs is created.

3.55 CONCLUSIONS AND SUMMARY

This chapter has discussed several spatial interpolation methods for point estimation and surface generation of precipitation data. Spatial interpolation for estimation of missing precipitation data is a crucial step in generation of serially complete data sets for climate change analyses. The methods discussed range from simple and naïve deterministic estimators to stochastic interpolators. Advantages and disadvantages of each method for continuous estimation of precipitation (without gaps) are elaborated in this chapter. Emerging soft computing tools embedded in spatial interpolation methods and distance metrics based on numerical taxonomy are also discussed. It is important to note that each method has unique characteristics, and applicability of these methods is not universal. Some of the point estimation methods are not applicable to surface generation. A crucial element of all the interpolation schemes is the definition of neighborhood. Local and global interpolation schemes may provide different estimates. All interpolation schemes introduce artefacts into the estimated data and these are difficult to avoid. Modelers using interpolated precipitation data are cautioned about these artefacts and should be aware of the uncertainties associated with the assessments. Multiple imputation schemes can be used to obtain an assessment of the uncertainty involved from a single method or a combination of multiple methods. Conversion of point measurements from rain gages to gridded data sets is one of the most difficult tasks that needs to be carried out before the global precipitation variability and trends from historical data can be assessed.

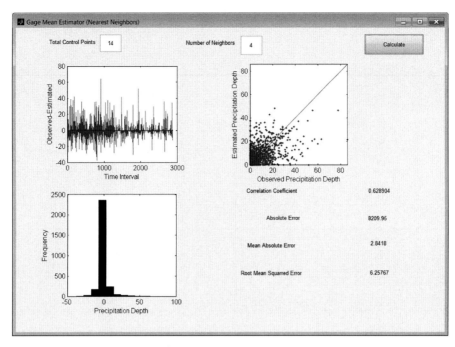

Figure 3.53 Visual and quantitative performance measures from RAIN.

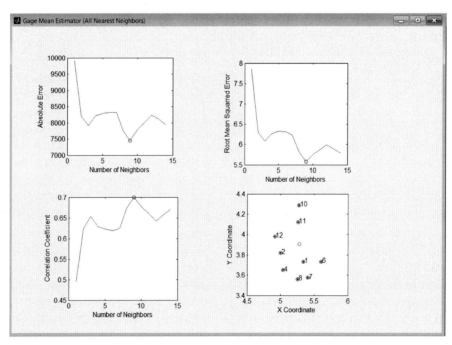

Figure 3.54 Results based on different neighbors using RAIN.

EXERCISES

Data sets available on the website and RAIN software can be used for solving some of these exercises.

3.1 Distinguish between point estimation of rainfall and surface generation.

3.2 Provide examples of exact and inexact surface interpolation schemes.

3.3 Use the RAIN software to estimate the missing precipitation values using the IDWM. Optimize the exponent value for the distance and compare the results. Long-term data provided on the website for 15 stations for Kentucky can be used for this specific exercise. Also evaluate different values (e.g., 1, 2, 4, 8) of exponents and observe the variations in the performance measures.

3.4 Use the RAIN software to obtain the local and global interpolated estimates of missing precipitation data at a point.

3.5 Using data provided in Table 3.22, estimate and evaluate four different semi-variograms discussed in this chapter. Develop a non-linear functional form to model the semi-variograms and check if the conditions of the authorized semi-variogram are met.

Table 3.22 *Observations at different rain gages*

X Coordinate	Y Coordinate	Precipitation depth (mm)
0	1	36.58
2	0	39.62
2	1	30.48
2	2	28.96
3	1	27.43
1	3	30.48
3	1	45.72
3	2	50.29
4	1	24.38
4	2	42.67

3.6 Discuss the limitations of deterministic interpolators and compare and contrast them with stochastic interpolation schemes.

3.7 What is multiple imputation and what are the advantages of this scheme of imputation? Using the observations in Table 3.10, provide estimates of missing precipitation data at point (1,1) using the different interpolation schemes discussed in this chapter. Consider these estimations as multiple imputations and evaluate the uncertainty in the estimate at the specified point.

3.8 Develop your own classifier system for rain or no rain conditions to improve the interpolation schemes for estimation of missing precipitation data.

3.9 Compare the inverse distance interpolation scheme with naïve interpolation schemes such as gage mean estimator, climatologic mean estimator, and single best estimator.

3.10 Using mean extreme precipitation data for different durations and different locations given in Table 3.23 and any geo-spatial tool (e.g., geostatistical wizard or spatial analyst) available under the ArcGIS environment, generate surfaces using different interpolation schemes and evaluate the authenticity of the surface by selecting a specific

number of calibration and validation points in space. Use performance measures such as root mean squared error, absolute error, and correlation coefficient for evaluation.

Table 3.23 *Extreme precipitation data (mm) for different durations*

Latitude	Longitude	Hours					
		1	2	6	12	24	48
30.417500	−84.985000	44	58	81	96	111	124
28.609833	−82.359500	49	64	85	102	122	146
28.475333	−82.434500	50	65	81	96	115	134
26.866667	−80.633333	47	60	77	85	101	122
26.738667	−80.937333	49	65	80	88	103	123
30.766667	−86.583333	48	66	93	107	119	135
30.774500	−86.520167	47	59	86	100	116	136
29.633333	−83.133333	41	53	68	83	100	130
29.643167	−83.159833	52	71	97	117	133	156
29.633333	−83.100000	44	59	76	84	96	118
29.722167	−85.019000	49	70	103	123	139	156
29.176333	−81.042333	46	62	80	93	109	126
29.017500	−81.306333	54	72	82	92	106	123
31.016667	−83.866667	42	53	73	86	98	104
30.243167	−83.255667	45	60	79	93	108	123
26.533333	−81.433333	47	67	76	80	89	99
27.524833	−80.816667	45	57	69	85	94	110
26.584333	−81.856833	58	76	92	104	120	138
30.687833	−82.558000	45	58	77	92	105	118

3.11 Use the data provided on the website and the RAIN software and estimate missing precipitation data at a specific rain gage using the coefficient of correlation weighting method and the non-linear least squares method. Compare the performance of the methods using different qualitative and quantitative measures.

3.12 Develop an ANN model in Microsoft Excel using a sigmoidal activation function to estimate missing data at a station based on the data supplied on the website. The GRG optimization solver available in the Microsoft Excel environment can be used to optimize the connection weights. Document the number of negative precipitation estimates obtained when ANN is used.

3.13 Waikato Environment for Knowledge Analysis (WEKA) is freely available data mining software. Download this software from the internet (http://www.cs.waikato. ac.nz/ml/weka/) and experiment with the daily precipitation data provided on the website to develop association rules. Use the exploratory data analysis option to obtain the correlation plot matrix to visualize the spatial correlations among different observation points.

3.14 Estimate and evaluate real and binary similarity measures using the daily precipitation data (mm) provided in Table 3.24 at different gages using rain gage 11 as base station.

3.15 Use all the performance and error measures discussed in this chapter to evaluate different models based on observed and estimated precipitation values provided in Table 3.25.

Table 3.24 *Rain gage observations*

Rain gage										
1	2	3	4	5	6	7	8	9	10	11
0	0	0	0	0	0	5	9	0	0	10
0	12	24	16	38	23	25	40	15	6	8
0	0	5	0	0	5	0	0	0	0	0
13	19	9	21	0	10	20	20	9	13	0
2	10	25	10	18	27	5	16	23	1	12
1	0	0	0	0	0	0	0	0	0	0
0	0	0	0	0	0	0	0	0	0	0
0	0	0	0	0	0	0	0	0	0	0
46	28	0	28	4	0	20	20	3	10	13
0	1	21	0	18	17	1	1	22	1	1
0	0	14	0	1	0	7	0	4	0	0
0	0	1	0	1	0	0	0	0	0	0
0	0	0	0	1	0	2	0	0	4	0
0	0	3	0	0	0	0	0	6	0	1
0	0	0	0	0	0	0	0	0	0	0
0	0	0	0	0	0	0	0	0	0	0
0	0	0	0	0	0	0	0	0	0	0
0	0	0	0	5	0	0	0	0	1	0
9	2	0	5	3	2	0	3	2	0	5
14	14	4	25	32	2	32	40	8	2	3
0	13	30	13	19	47	1	15	24	0	0
0	0	5	2	1	0	0	0	0	1	0
0	2	0	2	1	0	0	0	0	2	3
0	0	0	0	0	1	6	0	6	3	1
9	0	9	16	14	14	5	8	17	0	0
0	0	0	0	3	0	0	0	0	0	4
4	9	0	19	14	0	36	20	0	6	5
0	11	34	40	28	31	9	40	12	1	6
0	0	0	1	0	0	0	0	0	0	0
0	8	0	0	0	0	0	0	0	0	0

Table 3.25 *Rain gage observations*

	Estimated and observed precipitation depths (mm)								
Observed	64	71	203	163	188	203	279	305	254
Model A	51	64	114	152	170	203	279	254	229
Model B	51	76	127	127	127	152	229	279	203
Model C	51	127	178	178	178	203	262	279	102
Model D	25	51	178	178	229	102	279	229	127

4 Extreme precipitation and floods

4.1 INTRODUCTION

The timing and spatial distribution of precipitation in a river basin are the key components that determine where a flood occurs. To fully understand the origins of hydrometeorological floods, the temporal and spatial scales of atmospheric and hydrologic processes must be addressed for the basin in any study. It is crucial to adequately interpret the spatial and temporal distribution of precipitation to consequently evaluate the streamflow response. Typically, the atmospheric causes of flooding are understood using the theory of flood hydroclimatology. Hydroclimatology is an approach to analyze floods through the temporal context of their history of development and variation and the spatial context of the local, regional, and global atmospheric processes and circulation patterns from which the terrestrial flood develops (Hirschboeck *et al.*, 2000). The antecedent conditions, regional relationships, large-scale anomaly patterns, global-scale controls, and long-term trends can also affect the flood variability (Hirschboeck, 1988; Hirschboeck *et al.*, 2000) in space and time. According to Hirschboeck *et al.* (2000), there are four scales of meteorological phenomena, which can be enumerated as shown in Table 4.1.

4.2 HYDROMETEOROLOGICAL ASPECTS OF PRECIPITATION

4.2.1 Smaller-scale precipitation systems

Stormscale and mesoscale features that produce heavy rains are almost always convective in nature. Convective storms form within an atmosphere that is conditionally unstable. These systems are generally described (Hirschboeck *et al.*, 2000) as:

- **Stormscale systems:** Isolated thunderstorms can deliver localized rainfall rates and amounts sufficient to produce local flash flooding, which is likely if the intense rainfall produced by isolated thunderstorms is concentrated within a small drainage basin or if the storm moves slowly across the basin.

- **Mesoscale systems:** Convective clouds do not typically evolve as isolated entities; rather they tend to organize into narrow lines or bands or into clusters or complexes of individual storms. In a very broad sense, mesoscale systems in the mid latitudes can be classified as:
  - Precipitation bands, which occur most frequently during the cool season over the oceans and adjacent coastal regions, result when the circulations within the parent cyclone produce an unstable thermal stratification that will support convection.
  - Squall lines, which are most frequent over the continents and usually occur in the warm, moist air ahead of cold fronts, are basically long narrow lines of intense thunderstorm cells.
  - Mesoscale convective complexes (MCC) are large, organized convective cloud systems that lack the distinct linear structure of squall lines (Maddox, 1980; Fritsch and Maddox, 1981; Fritsch *et al.*, 1986). The precipitation structures embedded within MCCs can be very complicated, consisting of short squall lines, rainfall bands, a cluster of thunderstorms, and individual thunderstorms as well as widespread areas of steady, light-to-moderate rainfall.
  - Tropical systems are basically short-wave troughs in the easterly wind flow that manifest as convective cloud clusters.

4.3 LARGER-SCALE PRECIPITATION SYSTEMS

The macroscale and synoptic-scale events occur over large regions affecting vast geographic areas. The larger-scale processes that produce heavy rains are often associated with anomalous and persistent middle-level atmospheric wave patterns, slow moving or stagnant features such as blocking anticyclones and cutoff lows, or nearly stationary synoptic frontal zones. These systems are:

- **Synoptic-scale systems:** Heavy rains in these systems can occur along and ahead of cold fronts or along and to the cool

Table 4.1 *Different scales of meteorological phenomena*

System	Scale	Typical length scale	Typical time scale
Larger-scale precipitation systems	Macroscale	More than 10,000 km	More than 1 month
	Synoptic-scale	10,000 km to 2,000 km	Less than 1 month
Smaller-scale precipitation systems	Mesoscale	2,000 to 500 km	Approximately few days
	Stormscale	5 km to 50 km	Few hours

side of warm and stationary fronts. This type of situation can also indirectly contribute to heavy rains by setting up an environment very conducive to the development of smaller-scale precipitation systems. The floods associated with synoptic weather features are usually widespread, so flooding occurs over large river systems during an extended period.

- **Macroscale systems:** Processes in these systems can support the development, persistence, or sequential recurrence of smaller-scale precipitation systems that produce floods.

In subtropical areas, or in areas with extensive regions of warm oceans, the land–sea breeze circulations can be important in developing deep moist convection (DMC). This role for land–sea breeze circulations can be enhanced for islands and peninsulas (Pielke, 1974). For example, subtropical peninsular Florida is noted for its frequency of warm-season thunderstorms even though the synoptic flows in the region remain generally weak (Doswell and Bosart, 2000). The convective features over South Florida present a large variation in both space and time. Consequently, rainfall amounts and areal coverage can both vary by orders of magnitude and the location of convection may differ considerably from one day to the next (Blanchard and Lopez, 1985).

4.4 CONVECTIVE PATTERNS

Convective rainfall patterns are common in several parts of the world especially in tropical regions. Several factors can be responsible for the observed large day-to-day fluctuations in convective patterns:

- Variations in the large synoptic regimes should develop distinct temporal and spatial patterns of convection.
- Surface features in large water elements, such as Lake Okeechobee, water conservation areas, and the coastal configuration should influence the generation, maintenance, and decay of circulations on the peninsular and local scales.
- Synoptic- and regional-scale flow.

Convective storms or, more commonly, thunderstorms are produced by a phenomenon known as convective lifting, where air rises by virtue of being warmer and less dense than the surrounding air, resulting in the precipitation events. Convective storms

have been extensively investigated in several studies (Byers and Rodebush, 1948; Day, 1953; Gentry and Moore, 1954; Frank and Smith, 1968; Pielke, 1973) for organization and evolution of the convective field and attempts to show its relationship to the peninsular- and synoptic-scale forcing.

4.5 PRECIPITATION AND RIVER REGIMES

Precipitation regimes describing the distribution of precipitation totals on an annual basis were described by Shaw (1994) for eight regions around the world. Some of the regions were selected to coincide with the global precipitation regimes described by Petterssen (1958). According to Petterssen, the precipitation regimes can be divided into seven classes: (1) Equatorial (within a few degrees north and south of the Equator; (2) Tropical (between the Equator and tropics of Cancer and Capricorn); (3) Monsoon (within and outside the tropics in the Indian sub-continent on the east side of continents); (4) Subtropical (tropics to 30° north and southwest side of continents); (5) Mediterranean (within 30° to 40° north and southwest side of continents); (6) Mid latitudes; and (7) Polar.

River regime refers to the expected pattern of river flow during a year (Shaw, 1994). Definition of river regimes requires long-term flow data to obtain the general characteristics of the river flows. The river regime is the direct consequence of the climatic factors influencing the catchment runoff, and can be estimated from knowledge of the climate of a region. Classifications (Shaw, 1994) are possible based on two factors: temperature and rainfall. The temperature-dependent regimes are further categorized into glacial, mountain snowmelt, and plains snowmelt. The rainfall-dependent regimes are divided into four types: equatorial, tropical, temperate oceanic, and Mediterranean. Extensive details and the origins of classification schemes are provided by Shaw (1994).

4.6 HYDROMETEOROLOGICAL ASPECTS OF FLOODS: REVIEW OF CASE STUDIES

The meteorological systems described in the previous sections are not enough to predict flooding. Other key components that must be included in the analysis are rainfall–runoff processes, surface and subsurface hydrologic conditions, and land-use properties. Also, the antecedent climate, the hydrologic factors, and the available reservoir storage affect the possible flooding in a basin (Dunne, 1983). The soil moisture, soil saturation, and groundwater levels (shallow groundwater tables) are determinant factors in flooding events. Saturated soils are not a required precursor to severe flooding. For example, flash floods commonly occur when runoff is conveyed to stream channels rapidly as Hortonian overland flow, regardless of soil moisture content. In areas other than South Florida, snow and snowmelt also have a strong influence in flooding.

Floods in a single drainage basin can be generated by different types of precipitation events. The type of event that occurs depends in part on the season, but also may depend on factors related to the large-scale circulation environment. It is possible to forecast or monitor a developing flood situation over meteorological and short climatic time scales. Hydrologists have long used time series of gage flood-peak data to evaluate the probability of occurrence of a flood of a given magnitude. Precipitation systems have different intensities and durations and the flood peaks in a given stream may reflect this, resulting in hydroclimatically defined mixed distributions in the overall probability density function of flood peaks (Hirschboeck, 1987a, b, 1988). For example, two rivers in the Arizona portion of the Lower Colorado River basin that experience flooding from different precipitation systems have been grouped into three categories: stormscale precipitation systems, tropical storm-related precipitation systems, and synoptic-scale extratropical cyclone–frontal precipitation systems (House and Hirschboeck, 1997). This approach can supplement a purely statistical analysis of the flood peaks by revealing the kinds of weather systems that produce different magnitudes of floods. Also, it can be used to evaluate the causes of flood variations over time by different types of storms in several streams throughout a region or by evaluating the long flood record of a single stream in detail (Hirschboeck, 1987b; Webb and Betancourt, 1992).

In British Columbia, a mountainous watershed has been studied by Loukas and Quick (1996) by analyzing 175 storms and classifying them into four categories: winter rainfalls, snowstorms, summer rainfalls, and storms (mixture of rain and snow). They analyzed the spatial distribution by comparing the statistical correlation coefficients of the values of various storm features (precipitation, duration, average intensity, maximum hourly intensity, and relative start time of the storm) at each station to values at a selected base station in the area. Temporal distributions of the storms were evaluated using the well-known Huff's methodology (Huff, 1967), where, by recognizing the variability of hyetographs, the temporal distribution of precipitation can be expressed as isopleths of probabilities of dimensionless accumulated storm depths and durations. An analysis of historical large storms was performed. The results from this analysis indicate that

the 175 storms were comparable with large storms that occurred in the same area of study (Loukas and Quick, 1996). The potential effect of ENSO on the inter-annual variations in regional precipitation and streamflow has been widely studied (Ropelewski and Halpert, 1986, 1987; Redmond and Koch, 1991). In some areas of the world the ENSO–precipitation–flood relationship is straightforward, but in other areas a corresponding relationship to floods is not evident. Most regions are affected by multiple hydroclimatic mechanisms for generating floods, and this complicates the relationship of floods with the inter-annual variations of a single regional climate component. Researchers report that an analysis of variations in the position, intensity, or frequency of the large-scale atmospheric circulation patterns can increase understanding of the hydroclimatic causes for the spatial distribution of flooding (Hirschboeck et al., 2000). Also, anthropogenic climatic changes and natural shifts in climatic conditions are capable of affecting hydrologic systems.

A correlation has been made by paleohydrologic researchers between small changes in mean temperature and precipitation over the past few thousand years and large changes in flood frequency and magnitude on river systems (Chatters and Hoover, 1986; Knox, 1993; Ely et al., 1993; Ely, 1997). In the Lower Colorado River basin, a relationship between floods and climatic variations appears to hold true for at least the past 5,000 years (Ely, 1997).

A study of national trends and variation in UK floods was performed by Robson et al. (1998). Using extensive peaks over threshold and annual maxima data from 890 gaging stations, two annual series representing flood size and flood occurrence were examined by applying three main tests for trend: linear regression, normal scores linear regression, and Spearman's correlation. The study showed evidence that urbanization can affect flood regime, and also progressive changes in the UK flood regime can be observed. The rare and extreme, catastrophic flooding events are the most difficult to understand by traditional methods or direct observation, since the gage record from a river does not provide enough data to characterize the frequency of the event. According to Hirschboeck et al. (2000), examining the likelihood of clusters of floods within multiple-year periods dominated by a particular set of climatic conditions conducive to flooding is more successful than attempting to assess the occurrence of floods on an annual scale. Differences in paleoflood patterns in the magnitude and frequency of the largest floods in the USA show enormous implications for flood-frequency forecasting and river management (Hirschboeck et al., 2000).

The shape, timing, and peak flow of a streamflow hydrograph are significantly influenced by spatial and temporal variability in rainfall and watershed characteristics such as land use. Studies have been carried out to investigate linkages between climate indices, streamflows, and precipitation. A composite analysis approach was used to identify linkages between extreme flow and timing measures and climate indices for a set of 62 hydrometric stations in Canada (Sharif and Burn, 2009). The methodology involves examining the data record for the extreme flow and timing measurements for each station for the years associated with the ten largest values and the ten lowest values of various climate indices from the data record. In each case, a re-sampling approach was applied to determine if the ten values of extreme flow measures differed significantly from the series mean. The results show that several stations are influenced by the large-scale climate indices. The trends in extreme hydrologic events, both high- and low-flow, were analyzed by investigating four extreme flow measures: high-flow magnitude (HFM), high-flow timings (HFT), low-flow magnitude (LFM), and low-flow timings (LFT). In Northern Italy, a compound measure was created to describe the mean delay between combined flood and rainfall seasonality. The relative performance of four combined flood and rainfall seasonality descriptors was evaluated in the research (Castellarin et al., 2001).

As is known, the hydrologic response of a watershed varies with the dynamics of the storm precipitation. Several studies have investigated the linkage between precipitation and the streamflow data by correlating the data using techniques such as the Mann–Kendall test and linear regressions (Lettenmaier et al., 1994; Lins, 1997; Lins and Slack, 1999). The Mann–Kendall test was applied for the Mackenzie Basin in northern Canada to identify the trends and variability in the hydrologic regime by collecting data from 54 flow gaging stations and 10 meteorological stations. Trends were evaluated for different seasons, and winter month flows show strong increasing trends (Burn and Elnur, 2002; Abdul Aziz and Burn, 2006). Streamflow records were collected for five major river basins of Minnesota, USA, from 36 gaging stations (Novotny and Stefan, 2007). Discharge trends and the dynamics of South American rivers flowing into the South Atlantic were analyzed using this methodology (Andrea and Depetris, 2007). The annual and seasonal precipitation in the Hanjiang Basin as well as the water levels and the streamflow in the Yangtze River basin in China were investigated following these techniques (Hua et al., 2007; Zhang et al., 2006). Trends in precipitation and streamflow at different time scales for 50 years of data and the linkage with recent watershed and channel changes were investigated in the semi-arid Rio Puerco Basin in northwestern New Mexico. The Rio Puerco Basin was divided into five hydrologic units. For a 50-year period, hydroclimatic data were collected and analyzed in order to identify seasonality, variability, trends, and other characteristics using the non-parametric Mann–Kendall test (Molnar and Ramirez, 2001).

Research conducted on Hudson Bay rivers in the Canadian Arctic shows a decreasing trend in the streamflow that was linked to various large-scale atmospheric phenomena (Dery and Wood, 2004, 2005). Seasonal flood peaks, annual runoff volumes, and trends from 1807 to 2002 were analyzed in a study conducted in Sweden (Lindstrom and Berstrom, 2004). A statistical methodology was applied in a study that was conducted to explore the seasonality and multi-modality of floods as well as to identify any regularity in the spatial distribution of the corresponding extreme rainfall event. Non-parametric tests were used to link rainfall and flood events, and a pooling framework based on the rainfall regime was applied to identify hydrologically homogeneous pooling groups (Cunderlik and Burn, 2002).

Seasonal linkage between streamflow and precipitation has been studied in previous research in the USA. The increments in winter and spring streamflow were investigated (Lettenmaier et al., 1994), as well as an analysis of trends in selected quantities of flow discharge for 295 stations (Lins and Slack, 1999). One of the results found in these works was that the conterminous USA is getting wetter but less extreme (Lins and Slack, 1999; Douglas et al., 2000; Burn and Elnur, 2002). The two most frequently used statistical methods to capture flood seasonality are the annual maximum model and the peaks over threshold sampling model. An evaluation of these techniques has been performed by Cunderlik et al. (2004), demonstrating that the peaks over threshold sampling model outperforms the annual maximum model in most analyzed cases, providing more information on flood seasonality.

To recapitulate, in short, it is fundamental to explore the synergism between meteorological, climatic, hydrologic, and drainage basin factors as well as both the short-term meteorological and the long-term climatic processes in order to fully understand the causes of flood. Streamflow data have the advantage over rainfall data in that the complex variability of precipitation, evaporation, transpiration, land use, topography, and other physical characteristics of the region are reflected in the streamflow records (Sharif and Burn, 2009). It can be concluded that the common approach used by different authors relies on the use of statistical methods to find the existing links between hydrologic events and flooding in populated and agricultural areas. In that sense, the authors are proposing a methodology that starts with wide data collection, continues with detailed processing, and ends with an analysis that delivers a better understanding of the linkage between rainfall and flooding events, thus improving the warning systems currently in use.

4.7 PROBABLE MAXIMUM PRECIPITATION

The probable maximum precipitation (PMP) is defined by WMO (1994) as the theoretically greatest depth of precipitation for a given duration that is physically possible over a storm area of a given size under particular geographical conditions at a specified time of the year. It is widely used in the design of dams and other large hydraulic systems, for which a very rare event could have disastrous consequences. Probable maximum precipitation refers to the quantity of precipitation that approximates the physical upper limit for a given duration over a particular basin (WMO, 1994).

4.7.1 Estimation methods

The estimation of PMP is possible using one of the methods provided by WMO (1986a). These methods include (1) a storm model approach; (2) maximization and transposition of actual storms; (3) generalized depth–area–duration relationships; and (4) statistical analysis of extreme rainfall totals. PMP is commonly estimated by using storm transposition and maximization. This approach is based on the assumptions (1) that precipitation can be expressed as the product of the available moisture and the combined effect of the storm efficiency and inflow of wind, and (2) that the most effective combination of storm efficiency and inflow wind can be estimated from outstanding storms on record (WMO, 1994).

4.7.2 Preliminary estimates and parameters

Probable maximum precipitation estimates may provide sufficient information to guide the design of hydraulic and hydrologic structures in the earliest planning stages. For various regions around the world, generalized PMP estimates are available in the form of maps and diagrams (WMO, 1994). To determine a PMP for a given project basin, depth–area duration analyses for the region of interest are required. In the absence of this information, individual storm studies should be undertaken and the selection of the likely critical rainfall duration for the project design of interest should be reasonably determined (WMO, 1994). For relatively large drainage basins, it may be necessary to divide the region into sub-basins and to determine the maximum flood hydrographs for each sub-basin before conducting storm analyses (WMO, 1994).

4.7.3 Storm transposition

By considering large storms occurring within a meteorologically homogeneous region, it is possible to overcome the obstacle of shortness of rainfall records; however, it is important to consider that individual storms do not always have the same probability of occurrence over all sections of their transposition zone, especially if within the transposition zone there exist variations in orographic features. The US Weather Bureau (1976) and Kennedy *et al.* (1988) report advances in the evaluation of orographic effects for use in storm transposition and synthesis (WMO, 1994, 2009a).

4.7.4 Selection and analysis of major storms

A common method for selecting storms for analysis involves determining the meteorologically homogeneous region that encompasses the project basin and extracting the dates of the occurrence of major rainfalls from available rainfall station records (WMO, 1994, 2009a). Next, the synoptic features of the major storms are examined to determine whether a storm may be transposed to the project basin (WMO, 1994); this may be accomplished via methods proposed by the WMO (1986a).

After the selection of the major storms, the storms are maximized such that the percentage by which a particular storm's rainfall would have been increased if the meteorological characteristics of the storm approached estimated upper physical limits (WMO, 1994) is determined. The reader is referred to Weisner (1970) and the US Weather Service (1986) for published methods of storm maximization and to WMO (1986a) and Hansen *et al.* (1982) for studies relevant to the orientation of maximized storms. With maximized storm data, it is possible to estimate the highest rainfall depth for any selected duration for the project basin or relevant sub-basins (WMO, 1994).

4.7.5 Data scarcity: dealing with missing data

If critical meteorological and streamflow data are missing, then PMP estimates should be determined by way of analogy to PMP in climatologically similar regions with available data (WMO, 1994). However, special care should be taken to account for significant topographic, orographic, and meteorological differences between climatologically comparable regions. Statistical procedures may be of utility in determining PMP in such scenarios and are described in the WMO *Manual for Estimation of Probable Maximum Precipitation* (WMO, 1986a).

4.7.6 Statistical estimation method

Hershfield (1965) provided a statistical method to determine PMP if times series data for observed rainfall of a given duration are available:

$$PMP = \bar{X}_n + K_m S_n \qquad (4.1)$$

where $\bar{X}_n$ refers to the times series mean of observed precipitation for a specified time interval, K_m is a modification coefficient, and S_n is the standard deviation of the time series. The variable K_m may be defined by (Hershfield, 1965):

$$K_m = \frac{X_{max} - \bar{X}_{n-1}}{S_{n-1}} \qquad (4.2)$$

where X_{max} is the largest (first ranked) item in the ranked observed series, $\bar{X}_{n-1}$ is the mean of the series excluding X_{max}, and S_{n-1} is the standard deviation of the series excluding X_{max}.

The following example is used to illustrate the calculation for PMP using a statistical method. Annual time series of observed

Table 4.2 *Annual extreme precipitation totals for 24-hour duration*

Year	Depth	Year	Depth	Year	Depth	Year	Depth
1942	387	1958	179	1974	129	1990	114
1943	74	1959	162	1975	110	1991	215
1944	84	1960	217	1976	139	1992	124
1945	234	1961	56	1977	134	1993	142
1946	167	1962	91	1978	135	1994	155
1947	224	1963	169	1979	186	1995	161
1948	180	1964	180	1980	88	1996	61
1949	150	1965	243	1981	106	1997	98
1950	106	1966	122	1982	224	1998	232
1951	190	1967	122	1983	132	1999	181
1952	147	1968	117	1984	195	2000	131
1953	77	1969	124	1985	80	2001	148
1954	118	1970	124	1986	105	2002	134
1955	116	1971	138	1987	102	2003	140
1956	88	1972	157	1988	171	2004	278
1957	162	1973	71	1989	99		

extreme precipitation (mm) data for 24-hour duration for an observation site in South Florida, USA, are provided in Table 4.2. Calculations for PMP using Hershfield's method are shown in this section.

The variables, X_{max}, $\bar{X}_{n-1}$, and S_{n-1} after ranking data by depth of precipitation from largest to smallest are calculated.

$$X_{max} = 387\,mm \tag{4.3}$$

$$\bar{X}_{n-1} = 143\,mm \tag{4.4}$$

$$S_{n-1} = 48.7\,mm \tag{4.5}$$

The value of K_m is estimated by:

$$K_m = \frac{X_{max} - \bar{X}_{n-1}}{S_{n-1}} = \frac{387 - 143}{48.7} = 5.0 \tag{4.6}$$

Variables $\bar{X}_n$ and S_n are:

$$\bar{X}_n = 146\,mm \tag{4.7}$$

$$S_n = 57.3\,mm \tag{4.8}$$

The value of PMP is calculated using:

$$PMP = 146 + 5.0\,(57.3) = 433\,mm \tag{4.9}$$

The PMP value for a 24-hour duration rainfall is estimated at 433 mm.

When the probable maximum precipitation is to be applied to an area larger than about 25 km^2, the point value is generally reduced by means of depth–area or area reduction curves provided by WMO (2010).

4.8 PRECIPITATION-BASED DRIVERS AND MECHANISMS INFLUENCING EXTREME FLOODS

Understanding of precipitation-based drivers and mechanisms influencing flooding events is critical for hydrologic modeling and design of structural and non-structural measures for infrastructure protection.

Often statistical methods are used to find the existing links between hydrologic events and flooding in populated and agricultural areas. In general, any methodology used for evaluation of extreme floods should start with a wide range of data collection with detailed processing of data. Analysis that provides a clear understanding of the behavior of the pilot watersheds and the linkage between rainfall extremes and any other hydrological variables that cause flooding events should be carried out. Short-term forecasting methods can be developed for warning systems that are currently in use.

4.9 FLOODING MECHANISMS

In general, floods fall into three major categories: (1) riverine flooding, (2) overland flooding, and (3) flash flooding. Riverine flooding is the most common of the flooding events. When a channel receives too much water during high storm events the excess water flows over its banks and into the adjacent area. Flooding that occurs along a stream or channel is called riverine flooding. In general, the terrain of the watershed determines the dynamics of riverine flooding. In relatively flat areas, floodwater may cover large areas of land for long periods of time. Flash floods usually occur in hilly areas and urban areas. Overland flooding or basin flooding is common in areas dominated by flat terrain. Overland flooding is defined as the increase in volume of water within a river channel and the overflow of water from the channel onto the adjacent floodplain. In many parts of the world, flooding occurs in combination with heavy rainfall on a large expanse of flat terrain during periods of high groundwater table.

4.10 FLOODING AND SHALLOW GROUNDWATER LEVELS

The patterns and mechanisms of flooding in areas dominated by surface and groundwater interactions due to a shallow groundwater table are different from traditional overbank flooding. The literature is replete with hydrologic simulation models that consider subsurface water interactions on surface or overland flooding or vice versa. However, many studies focus on assessment of groundwater table variations due to overland flooding. While it is essential to assess those interactions or influences, the impact of high water tables on floods needs to be evaluated. Also, peak flood discharges are a function of several factors that include flat terrain, highly permeable soils, and high water tables. Models capable of considering the interactions in the hyporheic zone and relevant processes should be adopted for modeling the groundwater–surface water interactions. The variation of soil moisture due to changes in high groundwater table conditions can be addressed by explicitly characterizing and modeling soil moisture zones and the groundwater table conditions. Complexity of the soil moisture variations linked to groundwater table conditions warrants changes to soil infiltration equations. A modified Green–Ampt equation that considers changing water table heights can be adopted in the modeling process. Single event models may not be sufficient to characterize the groundwater–surface water interactions and soil moisture exchanges. However, valuable insights can be gained into the influences of groundwater table conditions on peak discharges. Extensive spatial and temporal hydrologic data are required for characterizing the relationship between soil moisture capacities and groundwater table conditions. Non-linear relationships linking the soil moisture capacity and groundwater table heights can be adopted. Single and continuous simulation models may provide different results in terms of increases in peak discharges.

In many hydroclimatic settings, saturated and shallow groundwater table conditions play a critical role in causing and exacerbating floods. Models capable of characterizing the saturation overland flow explicitly are needed to fully understand the influence of shallow groundwater tables on soil moisture conditions. Continuous simulation models for extended periods are recommended for characterizing the subsurface soil moisture capacities related to changing groundwater levels. Detailed study of the patterns and mechanics of flooding (generally defined as inundation hydrology) needs to be carried out to establish the dominant nature of flooding in a region. Overbank and basin flooding occurring simultaneously needs to be considered in many situations. A distributed hydrologic simulation model capable of modeling the complete channel–floodplain system is required in such situations. Different antecedent moisture conditions linked to the varying groundwater table conditions can be modeled using single event and continuous simulation models. However, the results from such simulations for design rainfall events should serve only as preliminary assessments of expected changes in peak discharges. A hydrologic model used for simulating infiltration and percolation losses should account for all the flows entering, moving within, and leaving the system, as well as storage changes within the system.

Soil moisture accounting (SMA) is a viable approach in a hydrologic simulation model that is capable of accounting for the surface and subsurface processes that affect the peak flooding rates. Models adopted for floodplain analysis in shallow groundwater table conditions require explicit consideration of groundwater table conditions and their influence on soil moisture capacities. This consideration requires extensive surface and subsurface soil characterization and clear understanding of surface flooding mechanisms. Data relevant to the hydrogeology of the region become crucial for accurate soil characterization. Floodplain analysis and peak discharge estimation using design rainfall events along with different antecedent moisture conditions (AMC) alone are not sufficient to ascertain flood risk. Detailed studies are required to evaluate the influence of shallow water tables combined with already high AMC. Detailed modeling studies are needed to direct the development of simplified modeling approaches for modeling groundwater–soil moisture interactions. Defensible modeling approaches that are conceptually sound, simple, and robust are essential when the task of exhaustive modeling of groundwater–soil moisture interactions becomes exorbitantly expensive due to extensive spatial and temporal data requirements.

4.11 SOIL MOISTURE CONTRIBUTIONS TO FLOODING

Shallow groundwater table conditions can increase the available soil moisture and therefore lead to decrease in soil storage capacity and increase in peak flooding events due to wet AMC. Fleming and Neary (2004) developed model parameters and calibration methods that are used in a SMA algorithm. Model performance indicates that the developed parameters and calibration methods work well when applied to the test watershed. In areas of high water table, there is a limit to the soil storage capacity, and infiltration cannot continue indefinitely without complete saturation of the soil. In such cases infiltration ceases, rainfall abstraction becomes zero, and rainfall excess intensity equals rainfall intensity. Soil storage capacity, S, can be calculated using the following equation (Bedient $et\ al.$, 2009):

$$S = L(\theta_s - \theta_i) \qquad (4.10)$$

where L is depth of water table, θ_s is the saturated moisture content, and θ_i is the initial moisture content. In some areas, regional information on available soil storage can be used for modeling purposes. The following equation is used to calculate the soil storage capacity using the curve number (CN):

$$S = \frac{1000}{CN} - 10 \qquad (4.11)$$

Initial abstraction is calculated using $I = 0.2S$. The Green–Ampt infiltration model can also be used for prediction of runoff (e.g., Manivannan $et\ al.$, 2001). The Green–Ampt model predicts runoff volume more accurately than the Soil Conservation Service (SCS) CN method.

4.11.1 Soil moisture and links to groundwater levels

Groundwater table levels can influence the amount of soil moisture and ultimately the storage capacity. Gregory *et al.* (1999) used different methods to estimate soil storage capacity for stormwater modeling applications. They used the following equation to calculate the soil storage capacity:

$$S = aH^b \qquad (4.12)$$

where H is the high water table depth (groundwater table) in feet and S is the soil storage capacity in inches. Gregory *et al.* (1999) indicate that soil storage capacities may be modified to better reflect development activity, whereby native soil values were reduced by 25% to account for the reduction in void spaces due to the compaction of earthwork operations. Based on the soil information and high water table (HWT) conditions, a functional relationship between soil storage capacity and HWT can be established. The relationship can then be used to calculate soil moisture capacity at different water table depths and the adjusted CNs are calculated. These adjusted numbers can then be used in methods that use the SCS CN approach for estimation of peak discharges.

4.11.2 Adjusted curve numbers and antecedent moisture conditions

This section provides an example that helps to model and understand the influence of high groundwater table conditions on peak discharge. Based on the soil information and high groundwater table conditions, a functional relationship between soil storage capacity and HWT or groundwater table is first established. The relationship is then used to calculate soil moisture capacity at different water table depths and the adjusted CNs are calculated. These CNs are used in a single event hydrologic simulation model (e.g., HEC-HMS) for simulation of peak discharges in the event-based model that uses an SCS CN approach. A watershed in Polk County, Florida, USA, is used as a case study region for estimating

Table 4.3 *Basin and meteorological parameters*

Parameters	Description
Basin	
Basin area	153 km^2
Loss method	SCS CN
Transform method	SCS unit hydrograph
Graph type	Standard
Lag time	9.93 hours
Meteorological	
Precipitation type	SCS storm
Storm type	SCS Type III
Precipitation depth	11 inches
Design storm	100 years

Table 4.4 *Curve number and peak discharge variations based on groundwater table (GWT) conditions*

GWT (mm)	Soil storage capacity (mm)	CN	Peak discharge (m^3/s)
152.4	34.622	88	1,335.25
304.8	67.818	79	1,253.59
457.2	100.497	72	1,166.88
609.6	132.845	66	1,080.17
762.0	164.948	61	998.71
914.4	196.858	56	927.29
1,066.8	228.608	53	855.85
1,219.2	260.221	49	796.44
1,371.6	291.717	47	738.50
1,524.0	323.107	44	683.03
1,676.4	354.403	42	638.32
1,828.8	385.613	40	596.24

peak discharge values and the influences of high seasonal groundwater table. The peak discharges are estimated using HEC-HMS with basin and meteorological parameters as shown in Table 4.3. Details of the HEC-HMS and model simulations can be found in the study by Teegavarapu *et al.* (2010). The SCS CN used for this case study example is 65.

The values 2.67 and 0.97 are used for coefficient a and exponent b, respectively, in the relationship $S = aH^b$ and the CNs are calculated based on different groundwater table conditions. The GWT condition refers to water table height below the soil surface.

The percentage increases in the CNs in Table 4.4 are used to calculate the adjusted CNs and corresponding peak discharges. These adjusted numbers and peak discharges are given in Table 4.5.

Table 4.5 *Adjusted CNs and peak discharge increases based on groundwater table (GWT) conditions*

GWT (mm)	Adjusted CN	Soil storage capacity (mm)	Peak discharge (m^3/s)	Increase in peak discharge (%)
152.4	98	5.2	1,378	29
304.8	98	5.2	1,378	29
457.2	98	5.2	1,378	29
609.6	98	5.2	1,378	29
762	98	5.2	1,378	29
914.4	92	21.4	1,359	28
1,066.8	86	40.8	1,320	24
1,219.2	81	60.2	1,275	20
1,371.6	76	79.4	1,219	14
1,524	72	98.6	1,168	10
1,676.4	68	117.7	1,112	4
1,828.8	65	136.8	1,065	–

Table 4.6 *Conversion table for CN from AMC Class II (normal) to AMC Class I (dry) or Class III (wet)*

AMCII	AMCI	AMCIII	AMCII	AMCI	AMCIII
100	100	100	58	38	76
98	94	99	56	36	75
96	89	99	54	34	73
94	85	98	52	32	71
92	81	97	50	31	70
90	78	96	48	29	68
88	75	95	46	27	66
86	72	94	44	25	64
84	68	93	42	24	62
82	66	92	40	22	60
80	63	91	38	21	58
78	60	90	36	19	56
76	58	89	34	18	54
74	55	88	32	16	52
72	53	86	30	15	50
70	51	85	25	12	43
68	48	84	20	9	37
66	46	82	15	6	30
64	44	81	10	4	22
62	42	79	5	2	13
60	40	78	0	0	0

After Soil Conservation Service (1972)

The adjusted CNs based on the HWT conditions can be compared with the standard CNs published by SCS (1972) for different AMC, as provided in Table 4.6. For a given groundwater table height, the adjusted CN equates to the CN for AMC III (wet condition) based on the CN for AMC II (normal or average condition) as shown in Table 4.7. This also suggests an obvious conclusion of increase in soil moisture with a decrease in groundwater table measured from the surface.

It is evident from this example that the groundwater table condition greatly influences the soil moisture conditions. Peak discharge values are extremely sensitive to soil moisture

Table 4.7 *Relationship between adjusted CNs and CNs for normal and wet conditions*

HWT (mm)	Adjusted CN	AMC II	AMC III
63.5	98	94	98
76.2	92	82	92
88.9	86	72	86
101.6	**81**	**64**	**81**
114.3	76	60	78
127.0	72	54	71
139.7	68	48	68
152.4	65		

conditions. Variability of these conditions and corresponding peak discharges can be evaluated using sensitivity and uncertainty analysis approaches. Uncertainty assessment of hydrologic model inputs and outputs is generally achieved by using Monte Carlo perturbation schemes adapted for specific input parameters. Several variants of perturbation schemes ranging from simple random sampling to Latin-hypercube sampling schemes can be used in hydrologic modeling for uncertainty estimation purposes. Teegavarapu *et al.* (2010) used a hydrologic simulation model with a SMA approach to model soil moisture interactions on peak flooding events for the watershed discussed in this section. These comparisons provided valuable insights into model parameter uncertainties and their influences on the ensemble of outputs generated from the simulation model.

4.12 SPATIAL AND TEMPORAL OCCURRENCE OF EXTREME EVENTS: DEPENDENCE ANALYSIS

The essential elements of a joint extremes assessment are the *distribution* of each variable, the *extreme values* of each variable, and the *dependence* for each variable-pair (Hawkes, 2008). The dependence between any two variables can be established by statistical measures (e.g., correlation) and subjective assessments of dependence based on scatter plots made before any advanced joint probability analyses are carried out.

The steps for dependence analysis are described as follows:

1. Collection of long-term rainfall data, streamflow data, water flow stages data, and time of occurrence for the storm events in the selected areas of interest.

2. Development of relationships among peak flooding events and spatial and temporal characterization of storms to analyze seasonality (dry and wet season), variability, trends, shifts in the distributions of rainfall and streamflow events, and other properties. Evaluate if there is a relationship between the annual rainfall and the flood peaks at each of the selected control structures.

3. Based on the analysis of the basin response time and knowing that it is one of the main factors in the flood generating processes, an analysis of the correlations between the available rainfall and peak flow historical data (short- and long-term data sets) should be evaluated in the region of study for different lag times. The peak flow for each year at critical points of interest (e.g., outlet and conveyance structures) should be obtained. Then the accumulated rainfall needs to be calculated from several days back in time (lag time) from the time of occurrence of the annual peak flow. The number of days or lag times can be selected from zero days (the same day that the peak flow occurred) to a specific number of days (e.g., 10 days).

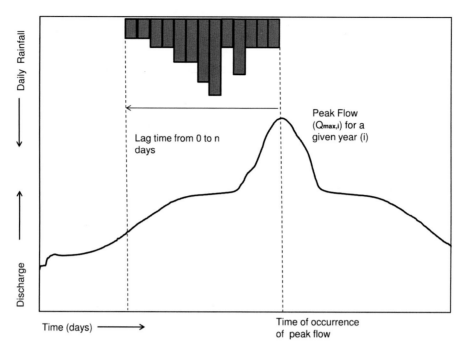

Figure 4.1 Schematic representation of changes in peak flow and accumulated rainfall.

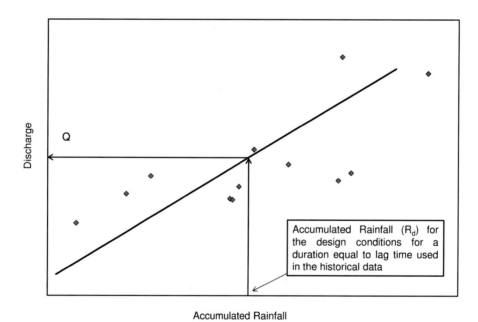

Figure 4.2 Criteria for the evaluation of design conditions for each structure using discharge as the comparison parameter.

4. Using statistical distributions (normal, log-normal, etc.), the short- and long-term rainfall data sets should be analyzed. A check for possible statistical distribution fits should be evaluated for rainfall and peak discharges.

5. Watershed-based analysis of precipitation and stream-flow trends falling under meteorologically homogeneous precipitation areas can be used to understand the variability of these storms occurring with a specific duration. Extreme storm events (e.g., hurricane events) can be also related with peak flows for each meteorologically homogeneous precipitation area. The number of flow peaks and the time of recurrence can be linked with extreme rainfall events.

6. Investigate the effects of maintenance operations, dredging, seasonal variations, and general works at different hydraulic structures on the variation of peak discharges, water levels, and ultimately in the conveyance capacity of the canals (see Figures 4.1, 4.2, and 4.3). Development of a methodology to

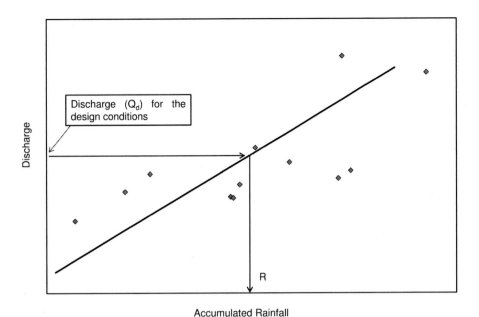

Figure 4.3 Evaluation of design conditions for each structure using rainfall as the comparison parameter.

generate relationships among extreme rainfall, peak flooding events, and stage discharge curves is included in this project. Events exceeding a threshold condition could produce a critical condition for outlet structures exposed to extreme rainfall events. The procedure for the definition of critical parameters is based on the evaluation of the relationships among the annual extreme rainfall amounts and the flood peaks and the water surface elevation at each of the selected control structures.

4.12.1 Peak discharge and accumulated rainfall

Peak discharge and accumulated rainfall based on a specific number of time intervals until the occurrence of the maximum flow can be evaluated, and their dependencies can be studied. The following section provides details of dependence analysis that links peak discharge and accumulated rainfall amounts.

Accumulated rainfall preceding a peak discharge event is calculated based on different lag times using historical data for each year (see Figure 4.1) based on the following equation:

$$R_i = \sum_{j=l}^{l-n} R_{j,i}, \left\{ Q_l = Q_{max,i} \right\} \forall i \qquad (4.13)$$

where:

i, year under analysis;
l, position where the peak is located in the 365 days of year i;
n, time from peak discharge to the beginning of rainfall (lag time);
R_i, accumulated rainfall for year i.

The lag time is calculated as the interval between rainfall occurrence and the peak discharge in response to such rainfall. The lag time can be significantly reduced as the watershed runoff characteristics change by urbanization, vegetation change, existence of control structures, or hydrometeorological processes.

Dependence analysis can be carried out by varying different conditions enumerated as follows:

• Peak discharge and individual rainfall from lag time zero to specific time interval.
• Peak discharge and accumulated rainfall from lag time zero to specific time interval.
• Peak discharge and annual extreme precipitation.
• Annual extreme precipitation and accumulated precipitation from lag time zero to specific time interval.
• Annual extreme precipitation and maximum stage per year.
• Peak discharge and maximum stage from lag time 0 to n days.
• Peak discharge and maximum stage per year.
• Maximum annual stage from lag time zero to n days and peak discharge.
• Peak discharge and average stage from lag time zero to specific number of time intervals (i.e., days).
• Correlations between peak flows, individual rainfall, accumulated rainfall, individual stage, and average stage data sets can be calculated for each hydrologic/hydraulic structure of importance in a basin. The behavior of the basins can be investigated based on the lag times by determining if they are dominated by rainfall-driven processes that are representative of natural watershed runoff processes instead of other control conditions.

4.12.2 Adequacy of design: extreme precipitation and discharges

The peak discharge and rainfall for the design conditions of outlet structures can be evaluated using available historical data. The rainfall from the design conditions (Q_d) is used to determine the discharge from the historical data. The agreement between discharge/rainfall from the historical data and discharge/rainfall from the design conditions can provide guidelines regarding the adequacy of the existing design conditions of critical structures. The criterion for the evaluation of the design conditions is based on the following cases.

Case A: Rainfall used for design conditions and the discharge versus accumulated rainfall from historical data are known. Using the discharge versus accumulated rainfall plot for a given lag time equal to the duration used for the rainfall from the design conditions, the design discharge is obtained from the graph and compared with the discharge for the design conditions. The following conditions may apply:

$Q > Q_d$: discharge (Q) from historical data is greater than design discharge (Q_d) (relevant to design conditions). The design condition is exceeded. Rainfall abstractions are overestimated. The condition may suggest inadequate design for discharge and revision of design is required.

$Q < Q_d$: discharge (Q) from historical data is less than design discharge (Q_d). The design condition is not exceeded. The design is conservative. Rainfall abstractions are underestimated. This condition may suggest adequate design.

$Q = Q_d$: discharge (Q) from historical data is equal to design discharge (Q_d). The design condition agrees with historical data. This condition may suggest adequate design.

Case B: Rainfall for design conditions and discharge versus accumulated rainfall from historical data are given. Use the peak discharge versus accumulated rainfall plot for a given lag time equal to the duration of rainfall from the design conditions. The following conditions may apply:

$R > R_d$: rainfall (R) from historical data is greater than design rainfall (R_d) from design conditions. The design condition is not exceeded. The rainfall required to produce design discharge is greater than the rainfall from design conditions. Rainfall abstractions are underestimated. This condition may suggest a need for modifications to design storm values.

$R < R_d$: rainfall (R) from historical data is less than design rainfall (R_d) from design conditions. The design condition is exceeded. The rainfall required to produce design discharge is less than the rainfall from design conditions. Rainfall abstractions are overestimated. This condition may suggest a need for modifications to design storm values.

$R = R_d$: rainfall (R) from historical data is equal to design rainfall (R_d) from design conditions. The design condition agrees with historical data. The design storm values are appropriate. This condition may rarely happen.

Conclusions regarding the likelihood of exceedence of the design conditions based on the analysis of the historical data can be made. Critical structures are thus evaluated to determine if they are functioning at their assumed design conditions. The general recommendation is to conduct a detailed analysis to determine new design conditions based on the existing drainage basin and canal system configuration.

Definition of the probability of exceedence becomes more complicated when two or more variables are considered. In this case, the likelihood that rainfall and discharge are exceeded at the same time requires the establishment of joint probability. The probability of exceedence was calculated as a probability of a joint critical value that is going to be exceeded based on the collected data available. Joint cumulative probability functions for accumulated rainfall from a given lag time corresponding to the design conditions (durations of 1 day and 3 days) versus peak discharges were generated for all structures. The probability of exceedence was calculated and reported for different critical scenarios, providing the chance of flooding based on the historical data analysis.

The analysis is performed based on the following expressions:

$$P_{occurrence} = P\left(R \leq R_d, Q \leq Q_d\right) \tag{4.14}$$

$$P_{exceedence} = 1 - P_{occurrence} \tag{4.15}$$

$$P_{exceedence} = 1 - P\left(R \leq R_d, Q \leq Q_d\right) \tag{4.16}$$

$$P_{exceedence} = P\left(R \geq R_d, Q \geq Q_d\right) \tag{4.17}$$

where:

$P_{occurrence}$, probability of occurrence of rainfall and discharge less than a given design condition;

$P_{exceedence}$, probability of exceedence of rainfall and discharge greater than a given design condition;

R, accumulated rainfall from historical data;

Q, discharge from historical data;

R_d, rainfall for design conditions;

Q_d, discharge for design conditions.

If $P_{exceedence} > P_{occurrence}$, the design parameters of structures need to be reviewed.

Individual probabilities for a given rainfall and discharge were determined using the cumulative joint distribution function for each structure under scrutiny. Recommendations to evaluate the existing conditions of a basin were given based on the calculated individual probabilities. For rainfall and peak discharge values

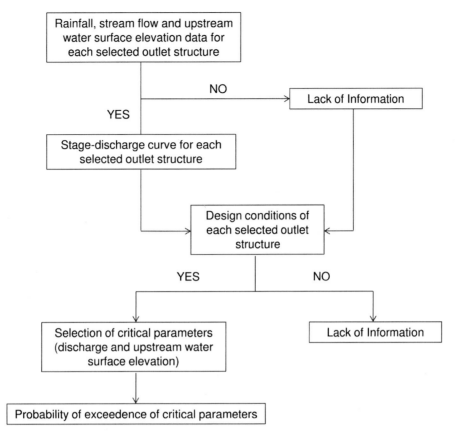

Figure 4.4 Probability of exceedence of discharge and stage knowing the design conditions for each selected outlet structure.

in the upper limits (extreme events) of the historical data, the individual probability should be low, otherwise the drainage condition of the basin (canal and outlet structures) would have to be reviewed (see Figure 4.4).

Discharge and stage for the design conditions of each selected outlet structure can be evaluated. A stage–discharge curve per year was plotted for each outlet structure and the probability of exceedence was calculated as the probability of a critical value that is going to be exceeded based on the collected data available. Comparing the discharge and water surface elevation for the design conditions with the stage–discharge curves, conclusions were made defining the threshold values for a given structure. The collection of events that exceed the design conditions and the analysis of the stage–discharge curve convey information regarding the reduction of any conveyance capacity and the threshold rainfall amounts that lead to flooding (see Figure 4.4).

The existing land-use patterns in the upstream drainage area for each basin should be investigated to assess the pervious and impervious area ratios. The area that is contributing to the runoff and its link with the lag time and the peak discharges for the selected control structures needs to be evaluated. Geo-spatial analysis showing patterns, trends in land use and their connections to flooding for each basin should be investigated.

4.13 JOINT PROBABILITY ANALYSIS

Joint probability analysis can be used to analyze the joint distribution of two correlated random variables. In the current context the two variables are annual peak discharges and accumulated rainfall amounts preceding the occurrence of annual peaks. In general, peak discharges and accumulated rainfall amounts may follow different probability distributions. The multivariate distribution of correlated random variables with different marginals is applicable in theory. In practice, an alternative is to use the multivariate normal distribution to describe the joint probability distribution of correlated random variables (Yue, 2000). This method first normalizes the different marginal distributions using a transformation technique. Then one only needs to deal with a joint distribution of the normalized variables using the multivariate normal distribution.

Bivariate normal distributions can be used to develop joint probability distributions. Transformation of the data (e.g., Box–Cox transformation method) may be required to transform the original data to near the normal distribution. The transformation parameters can be estimated using optimization approaches by maximizing the maximum likelihood function. Once joint probability distributions are obtained, the conditional

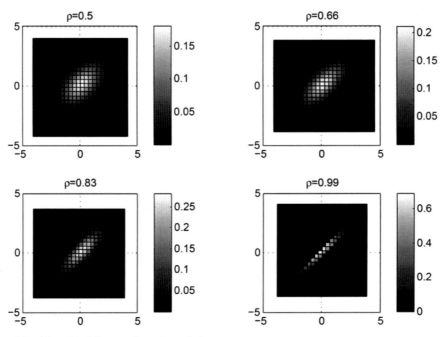

Figure 4.5 Bivariate normal densities with different values of correlation.

distributions can be used to obtain the associated return periods of annual storm peaks and accumulated rainfall amounts preceding the peak event. A brief description of the bivariate normal distribution is given next. The description and derivation of the probability density function are based on two variables (e.g., X and Y). In the current context, the first variable X is the accumulated rainfall and the variable Y is the annual peak discharge. The notation and the description of the equations for the bivariate normal distribution that is described next are adopted from Ugarte et al. (2008).

The bivariate normal distribution is a joint distribution of the random variables X and Y, when its joint density takes the form:

$$f_{X,Y}(x, y) = \frac{1}{2\pi\sigma_X\sigma_Y\sqrt{1-\rho^2}} \exp\left\{-\frac{1}{2(1-\rho^2)}\left[\left(\frac{x-\mu_x}{\sigma_x}\right)^2\right.\right.$$
$$\left.\left. - 2\rho\left(\frac{x-\mu_x}{\sigma_x}\right)\left(\frac{y-\mu_y}{\sigma_y}\right) + \left(\frac{y-\mu_y}{\sigma_y}\right)^2\right]\right\} \quad (4.18)$$

for $-\infty < X, Y < +\infty$, where $\mu_X = E[X]$, $\mu_Y = E[Y]$, $\sigma_X^2 = \text{Var}[X]$, $\sigma_Y^2 = \text{Var}[Y]$, and ρ is the correlation coefficient between X and Y.

An equivalent representation of (4.18) is given in Equation 4.20, where $X = (X, Y)^T$ is a vector of random variables where T represents the transpose; $\mu = (\mu_X, \mu_Y)^T$ is a vector of constant;

and Σ is a 2×2 non-singular matrix such that its inverse Σ^{-1} exists and the determinant $|\Sigma| \neq 0$, where:

$$\Sigma = \begin{pmatrix} \text{Var}[X] & \text{Cov}[X, Y] \\ \text{Cov}[Y, X] & \text{Var}[Y] \end{pmatrix} \quad (4.19)$$

$$f(x) = \frac{1}{(\sqrt{2\pi})^2}\left|\sum\right|^{-\frac{1}{2}} \exp\left\{-\frac{1}{2}(X-\mu)^T - \sum^{-1}(X-\mu)\right\} \quad (4.20)$$

The shorthand notation used to denote a multivariate (bivariate being a subset) normal distribution is $X \sim N(\mu, \Sigma)$. In general, Σ represents what is referred to as the variance covariance matrix. When $X = (X_1, X_2, \ldots, X_n)^T$ and $\mu = (\mu_1, \mu_2, \ldots, \mu_n)^T$ it is defined as:

$$\Sigma = E(X - \mu)(X - \mu)^T \quad (4.21)$$

Bivariate densities for different values of correlations are shown in Figure 4.5.

The results from the proposed method provide additional information that cannot be obtained by single variable storm frequency analysis, such as the joint return periods of the combinations of annual discharge peaks and accumulated rainfall amounts, and the conditional return periods of one variable given the other. Also, the results can contribute meaningfully to the design of flood control structures. For a given storm event return period, it is possible to obtain various occurrence combinations of storm peaks and amounts, and vice versa (Yue, 2000). The return periods exceeding threshold values of variables X and Y can be calculated if cumulative distribution functions ($F(.)$) are available for X and Y variables individually or jointly.

$$F(x) = P_r(X \leq X_{th}) \qquad (4.22)$$

$$F(y) = P_r(Y \leq Y_{th}) \qquad (4.23)$$

The return periods based on cumulative distributions for X and Y are given by:

$$T_x = \frac{1}{1 - F(x)} \qquad (4.24)$$

$$T_y = \frac{1}{1 - F(y)} \qquad (4.25)$$

The return periods for two variables for X and Y exceeding specific threshold values can be calculated using:

$$F(x, y) = P_r(X \leq X_{th}, Y \leq Y_{th}) \qquad (4.26)$$

$$T_{x,y} = \frac{1}{1 - F(x, y)} \qquad (4.27)$$

The adequacy of the design criteria can be evaluated using the return periods obtained from historical data using joint probability distributions.

The exceedence probability is estimated using the following equation:

$$P_e(X \geq X_{th}, Y \geq Y_{th}) = 1 - P_r(X \leq X_{th}) - P_r(Y \leq Y_{th})$$
$$+ P_r(X \leq X_{th}, Y \leq Y_{th}) \qquad (4.28)$$

4.13.1 Requirements of bivariate normal distribution

One of the main requirements of the bivariate normal distribution is that the distributions fitted to the two variables should be normal. Different data transformation methods are available to achieve this if a required condition of the bivariate normal distribution is not satisfied.

4.13.2 Box–Cox transformations for normality

The Box–Cox transformation (or power transformation) (Box and Cox, 1964) is applied to normalize the sample data. In the current context, the sample data are annual peak discharge and accumulated lagged rainfall values. The Box–Cox transformations applied to rainfall and discharge are given by following equations:

$$R_{\lambda,i} = \log(R_i)(\lambda = 0) \qquad (4.29)$$

$$R_{\lambda,i} = \frac{R_i^\lambda - 1}{\lambda}(\lambda \neq 0) \qquad (4.30)$$

$$Q_{\lambda,i} = \log(Q_i)(\lambda = 0) \qquad (4.31)$$

$$Q_{\lambda,i} = \frac{Q_i^\lambda - 1}{\lambda}(\lambda \neq 0) \qquad (4.32)$$

The value of the original variable values for rainfall can be obtained using the following back transformations:

$$R_i = \left(R_i^\lambda \lambda + 1\right)^{\frac{1}{\lambda}}(\lambda = 0) \qquad (4.33)$$

$$R_i = e^{R_{\lambda,i}}(\lambda \neq 0) \qquad (4.34)$$

The transformation parameter (λ) needs to be estimated for each sample data to achieve the expected transformation. Similar back transformation can be applied to discharge data. It is important to note that oftentimes the power-transformed data may not fit the anticipated normal distribution.

4.13.3 Box–Cox transformation parameter estimation

The parameter (λ) can be estimated using an optimization approach. The formulation for the optimization of the parameter is given by the following equations (4.35–4.39). The objective is the log-likelihood function (Equation 4.35) that needs to be maximized.

Maximize

$$\log(\lambda) = -\frac{n}{2}\log(2\pi) - \frac{n}{2}\log\left(\sigma_\lambda^2\right) + (\lambda - 1)\sum_{i=1}^{n}\log(Q_i)$$
$$- \frac{1}{2\sigma^2}\sum_{i=1}^{n}\left(Q_{\lambda,i} - \mu_\lambda\right)^2 \qquad (4.35)$$

subject to:

$$\sigma^2 = \frac{1}{n}\sum_{i=1}^{n}\left(Q_{\lambda,i} - \mu_\lambda\right)^2 \qquad (4.36)$$

$$\mu_\lambda = \frac{1}{n}\sum_{i=1}^{n}Q_{\lambda,i} \qquad (4.37)$$

$$\lambda \leq \lambda_u \qquad (4.38)$$

$$\lambda \geq \lambda_l \qquad (4.39)$$

A non-linear optimization solver is required for optimal solution (i.e., the optimal value of transformation parameter, λ, with upper and lower limits specified. Optimal estimation of the transformation parameter can be obtained using genetic algorithms (GA). Gradient-based optimization solvers may not always provide global optimal solutions and have a tendency to produce local optimal solutions.

4.13.4 Normality tests

The transformed values of annual peak discharges and accumulated rainfall values using optimal Box–Cox transformed parameters are tested using several goodness-of-fit tests. Visual assessments of probability plots give a preliminary indication of the normality of the sample data. However, several other tests such as Chi-square, Kolmogorov–Smirnov (KS), Anderson–Darling,

Lilliefor, Shapiro–Wilk, and D'Agostino–Pearson goodness-of-fit tests must be carried out to assess the normality of the sample. One major limitation of the KS method is that population parameters are required for the test as opposed to sample parameters. Several modifications to the KS test have been made recently, which overcome this limitation of the test. The most preferred tests are D'Agostino–Pearson, Anderson–Darling, and Lilliefor.

4.13.5 Conditional probability densities

Given a bivariate normal distribution of annual peak discharge, Q, and accumulated rainfall, R, the marginal distribution of Q is $N(\mu_Q, \sigma_Q)$ and of R is $N(\mu_R, \sigma_R)$; the conditional density of Q, given $R = R_{th}$, is a normal distribution with mean $(\mu_{Q|R})$ and variance $(\sigma_{Q|R}^2)$ given by:

$$\mu_{Q|R} = \mu_Q + \rho \frac{\sigma_Q}{\sigma_R}(R_{th} - \mu_R) \qquad (4.40)$$

$$\sigma_{Q|R}^2 = \sigma_Q^2(1 - \rho^2) \qquad (4.41)$$

The conditional distributions of Q and R are also normally distributed with modified mean and standard deviation values as given by:

$$P_e(Q \geq Q_{th}, |R = R_{th}) = 1 - P_r(Q \leq Q_{th}|R = R_{th}) \quad (4.42)$$

$$P_e(Q \geq Q_{th}, |R = R_{th}) = 1 - P_r\left(\frac{Q - \mu_{Q|R}}{\sigma_{Q|R}} \leq \frac{Q_{th} - \mu_{Q|R}}{\sigma_{Q|R}}\right) \qquad (4.43)$$

The conditional period, $T_{X|Y}$, is estimated using:

$$T_{Q|R} = \frac{1}{P_e(Q \geq Q_{th}|R = R_{th})} \qquad (4.44)$$

The conditional densities and the return periods can be estimated using the Box–Cox transformed data.

4.13.6 Validity of joint distributions

The observed and theoretical joint probabilities based on the peak discharges and accumulated rainfall amounts can be evaluated to assess the validity of the bivariate normal distribution model fitted. Theoretical probabilities from the fitted model and observed non-exceedence probabilities can be easily obtained and compared. A high correlation between the observed and the theoretical values generally indicates the validity of the bivariate normal distribution model used for fitting the joint distribution of the sampled data. Alternatively, a conceptually simpler approach would be to use random vectors (joint variables) generated based on the fitted bivariate normal distribution parameters, and the generated values can be compared with observed data. The summary statistics from the observed data can be compared with generated values.

4.14 PARTIAL DURATION SERIES ANALYSIS: PEAKS OVER THRESHOLDS

There are two basic approaches for extracting extreme data when a univariate extreme value analysis is performed. These approaches can be described as:

1. Epochal method: Extract the maximum or minimum value for each year or in equal length intervals as seasons.
2. Peaks over threshold (POT) method: Extract all points above a given threshold. In this case the numbers of extracted points are usually not equal in each interval.

The POT approach enables the analyst to use all the data exceeding a sufficiently defined threshold and is more effective than the classical approach, which uses only the largest value in each of a number of comparable sets. In the POT approach it is important to choose a threshold that maintains the independence of the events. Lower thresholds would introduce a strong bias in some instances. One of the main difficulties in a study of extreme values is the limited length of the available records. When an annual maximum time series is used, the fitting of the parameters for a good analysis should rely on at least a 30-year-long time series; in most cases, the available length of the data set is much shorter. When an analysis is performed with a short set of data, the uncertainty associated with statistical analysis may increase. In order to reduce this uncertainty, daily streamflow data are used within a POT framework in which the basic idea is to use more than one extreme event per year, increasing the available information with respect to the epochal method, which uses annual maximum flood data.

Annual maximum series or partial duration series (PDS) (Langbein, 1949; Stedinger, 2000) are generally used in flood frequency analysis. In the case of annual maximum series, one extreme value is selected each year. Selection of extreme events in such a setting ensures that the events selected are independent. The PDS is also referred to as the POT approach. This approach requires identification of events that are independent and the values associated with all the extreme events within a year exceed a specific pre-fixed threshold. Figure 4.6 shows a typical hydrologic time series with extremes separated by specific time intervals. The selection of threshold is based on a criterion such as:

$$M_{th} = \min\{M_i, M_{i+1}, M_{i+2}, \ldots, M_{ny}\} \, \forall i \qquad (4.45)$$

where M_{th} is the threshold value obtained based on the minimum value of all annual extreme values obtained each year, i.

$$M_{th}^i = \varepsilon(M_i) \, \forall i \qquad (4.46)$$

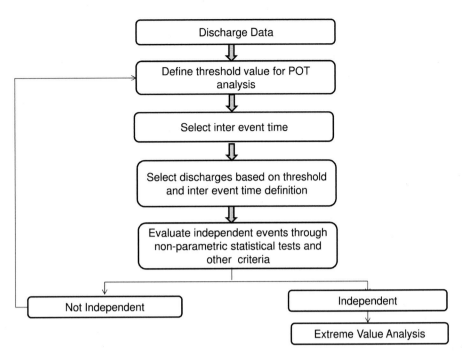

Figure 4.6 Schematic representation of the selection process for POT in the PDS analysis.

The variable ε is referred to as the threshold factor, which can vary between 0 and 1. Historically, the notion of independence has played a prominent role in the determination of POT events. If events form an independent class, much less information is required and calculations are correspondingly easier. To decide whether a given sequence is truly random, or independent and identically distributed, we need to resort to non-parametric tests for randomness. In statistical literature, a truly random process refers to a process that can produce independent and identically distributed samples. If an observed value in the sequence is influenced by its position in the sequence, or by the observations that precede it, the process is not truly random. The randomness is related to the property of the data, and it is essential to fulfill the theoretical basis of many classical statistical tests. Even when the observations are not truly random, which is true in many practical applications, we can still perform those simple and neat theoretical results with a certain degree of confidence if we can tell how close to random the data can be.

Tests can be carried out to ensure the independence of events selected for the POT analysis. A non-parametric test referred to as the "runs test" is one of the commonly used tests to assess serial randomness: whether or not observations occur in a sequence in time. Tests can be performed for the selected POT. The test is based on the null hypothesis that the values come in random order, against the alternative that they do not. The test is based on the number of runs of consecutive values above or below the mean of the selected data. The runs test can be performed for PDS selected for different lag times between them.

The selection process of the POT is described in Figure 4.7.

4.15 BASEFLOW SEPARATION METHODS

Determination of baseflow value for a streamflow is essential for any POT events. Once the baseflow is known, the events (i.e., extreme events) can be separated easily. Baseflow is defined as the proportion of streamflow that comes from the subsurface flow. There are different methodologies to estimate the baseflow of a stream in a watershed. It is necessary to separate out the baseflow from stream gage data. Baseflow separation carried out by different hydrologists manually is not likely to result in the same baseflow for a particular area. This suggests that there is a need for a set of procedures to ensure that baseflows are calculated objectively. Different types of methodologies are available and the most common methods are divided into two groups that can be summarized as:

- Graphical methods, which tend to focus on defining the points where baseflow intersects the rising and falling limbs of the quickflow response. Typical graphic strategies to estimate the baseflow are:
  - Constant discharge (Linsley *et al.*, 1958), which assumes that baseflow is constant during the storm hydrograph; typically the constant baseflow value is selected as the minimum streamflow value prior to the rising limb.
  - Constant slope, which assumes that the baseflow has an instant response to the precipitation, and the baseflow increases immediately during a rainfall event. Baseflow is calculated by connecting the rising limb of the hydrograph with the inflection point on the receding limb.

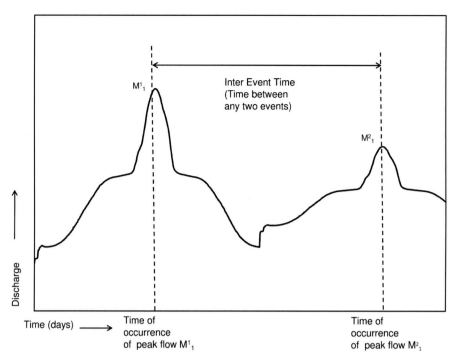

Figure 4.7 Steps for the selection of POT in the PDS analysis.

- Concave method, which also assumes that the baseflow has an instant response to the precipitation by decreasing immediately during the rising limb by projecting the declining hydrographic trend evident prior to the rainfall event to directly under the crest of the flood hydrograph (Linsley *et al.*, 1958), then it is connected to the inflection point on the receding limb of the hydrograph increasing the baseflow value.
- Typical methods of continuous hydrographic separation techniques that involve data processing or filtering where streamflow data is processed for the entire hydrograph resulting in a baseflow hydrograph are:
  - Smoothed minima techniques: Uses the minimum of *n* days of non-overlapping periods derived from the hydrograph to estimate the baseflow hydrograph by connecting a subset of points selected from this minima series (Institute of Hydrology, 1980; FREND, 1989; Sloto and Crouse, 1996).
  - Fixed interval method: Discretizes the hydrographic record into increments of fixed time where the baseflow component of each time increment is assigned as the minimum streamflow recorded within the increment (Pettyjohn and Henning, 1979).
  - Streamflow partitioning method: Uses both the daily record of streamflow and rainfall to calculate baseflow. The baseflow equates to streamflow on a given day, if

rainfall on that day and a set number of days previous is less than a defined rainfall threshold value (Shirmohammadi *et al.*, 1984).
- Antecedent precipitation index (API): Calculates the API for each storm. It is based on the daily precipitation for 30 days preceding a storm, using a decay coefficient that has been estimated in other studies as 0.85 (best fit curve for rainfall–runoff stations in Wisconsin) (Conger *et al.*, 1978). The equation for API can be described as follows:

$$API = 0.85^{t_o} P_o + 0.85^{t_1} P_1 + \cdots + 0.85^{t_{30}} P_{30}$$

(4.47)

where i is the number of days preceding the storm, t_i is time in days and P_i is the corresponding precipitation for day i. The API is plotted versus baseflow for each storm and a best fit straight line is drawn. The best fit straight line can be represented by a straight line equation where the ordinate intercept is the baseflow and the slope is the API. Using the best fit straight line equation, the long-term precipitation records can be substituted in the equation to determine the baseflow for any storm period desired.
- Recursive digital filters: Tools in signal analysis and processing that are used to remove the high-frequency

quickflow signal to derive the low-frequency baseflow signal (Smakhtin, 2001; Nathan and McMahon, 1990; Hughes *et al.*, 2003). The digital filter that has been applied to smooth hydrographic data in this project is based in the algorithm developed by Lyne and Hollick (1979) with slight modifications introduced by Smakhtin (2001) and later used by Hughes *et al.* (2003). The algorithm can be summarized as follows:

$$q_i = \propto q_{(i-1)} + \beta \, (1 + \propto)(Q_i - Q_{(i-1)}) \quad (4.48)$$

$$QB_i = Q_i - q_i \quad (4.49)$$

where q_i is the high-flow time series component (baseflow) at time step index i; $Q_{(i-1)}$ is the baseflow at the previous time step $i-1$; Q_i is the total flow at time step i; $\propto$ (recession constant) and β are known as the separation parameters, with recommended values of $0 < \propto < 1$, $0 < \beta < 0.5$; QB_i is the baseflow time series component. According to this methodology, the baseflow component (QB) for each time step is constrained to be never less than zero or greater than the total flow (Q) (Hughes *et al.*, 2003).

Some of the explained methodologies for the estimation of baseflow tend not to have any hydrologic basis but aim to generate an objective, repeatable, and easily automated index that can be related to the baseflow response of a catchment (Nathan and McMahon, 1990; Smakhtin, 2001; Wahl and Wahl, 1995). In this project, three different methodologies have been developed and applied to estimate the baseflow index (BFI), which is the long-term ratio of baseflow to total streamflow and the mean baseflow for the entire streamflow data set.

4.16 EXTREME PRECIPITATION AND FLASH FLOODS

A flash flood is sudden local-scale flooding generally produced by a meteorological-scale precipitation event of limited time and space scale that produces rapid runoff (Shelton, 2009). Extreme precipitation, possible due to convective systems (Doswell *et al.*, 2001), is mainly responsible for flash flood conditions. However, other factors that relate to watershed topography, land-use patterns, and several aspects of localized intense short duration precipitation events influence the onset of flash flood. Universally applicable magnitudes of precipitation and extreme precipitation event durations that lead to flooding cannot be defined. WMO (2007) defines flash flood as a flood of short duration with a relatively high peak discharge. Sene (2008) defines a flash flood as a flood that threatens damage at a critical location in the catchment,

where the time for the development of flooding from the upstream catchment is less than the time needed to activate warning, flood defense, or mitigation measures downstream of the critical location.

4.17 PRECIPITATION THRESHOLDS AND FLOODS

Knowledge of rainfall totals from rain gages, satellite observations, and weather forecasting models above a specific threshold provide early indications of the onset of flooding. Rainfall amounts can be used directly to initiate flood warnings. However, uncertainties related to transformation of rainfall amounts into flooding events remain. Sene (2008) indicates that reliance on only rainfall threshold-based warnings may lead to false alarms in many situations. Precipitation thresholds are often expressed as rainfall depths that are expected to cause flooding in a region based on past experiences. These thresholds are also referred to as triggers, warning levels, and alarms. The following section describes models that incorporate lead times for flood forecasting.

4.17.1 Models with lead times

The joint probability approaches previously discussed in this chapter can be extended to develop models with lead times for real-time use. Accumulated rainfall preceding a peak discharge event is calculated based on different lag times using historical data for each year (see Figure 4.8) based on the following equation:

$$R_i = \sum_{j=l-lt}^{(l-lt)-n} R_{j,i}, \left\{ Q_l = Q_{max,i} \right\} \forall i \quad (4.50)$$

where:

i, year under analysis;

l, temporal position where the peak is located in the 365 days of year i;

lt, lead time (accounted in time intervals backwards from the time of occurrence of peak);

n, time from peak discharge to the beginning of rainfall (lag time);

R_i, accumulated rainfall for year i.

The lag time is calculated as the interval between rainfall occurrence before the lead time and the peak discharge in response to such rainfall. The lag time can be significantly reduced as the watershed runoff characteristics change by urbanization, vegetation change, existence of control structures, or hydrometeorological processes.

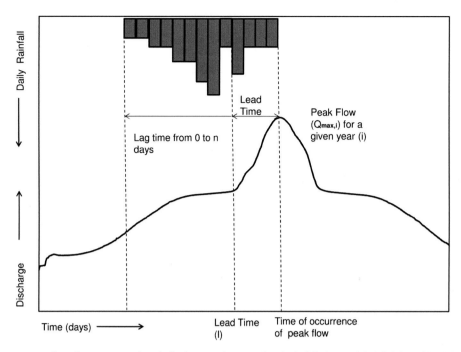

Figure 4.8 Schematic representation of occurrence of peak discharge and accumulated rainfall along with definition of lead time.

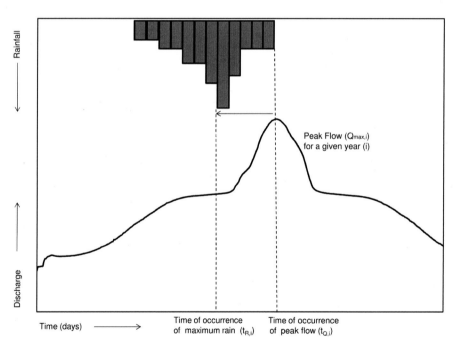

Figure 4.9 Time difference between occurrence of peak discharge and extreme rainfall event.

The incorporation of lead time in the analysis helps in developing flood control warnings/watches and protocols for operational changes based on joint exceedence conditional probabilities. The approach can also be used to evaluate the two critical responsible factors (high groundwater table and extreme precipitation) for flooding in a joint probability framework.

4.18 TEMPORAL DIFFERENCE IN OCCURRENCE OF PEAKS

Evaluation of the temporal difference in occurrence of peak discharges and extreme rainfall events is essential to assess the lag times. Figure 4.9 shows a schematic representation for

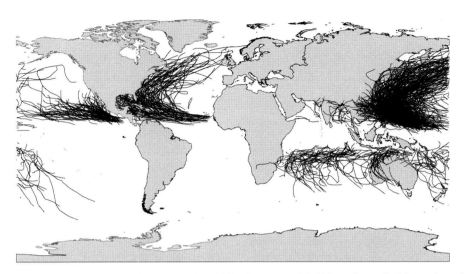

Figure 4.10 Tracks of hurricanes (category 4 and 5) from years 1853–2008 (data source: World Data Center for Meteorology, Ashville and NOAA).

calculation of temporal differences in occurrence of extreme values (i.e., annual extremes) of discharge and rainfall. The temporal differences in occurrence were calculated for two variables: maximum rainfall and peak discharge. The evaluation of these variables is fundamental to understanding how the basins are responding to a stressor such as rainfall. If stages are observed at any critical hydraulic structure, the stage values can also be used for this evaluation. Temporal differences in occurrence can be evaluated for the following variables:

• maximum rainfall and peak discharge in a given year;
• maximum rainfall and stage in a given year;
• maximum stage and peak discharge in a given year.

4.19 CYCLONIC PRECIPITATION: EPISODIC EVENTS

Precipitation related to cyclonic landfalls (especially tropical cyclonic events) results in the greatest precipitation totals across the globe for durations ranging from 1 hour to 96 hours. Therefore, consideration of precipitation totals resulting from cyclonic precipitation events in dependence and joint probability analysis of peak discharges and extreme rainfall is essential. Increased cyclonic event activity is also linked with several teleconnections (oscillatory modes) across the globe. One example of increased hurricane activity in AMO warm phases is well documented and is discussed in Chapter 6. Tracks of hurricanes (category 4 and 5) based on long-term historical data are shown in Figure 4.10.

An example case study from the South Florida region in the USA, discussed later in this chapter, suggests that extreme values of discharges are linked to accumulated lagged rainfall amounts due to episodic events such as hurricanes. The intensity of the rainfall amounts may be small over the entire period of hurricane landfall. However, the occurrence of rainfall for prolonged periods of time before and after the landfall of hurricanes leads to saturated soil moisture conditions and eventually peak flows or runoff events. Statistical analysis of extreme precipitation events requires separation of events based on storm classification to preserve homogeneity assumptions.

4.20 DESK STUDY APPROACH

Joint probability analysis predicts the probability of occurrence of events in which two or more partially dependent variables simultaneously take high or extreme values. Several different variables are potentially important in design and assessment of flood protection. Key variables that directly affect the hydrology of the basins are: streamflow, rainfall, groundwater levels, soil moisture conditions, and storage capacity.

The simultaneous occurrence of extreme rainfall events and extreme discharges requires the evaluation of the possibility of these discharges exceeding a critical design discharge of hydraulic structures. The desk study approach (DSA) aims to evaluate the current and future design of hydraulic structures, by effectively understanding the extreme behavior of the involved variables in the failure of the hydraulic structure. Previous research work by Hawkes and Svensson (2005) considers thresholds on several pairs of variables, such as wave height, sea level, surge, tide, river flow, precipitation, and wind, to analyze the impact over coastal structures on the coast of Britain. Another study (Yeh *et al.*, 2006) used wave height and water level data extracted from the typhoon season to analyze the traditional frequency analysis and joint probability method. Infinite solutions were found for a specific failure condition, so designers could determine the important factor to obtain the appropriate solution in a particular region. Both studies confirm the value of the joint probability method, although the scarcity of available case studies in this area is evident.

Since 2005, the Department of Environment, Food and Rural Affairs, UK (DEFRA) has been evaluating the dependence between some hydrologic and hydraulic variables and how best to quantify their combined impact on coastal defenses. A simplified method referred to as the "desk study approach" for joint probability analysis was developed, extending a method originally published in CIRIA (1996). Methods for marginal extremes are discussed by DEFRA (2005b). The DSA provides project-specific joint probability tables, being able to work with different dependence measures, any dependence value, any number of records per year, and any joint exceedence return period (DEFRA, 2005b). The input needed is enumerated as follows:

1. marginal extremes for each of two variables;
2. number of records per year;
3. joint exceedence return periods;
4. dependence value (DEFRA, 2005b, c).

The magnitude of an extreme event is inversely related to its frequency of occurrence (DEFRA, 2005b). For the peak discharge and accumulated rainfall values, rainfall and peak discharge for a given return period using annual maximum quantities of streamflow and rainfall data can be estimated. The return period refers to the average period of time between occurrences of a particular high value of that variable (DEFRA, 2005b). The events that give critical flooding conditions are easily characterized as events caused by extreme values of the variable whenever it exceeds some level. The flooding can be defined when the streamflow and rainfall fall within the failure region. Curves of different return periods from 1 to 50 years can be developed. Any one of these curves can provide the joint exceedence probability information when the variable states are more severe than at the critical condition (DEFRA, 2005b). To apply DSA, discharge and rainfall data are matched up day-by-day over several years' data sets. From these daily joint data sets, one record is extracted at each peak discharge. The number of records extracted is usually one per year. During the application of the DSA to the coastal defense project (DEFRA, 2005a), a sensitivity analysis for the dependence parameter ρ was performed. A similar threshold value of between 0.8 and 0.9 can be used.

An analysis of the dependence parameter ρ was reported based on different dependence levels (HR Wallingford, 2000a):

$0 < \rho < 1$ corresponds to positive dependence;
$\rho = 0$ corresponds to independence;
$-1 < \rho < 0$ corresponds to negative dependence.

Based on the sensitivity analysis of the DSA to variable ρ performed by DEFRA (2005a), it can be concluded that:

high values of ρ tend to give the highest joint exceedence values;
low values of ρ tend to give the lowest joint exceedence results.

The reported analysis suggests that sensitivity to ρ is a small part of the overall uncertainties associated with prediction of extreme states and their impacts (DEFRA, 2005b). The main source of uncertainty in this analysis was the data due to the inaccuracy in the collection process. Another source of uncertainty is the estimation of the parameters (DEFRA, 2005b).

Joint exceedence probability combinations of peak discharges and rainfall with a given chance of occurrence are defined in terms of flooding conditions in which a given streamflow quantity is exceeded at the same time as a given rainfall threshold is being exceeded (DEFRA, 2005a). Figure 4.11 (DEFRA, 2005b) shows an example of plotted joint exceedence curves using the DSA. Green and yellow areas illustrate ranges of wave height and sea level with the given joint exceedence probability for DEFRA analysis. Red and blue areas are failure regions and the probabilities they represent (DEFRA, 2005b). The two-variable joint exceedence extremes (several records per year) are plotted as curves for different return periods. If a rectangle is drawn, with sides parallel to the x–y axes, then the probability density (p) of an event falling within that rectangle is (the sum of the exceedence probabilities (P) corresponding to the bottom-left and top-right corners) minus (the sum of the exceedence probabilities corresponding to the top-left and bottom-right corners). The resulting curves for different return periods are contour lines for equal joint exceedence probability.

Joint probability refers to two or more partially related environmental variables occurring simultaneously to produce a response of interest. The DSA is based on the estimation of the probability of failure expressed in terms of the return period. The probability assessment that is described by the DSA can be described as follows (Hawkes et al., 2002):

1. Identify types of failure.
2. Identify the combinations of the conditions for the selected variables that cause failure.
3. Estimate the probability of these combinations that give failure.

In this approach there is a need for statistical methods to detect and quantify dependence on the two variables and to extrapolate from the observed data to the behavior in the extreme tail of the distributions (DEFRA, 2005b).

There are several stages involved in the implementation of the DSA; the first stage is the compilation and preparation of the input data. Each input corresponds to the extreme annual value for a specific variable (e.g., peak discharge, rainfall, or groundwater

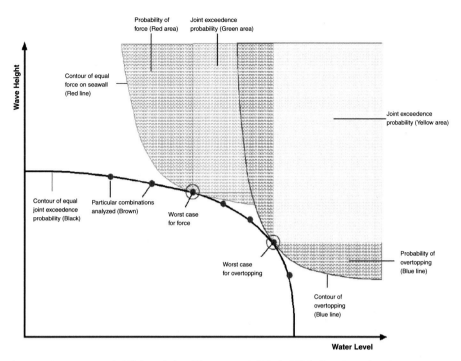

Figure 4.11 Joint exceedence and response probabilities relationship (source: DEFRA, 2005). See also color plates.

level). Each extreme value is taken as an independent record per year; in that sense, a 1-year and a 100-year return period have probabilities of occurrence of 1/1 and 1/100 respectively. Another stage in the DSA is the fitting of statistical distributions. The data sample represents the occurring conditions very well, although considerable uncertainty and irregularity in the upper tail of the data sample are present and can be reduced by using fitted probability distributions.

The third stage of the DSA involves conversion to normal scales and the fitting of a dependence function to the set of data. Simple diagnostic tests have been applied in previous projects to determine whether full dependence or independence models are adequate approximations for the variables involved in the analysis (HR Wallingford, 2000a; Bortot *et al.*, 2000). Other dependence models have been applied, such as the single bivariate normal (BVN) distribution and a mixture of two BVNs (HR Wallingford, 2000a; Coles *et al.*, 1999; Heffernan, 2000). The usual purpose of joint probability analysis is to estimate the probability of exceeding some critical value $P_r(Y_t > v)$:

$$P_r(Y_t > v) = \int\limits_{A > v} f(x)\, dx \qquad (4.51)$$

To simulate a large amount of conditions based on the extreme values, fitted distributions, and dependences, the DSA applies a Monte Carlo simulation to generate a large sample of synthetic records for the variables involved in the analysis. The DSA includes a refinement of the Monte Carlo simulation of data by re-scaling any simulated values (based on the fitted distributions)

above a specified return period threshold to new values with equal probabilities taken from the refined predictions (DEFRA, 2005b).

In such an analysis, the extreme values are extracted from existing collected data, and an estimated value of the dependence between these two variables is obtained. The combined two variables with their exceedence return period (T_r) is calculated. Then the map containing joint exceedence extremes and the associated joint exceedence return period curves is created, referring to both variables simultaneously exceeding their specified values. To construct the joint exceedence curves, parameters such as number of data per year and dependence measure ρ (DEFRA, 2005b) need to be selected. The joint exceedence curves represent the probability of two extreme variables occurring simultaneously and their joint exceedence extremes with the associated joint exceedence return period curves. Infinite solutions are possible for a specific failure condition, so that a designer can determine the important factor to obtain the appropriate solution in a particular region or structure. Joint probability density presents a better estimation toward the probability of flood risk.

4.21 REGRESSION ANALYSIS

Regression models based on the concepts of inductive modeling can be used for linking several hydrologic variables for flood peak estimation. In this section an example is discussed that links groundwater levels, rainfall, and peak discharges in a watershed.

The relationship between peak discharges and the corresponding groundwater table levels and accumulated rainfall values at a specific temporal lag can be evaluated to possibly predict peak discharge levels. A multiple regression model with positive weights can be conceptualized and implemented to analyze several variables, where the focus is on the relationship between the independent variable (peak discharge) and two or more dependent variables (groundwater table and accumulated rainfall). The regression analysis requires all the assumptions related to residuals to be satisfied.

The typical value of the dependent variable (peak discharge) changes when the independent variables (groundwater table level and accumulated rainfall) are varied. The available data set can be divided into two subgroups and those subgroups are used as a training or calibration data set and as a predicting or validation data set for the model. The model can be run several times for each data set, defining the distribution of the response variable in terms of the predictor variables. Peak discharges are predicted based on the least squares method; a standard approach is applied for data fitting, where the overall solution minimizes the sum of squared residuals, the difference between an observed value and the fitted value provided by the model.

The model is based on the following non-negative least squares problem:

$$\min_{x} \frac{1}{2} \|C_x - d\|_2^2 = \text{such that } x \geq 0 \qquad (4.52)$$

where the matrix C and the vector d are the coefficients of the objective function. The vector x of independent variables is restricted to be non-negative. A non-parametric regression is applied in this project referring to a technique that allows the regression function to lie in a specified function. The single predictor, Q^p, is determined based on the following multiple regression function:

$$Q^p = f(GW, R) \qquad (4.53)$$
$$Q^p = W_1 GW + W_2 R \qquad (4.54)$$

where Q^p is the peak discharge, GW is the groundwater table, R is the accumulated rainfall, and W_1 and W_2 are the corresponding weights. Once weights are obtained for the model using the training data set, the variation of the dependent variable is characterized around the regression function (probability distribution). Later on, the model is used to make predictions of the expected peak discharges based on the predicting data set. The peak discharges are predicted for a different set of data. An analysis of the relationship between peak discharge and rainfall, and peak discharge and groundwater table levels can be performed by defining a threshold in the peak discharge and calculating the regression function as well as the different weights. The goal of this analysis will be to determine the sensitivity of the dependent variable

(peak discharge) for different threshold values of discharges. The relation between peak discharge and rainfall, and peak discharge and groundwater table can be analyzed and the different weights for the multiple regression equation can be assessed.

4.22 EXTREME PRECIPITATION EVENTS AND PEAK FLOODING: EXAMPLE

South Florida is characterized by two predominant seasons: summer (wet) and winter (dry). They are determined more by precipitation than by temperature. The summer season is characterized by warm, humid conditions with frequent showers and thunderstorms. The winter season has cooler temperatures, lower humidity, and less frequent precipitation. The dry season in Florida can be identified from November to April and the wet season from May to October. In other words, convection in the form of short duration events, such as showers and thunderstorms, is almost a daily occurrence during the summer, occurring mostly in the late afternoon and early evening. Overall, the best indicators of the summer season in southeast Florida are dew-point temperatures and minimum temperatures remaining in the seventies (°F), and frequent daily rainfall (NOAA, 2010). This finding has been verified recently in a comprehensive investigation of summertime precipitation over the USA by Higgins et al. (1997).

Two heavily urbanized basins that are prone to frequent flooding in South Florida are considered for application of a joint probability analysis. A joint exceedence analysis for the dependence between key pairs of variables (i.e., precipitation and peak discharge) in basins C4 and the C6 in South Florida is carried out in this example. The locations of these basins are shown in Figure 4.12 and Figure 4.13. Extreme water levels in the man-made canals of South Florida are influenced by both streamflow and rainfall. Any dependence between them needs to be taken into account to understand the behavior of basins C4 and C6. The traditional design approach in hydrology and hydraulics tends to be essentially deterministic and empirical, a joint probability approach of extreme rainfall and high streamflow events helps in understanding the risk of urban flooding in these analyzed basins. Table 4.8 provides details of annual peak discharge and accumulated rainfall depths at a lag of 3 days excluding the day of occurrence of peak discharge. Rainfall data used for this example are from a nearby rain gage at Miami FS_R (shown in Figure 4.13) and the streamflow gage at C4_Coral site in C4 basin.

Summary statistics of the discharge (m³/s) and rainfall (in millimeters) are provided in Table 4.9.

Box–Cox-transformed peak discharge and rainfall values are obtained by optimizing the log-likelihood function. The values of transformation parameter (λ) are -0.699 and -0.109 for peak discharge and rainfall respectively.

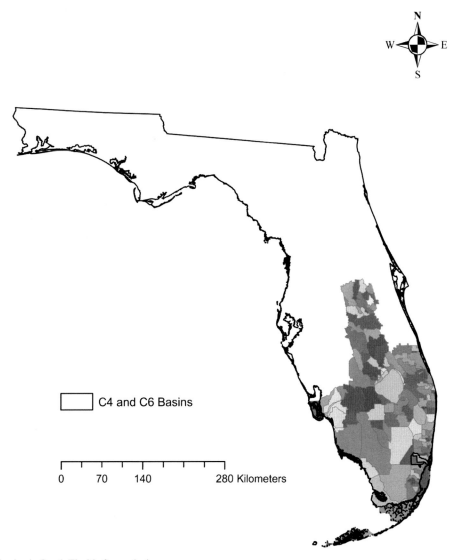

Figure 4.12 Selected basins in South Florida for analysis.

The bivariate cumulative probability distribution for peak discharge and accumulated rainfall is shown in Figure 4.15. The joint exceedence conditional probability density plots for different accumulated rainfall amounts are shown in Figure 4.16. The joint return period plots for peak discharge and accumulated rainfall (lag = 3) are shown in Figure 4.17.

4.22.1 Dependence analysis

The C6 basin shown in Figure 4.13 is selected for dependence analysis using peak discharge data at flow monitoring structure C6_NW36 and the same rain gage used in the previous example. Correlations for different lagged precipitation values and peak discharge are evaluated using Spearman's rho (ρ) and Kendall's tau (τ), which are alternatives to Pearson's correlation coefficient. Figures 4.18 and 4.19 show the two coefficients based on the

relationship between peak discharge and accumulated rainfall values at lag times of 0–3 days. Spearman's rho is calculated based on ordered ranks and the Kendall's tau is based on concordant and discordant pairs (Wilks, 2006). Both these measures of association are robust from the deviations of linearity and resistant to outliers.

Figures 4.20, 4.21, and 4.22 show variation of accumulated rainfall and peak discharge for different lag times from 0 to 3 days, 4–7 days, and 8–10 days for structure C6_NW36. The first set of coefficient of determination values reported in the figures relate to the entire data, the second set of coefficients relate to hurricane event data, and third set is for all events without hurricane events included.

It is evident that, at longer durations, accumulated rainfall amounts have higher correlations with peak discharges when hurricane events are considered. These stronger correlations are

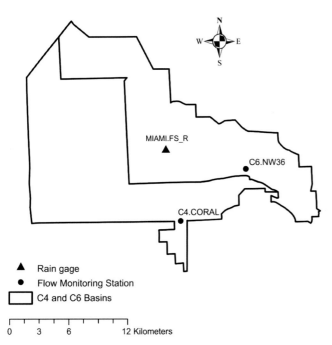

Figure 4.13 C4 and C6 basins and flow and rain monitoring stations.

Table 4.9 *Summary statistics for annual peak discharge and lagged rainfall values for C4_Coral–Miami-FS_R*

Variable	μ	σ	C_s	C_k
Discharge (Q)	12.49	5.47	1.90	6.75
R(lag 0 day)	32.82	39.83	1.70	6.57
R(lag −1 day)	66.68	67.34	2.24	8.91
R(lag −2 day)	93.52	85.39	1.96	6.84
R(lag −3 day)	104.01	85.98	1.70	5.83
R(lag −4 day)	116.23	88.29	1.54	5.16
R(lag −5 day)	125.81	89.69	1.45	4.75
R(lag −6 day)	134.75	89.13	1.31	4.52
R(lag −7 day)	141.94	89.92	1.21	4.23
R(lag −8 day)	147.93	92.61	1.11	3.80
R(lag −9 day)	153.72	91.34	1.01	3.75
R(lag −10 day)	161.26	89.71	0.87	3.66
μ	Mean		C_s	Skewness coefficient
σ	Standard deviation		C_k	Kurtosis

Table 4.8 *Peak discharge at C4_Coral and accumulated rainfall depth (lag = 3 days) at Miami FS_R (rain gage)*

Peak discharge (m^3/s)	Rainfall depth (mm)	Peak discharge (m^3/s)	Rainfall depth (mm)
11.89	139.45	9.91	199.90
8.95	81.53	6.91	102.87
9.63	118.87	27.61	119.38
11.58	58.42	12.43	125.48
7.87	52.83	14.38	6.60
11.36	150.62	15.77	203.71
8.50	85.34	10.93	104.39
7.53	19.05	10.93	68.33
6.40	38.35	13.93	64.26
7.87	23.62	10.99	92.46
10.68	66.29	31.71	353.82
7.87	31.75	27.72	389.13
13.88	314.20	18.04	188.21
9.83	89.92	8.75	27.18
19.11	52.32	12.40	47.24
15.01	221.49	15.15	18.03
15.55	68.58	15.12	129.54
10.48	95.76	12.03	80.26
9.71	83.82	13.45	73.91
10.31	119.38	11.47	40.64
7.45	105.41	8.72	38.35
13.39	85.34		

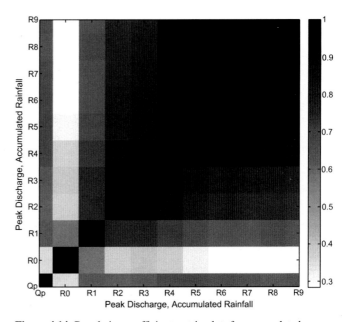

Figure 4.14 Correlation coefficient matrix plots for accumulated rainfall at different lag values and annual peak stage values for C4_CORAL basin.

appropriate for South Florida as landfall of several hurricanes in the wet season produces slow-moving storms and higher rainfall accumulations over time.

4.23 ASSESSMENT FROM DEPENDENCE ANALYSIS

- Dependence analysis using annual peak discharges and accumulated rainfall values provides a reasonable estimate of the

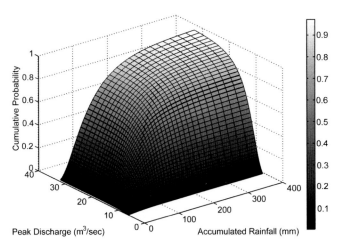

Figure 4.15 Joint cumulative probability plots for peak discharge and accumulated rainfall (lag = 3) for C4_CORAL.

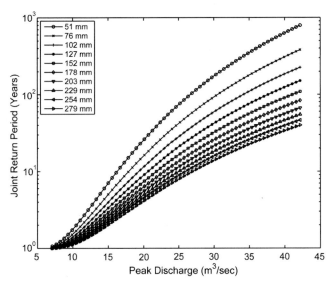

Figure 4.17 Joint return period plots for peak discharge and accumulated rainfall (lag = 3) for C4_CORAL.

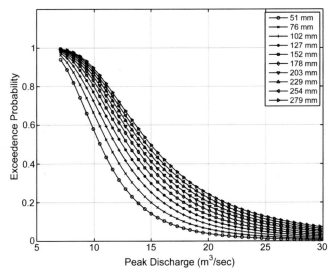

Figure 4.16 Joint exceedence conditional probability plots for peak discharge and accumulated rainfall (lag = 3) for C4_CORAL.

response time of the basins. Negative and positive dependencies may be observed in many basins at different dependence lag times. Negative or no dependence mainly suggests a rainfall–runoff response that is not typical of a natural watershed.

- Extreme rainfall events may not always be linked to the annual peak discharges in a watershed. This link may not be strong even at a lag time of several days from the day of maximum peak value of discharge.
- Dependence analysis and joint probability approaches can be helpful in assessment of existing design storms and peak discharges.
- The inclusion and non-inclusion of extreme precipitation events related to episodic extreme events (e.g., tropical

storms, hurricanes) may affect the conclusions of dependence analysis.

- Cumulative BVN functions can be used to assess the joint probabilities associated with extreme events and are helpful for assessment of adequacy of design storm and discharge values.
- Selection of an appropriate rain gage based on its location within a watershed (or a basin) is critical for the success of dependence analysis.
- Large time differences in occurrence and correlations between maximum rainfall and peak discharges demonstrate that maximum peak discharges in the basins may not necessarily relate to maximum precipitation events (i.e., annual extremes). It can be concluded this is expected behavior for basins where groundwater storage and soil moisture conditions play an important role in the runoff generation mechanisms.
- Smaller basin response times are suggestive of developed watersheds.

4.24 STATISTICAL ANALYSIS OF PEAK DISCHARGE AND PRECIPITATION DATA

Statistical analysis of available peak discharge and lagged accumulated precipitation data or annual maximum precipitation data can be conducted. The analysis need not necessarily be carried out in the joint probability analysis framework discussed in this chapter. Peak discharges and extreme rainfall amounts can be independently analyzed. These analyses of peak discharge and

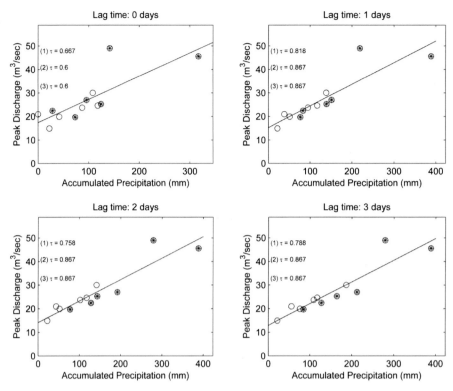

Figure 4.18 Variation of accumulated rainfall and peak discharge for different lag times from 0 to 3 days for structure C6_NW36, using Kendall's tau.

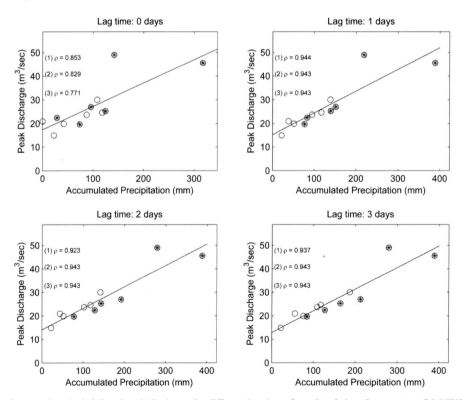

Figure 4.19 Variation of accumulated rainfall and peak discharge for different lag times from 0 to 3 days for structure C6_NW36, using Spearman's rho.

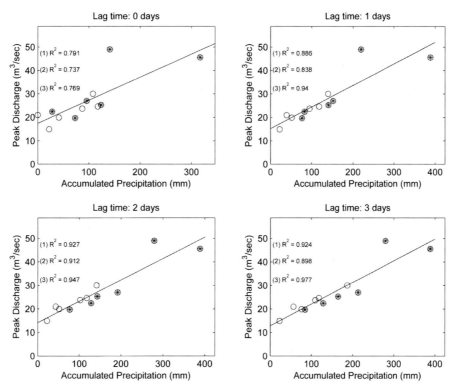

Figure 4.20 Variation of accumulated rainfall and peak discharge for different lag times from 0 to 3 days for structure C6_NW36.

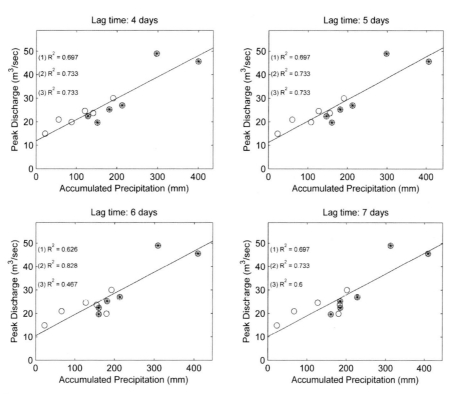

Figure 4.21 Variation of accumulated rainfall and peak discharge for different lag times from 4 to 7 days for structure C6_NW36.

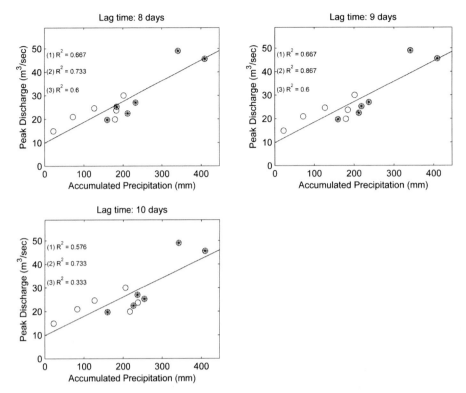

Figure 4.22 Variation of accumulated rainfall and peak discharge for different lag times from 8 to 10 days for structure C6_NW36.

extreme precipitation data are referred to as flood and precipitation frequency analyses. Several candidate statistical distributions are appropriate for characterizing peak discharge and accumulated precipitation data. These distributions are extreme value, generalized extreme value, normal, log-normal, three-parameter log-normal, Pearson, and log-Pearson. The distribution parameters can be estimated using maximum likelihood estimation (MLE) methods. The Chi-square test can be used as a goodness-of-fit method. The goodness-of-fit test can be conducted with varying bin (cell) sizes to assess the sensitivity of the test to bin size selection. Trend analyses to assess the stationarity of the peak discharge time series can be carried out using the Mann–Kendall tau-b with Sen slope method. The test is based on a non-parametric function that computes a coefficient representing strength and direction of a trend for time series data. Statistical analysis methods discussed in Chapter 7 can be used for analysis of peak flow data. Also, there are several textbooks that primarily focus on flood frequency analysis (e.g., Haan, 2002) from which the reader can obtain more detailed statistical procedures.

The generalized extreme value (GEV) distribution is one of the most used distributions for characterizing flood peaks in many flood frequency analysis studies. Villarini *et al.* (2011) document several studies that have used GEV distributions. Differences in flood generation mechanisms due to different meteorological processes and possible future climate change conditions require

evaluation of (1) "mixtures" of flood peak distributions; (2) upper tail and scaling properties of the flood peak distributions; and (3) the presence of temporal non-stationarities in the flood peak records (Villarini *et al.*, 2011). The temporal non-stationarity issue relevant to floods is discussed in the next section.

4.25 FLOODS IN A CHANGING CLIMATE: ISSUES

Flood frequency analysis based on the assumption of stationarity still continues to be the basis for estimation of future flood threats and risks. Changing hydrologic conditions, driven by the climate system or by human activities within the watershed, as well as concepts of multi-decadal climate variability, present a challenge to the assumption of stationarity (Kiang *et al.*, 2011). Assessment of non-stationarities in peak discharge series generally requires a series of statistical tests. Change-point tests using a non-parametric test such as the Pettitt test (Pettitt, 1979) can be used to assess abrupt changes in mean and variance of flood peak distributions. The presence or absence of monotonic trends can be assessed by the Mann–Kendall and Spearman tests (Villarini *et al.*, 2011).

Climate change as one of the drivers may alter flood generation processes when input variables (e.g., precipitation) to the hydrologic cycle change in future. Increases in temperature and precipitation extremes in future will affect the timing, magnitude, and frequency of flooding events. Timing shifts in the onset of spring snowmelt in some parts of the world may influence water availability and therefore affect the operation of water resources systems. Examples of these flooding events include snowmelt floods, which are not based on extreme short-duration precipitation events. It is important to recognize that the causes of non-stationarity may not always be related to climate change. Long-term groundwater developments, construction of water impounding structures (i.e., dams) and diversion structures may lead to reduction of flows in watersheds. An example of such causes of non-stationarity based on groundwater depletion in the San Pedro River is discussed by Hirsch (2011).

Projections of precipitation change from climate change models are less reliable than projections of temperature (Barsugli *et al.*, 2009). Barsugli *et al.* also indicate that long-term changes in large areas are less uncertain compared to short-term changes in small areas. Short-duration extreme rainfall (high intensity) may not always lead to peak flooding events. Long-duration rainfall events, especially due to slow-moving tropical cyclone systems (hurricanes) are often responsible for extreme flooding events. Short-duration rainfall events with high intensity are expected to increase in future (IPCC, 2007d). While these events are critical in some watersheds and may lead to flash floods (Shelton, 2009), long-duration rainfall events play a major role in watersheds where the response is not so quick. As is evident from the example presented for dependence analysis from South Florida, response of watersheds depends on several factors such as a groundwater levels, antecedent soil moisture, hurricane landfalls, and managed and controlled water conveyance systems. The availability of long-term extreme precipitation data and peak discharge data is critical for assessment of stationarity or non-stationarity of time series. The inclusion of a specific data set from a temporal window (most recent years) may lead to confirmation of stationarity or non-stationarity.

4.26 CONCLUSIONS AND SUMMARY

This chapter deals with hydrometeorological aspects of precipitation and relevant flood-causing drivers along with flood generation mechanisms. Evaluation of extreme precipitation events in combination with peak discharges in a watershed are addressed through joint probability approaches. Watershed conditions (e.g., soil moisture, water table levels) that influence runoff generation mechanisms are discussed. Detailed explanations of application of

the joint probability approach along with the desk study approach (DSA) are provided. Incorporation of lead time in forecasting occurrence of extreme flows is discussed in a joint probability framework. Dependence analysis considering extreme flows and precipitation events lagged by number of specific time intervals is described for understanding watershed response times. Issues related to flood frequency analyses under changing climate are discussed.

EXERCISES

4.1 Obtain long-term precipitation and peak flood values at a site in your region and assess the hydrometeorological conditions that have led to these extreme events. Quantify the occurrence of extreme events on a seasonal basis.

4.2 Obtain extreme precipitation data from your region either as a block maximum (i.e., yearly maximum values) or POT values and perform a statistical analysis on the data to characterize the data by any statistical distribution.

4.3 Using accumulated 10-day precipitation values, provided in Table 4.10, from the time of occurrence of peak discharge in several years at a structure, evaluate the normality of the time series. Use any transformation (e.g., Box–Cox) to normalize the data if required. Use normal probability plots as visual checks for normality.

Table 4.10 *Precipitation values*

Precipitation depth (mm)			
229	51	166	128
149	314	236	197
146	104	70	144
254	324	238	137
70	221	114	165
86	102	91	146
267	159	65	
129	161	104	
82	133	405	
40	107	401	
70	216	196	
2	231	40	
106	113	83	

4.4 Use the yearly peak discharge and lagged accumulated rainfall values for different durations provided in Table 4.11, for a hydraulic structure in a basin in South Florida, to perform a joint probability analysis described in this chapter.

Table 4.11 *Peak discharge and lagged accumulated rainfall values*

Peak Discharge (m³/s)	Daily precipitation depths						
	lag = 0	lag = 1	lag = 2	lag = 3	lag = 4	lag = 5	lag = 6
11.89	89	139	139	139	149	149	149
8.95	0	0	75	82	82	82	124
9.63	49	119	119	119	163	163	171
11.58	0	23	58	58	62	70	70
7.87	44	44	44	53	57	65	66
11.36	6	151	151	151	163	210	212
8.50	51	70	81	85	91	91	91
7.53	7	19	19	19	41	43	43
6.40	38	38	38	38	38	38	38
7.87	0	6	24	24	26	69	69
10.68	0	41	47	66	80	101	101
7.87	6	32	32	32	35	35	35
13.88	0	19	314	314	314	314	314
9.83	48	63	88	90	100	100	100
19.11	0	8	15	52	250	319	319
15.01	1	30	221	221	221	221	221
15.55	0	47	48	69	72	86	86
10.48	86	86	86	96	124	124	126
9.71	0	34	84	84	84	84	136
10.31	119	119	119	119	119	119	125
7.45	0	29	105	105	105	105	105
13.39	44	85	85	85	85	146	172
9.91	59	94	163	200	200	200	229
6.91	26	44	73	103	103	103	103
27.61	41	81	118	119	119	119	120
12.43	0	48	69	125	132	170	189
14.38	6	7	7	7	38	38	61
15.77	88	147	160	204	204	204	204
10.93	53	67	93	104	104	106	106
10.93	32	55	56	68	68	77	77
13.93	64	64	64	64	64	65	65
10.99	16	92	92	92	92	92	92
31.71	188	280	353	354	380	382	390
27.72	0	337	389	389	389	392	393
18.04	89	120	120	188	196	196	196
8.75	6	8	20	27	27	27	27
12.40	39	46	46	47	51	51	72
15.15	1	1	1	18	23	62	113
15.12	69	81	91	130	157	157	157
12.03	15	66	80	80	100	113	125
13.45	7	19	51	74	76	98	98
11.47	29	40	40	41	68	86	165
8.72	27	33	34	38	59	61	70

4.5 Perform a dependence analysis using peak discharge and rainfall data provided in Table 4.11 to comment on correlations at different lags and response of the watershed or basin.

4.6 Assuming a non-linear (power) relationship between groundwater table height (H) and soil moisture capacity (S) discussed in this chapter, evaluate the impact of increased soil moisture on peak discharge using the SCS CN method.

Table 4.12 *Annual 1-hour maximum precipitation data*

Year	Depth	Year	Depth	Year	Depth	Year	Depth
1942	112	1958	45	1974	47	1990	71
1943	52	1959	50	1975	54	1991	75
1944	31	1960	59	1976	80	1992	76
1945	63	1961	41	1977	45	1993	64
1946	67	1962	40	1978	41	1994	53
1947	58	1963	81	1979	65	1995	47
1948	54	1964	58	1980	67	1996	43
1949	56	1965	44	1981	36	1997	46
1950	34	1966	53	1982	48	1998	43
1951	51	1967	51	1983	39	1999	39
1952	63	1968	79	1984	71	2000	64
1953	35	1969	90	1985	40	2001	43
1954	37	1970	47	1986	56	2002	56
1955	47	1971	57	1987	58	2003	64
1956	32	1972	50	1988	65	2004	39
1957	64	1973	32	1989	60		

Use values of 2.67 for coefficient a and 0.97 for exponent b in the relationship $S = aH^b$.

4.7 Discuss different methods for selecting POT values and confirm that these are indeed independent and random samples.

4.8 Use the annual 1-hour maximum precipitation time series provided in Table 4.12 to obtain PMP using the method discussed in this chapter.

Table 4.13 *Annual (water year) peak discharge data*

Year	m³/s	Year	m³/s	Year	m³/s	Year	m³/s
1941	101.94	1961	14.78	1981	14.44	2001	97.41
1942	105.62	1962	71.92	1982	78.72	2002	108.45
1943	39.08	1963	46.44	1983	75.89	2003	116.67
1944	65.13	1964	174.43	1984	56.92	2004	228.52
1945	180.66	1965	22.85	1985	30.30	2005	120.06
1946	92.60	1966	122.33	1986	120.63	2006	162.82
1947	274.39	1967	98.83	1987	40.21	2007	35.68
1948	353.96	1968	132.81	1988	131.96	2008	192.55
1949	99.96	1969	79.57	1989	10.76	2009	78.72
1950	186.04	1970	314.60	1990	35.40		
1951	262.78	1971	16.28	1991	55.22		
1952	77.59	1972	41.63	1992	55.78		
1953	224.55	1973	38.23	1993	54.93		
1954	319.98	1974	71.36	1994	43.32		
1955	93.16	1975	51.82	1995	145.83		
1956	30.58	1976	68.24	1996	110.15		
1957	509.70	1977	17.75	1997	65.70		
1958	50.40	1978	79.85	1998	105.06		
1959	89.20	1979	149.80	1999	47.86		
1960	246.92	1980	25.03	2000	151.21		

4.9 Peak discharge values at a United States Geological Survey (USGS) station (identification number: 02232000) are provided in Table 4.13. Evaluate (a) any linear trends; (b) trends in variances; and (c) trends in different 30-year windows.

5 Climate change modeling and precipitation

5.1 DOWNSCALING PRECIPITATION

Spatial and temporal downscaling of precipitation from GCM outputs is one of the most challenging tasks encountered by climate change researchers. In general, the skill of any climate change model is lower for precipitation compared to temperature. Oftentimes the climate regime, the temporal resolution of downscaling, the downscaling schemes, and bias-correction procedures affect the final outcome of the downscaling process. The process also inherits some of the biases from the GCM simulations and correction of these biases is often difficult. This chapter provides a brief overview of downscaling methods. For a detailed description of statistical downscaling methods and their applications, readers are referred to the books by Mujumdar and Kumar, and Simonovic in the four-volume IFI series. Details and discussions of different downscaling methodologies reported in this chapter are mainly adopted from CICS (2011) and Wilby et al. (2004).

5.2 DOWNSCALING METHODS

Downscaling of GCM simulations is required for climate scenario development and impact analysis. The results from GCMs can be downscaled using either a nested high-resolution regional climate model (RCM) (Benestad et al., 2008; Christensen et al., 2001; King et al., 2011) or through empirical/statistical downscaling (von Storch et al., 1993; Rummukainen, 1997; Benestad et al., 2008). Puma and Gold (2011) provide guidance in selection of a specific downscaling approach for climate scenario development based on complexity of analysis and spatio-temporal resolution. For example, for low spatio-temporal resolution (e.g., month) and low complexity (i.e., single basic downscaling approach) they suggest using a change factor method with 50 km spatial resolution and monthly temporal resolution. Multiple dynamic and statistical downscaling methods at 15 km spatial resolution and daily temporal resolution are recommended by Puma and Gold (2011).

Downscaling of GCM simulations needs to be carried out at spatial and temporal scales. Spatial and temporal downscaling methods are discussed in the next sections.

5.3 DOWNSCALING AT SPATIAL LEVEL

Spatial downscaling (SD) refers to the techniques used to derive finer resolution climate information from coarser resolution GCM output. Relationships between local- and large-scale climate information can be established based on the assumption that these relationships remain the same for future climate. Site-specific information can be obtained from regional- or larger-scale GCM outputs. One of the simplest SD methods is the change factor method. The change factor method for obtaining higher spatial resolution scenarios involves use of high-resolution baseline data with coarse-scale information on climate change from GCMs (Puma and Gold, 2011).

The spatial resolution of most GCMs is between about 250 and 600 kilometers, and the temporal resolution is a month. The forcings and circulations that affect regional climate, however, generally occur at much finer spatial scales than these and can lead to significantly different regional climate conditions than are implied by the large-scale state. Spatial downscaling may be able to incorporate some of these regional climate controls and hence add value to coarse-scale GCM output in some areas – although its usefulness will be very much dependent on the region and climate data available. Each case will be different and may necessitate the investigation of a number of different downscaling techniques before a suitable methodology is identified – and in some cases it will not be possible to improve upon the coarse-scale scenarios of climate change by downscaling.

There are a number of general recommendations (CICS, 2011) concerning SD:

- The GCM being used for SD should be able to simulate well those atmospheric features that will influence regional climate, e.g., jet streams and storm tracks.

- The downscaling technique should be based on a climate variable that does not exhibit large sub-grid scale variations.
- The variables used in the downscaling process should also ideally be primary model variables; i.e., they are direct model output (e.g., sea level pressure) and are not based on parameterizations involving other model variables, as is the case with precipitation.

5.4 DOWNSCALING AT TEMPORAL LEVEL

Downscaling of GCM simulations at a temporal level is possible by the use of stochastic weather generators. Detailed discussion about the stochastic weather generators is provided in a later section in this chapter.

5.5 STATISTICAL DOWNSCALING TECHNIQUES

Spatial downscaling techniques can be divided into empirical/statistical methods, statistical/dynamic methods, and higher-resolution modeling referred to as regional climate modeling. The techniques mainly aim at modeling sub-grid scale changes in climate as a function of larger-scale climate. Three different types of classifications (CICS, 2011) are available under this set of techniques:

- Transfer functions (or regression models) – statistical relationships are established based on large-area and site-specific surface climate information.
- Weather typing – statistical relationships are established between specific atmospheric circulation types and local weather patterns.
- Stochastic weather generators – these statistical models may be conditioned on the large-scale state in order to derive site-specific weather.

5.5.1 Approaches using transfer functions

The first step in transfer function-based SD is the definition of the statistical relationships based on observed climate data. This requires identification of the large-scale climate variables (the independent or predictor variables) to be used in the transfer functions or regression models. These predictor variables may be large-scale variables (e.g., mean sea level pressure (MSLP)) that are required to calculate mean values for a region, appropriate to the size of the relevant GCM grid resolution. The approach using transfer functions is shown in Figure 5.1.

Wilby *et al.* (2002) suggest the following requirements for candidate predictor variables:

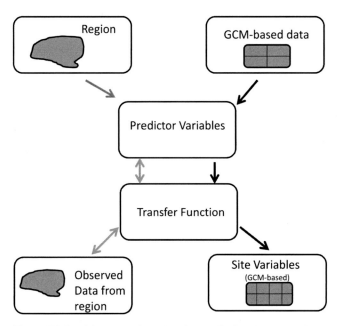

Figure 5.1 Spatial downscaling using the transfer function approach.

- variables that are physically and conceptually sensible with respect to the site variable (the dependent variable or predictand);
- variables that are strongly and consistently correlated with the predictand;
- variables that are readily available from archives of observed data and GCM output; and
- variables that are accurately modeled by GCMs.

Transfer functions can be established based on the selected predictor variables. Normalization of variables is the first step that is reported in the literature, where the variables are first transformed using mean and standard deviation information. This facilitates the use of GCM data in the climate change component of downscaling, since GCM-derived scenarios of future climate change are expressed as changes with respect to a particular baseline time period in an attempt to overcome any shortcomings a GCM may have in simulating current climate conditions (CICS, 2011).

There are a number of methods that can be used to calculate the transfer functions: (1) multiple linear regression (MLR), (2) principal components analysis (PCA), and (3) artificial neural networks (ANNs). Transfer functions are classified under regression models by Wilby *et al.* (2004). Linear downscaling assumes that local observations may be related to large-scale circulation patterns through a simple linear relationship (Benestad *et al.*, 2008). The selection of the method for SD depends on the availability of data. Limitations of the methods should also be realized. For example, use of MLR requires conditions relating to independence of predictor variables. Lack of independence may lead to multicollinearity. It is expected that the SD method should explain a large percentage of the variance in the observed site/station data.

The error associated with the statistical transfer functions, usually defined by the standard error, should also be less than the projected future changes in the variable in question, if these functions are to be used to determine future climate at the location in question. If this is not the case then it cannot be determined whether the statistical model sensitivity to future climate forcing is greater than the model accuracy (CICS, 2011).

Verification of the transfer functions or regression models is a crucial step in the development of SD models. A set of independent data, i.e., data withheld from the calibration process is used for the verification process. Data length is also crucial for the verification and calibration process. Many studies recommend at least 20 years of data to be used for calibrating and 10 years for verification of the SD models.

The determination of model performance is rather subjective and will depend on the study location and the data available. It is a possibility that SD of precipitation may not produce accurate results in spatially varying terrains (landforms defined by complex topography). Local forcing factors will play a much larger role in the determination of local climate characteristics (compared to the role of large-scale forcing factors) in these topographically complex regions than they do in areas of more homogeneous terrain.

If the model calibration and verification processes are accomplished with satisfactory results, then the statistical downscaling models may be used for climate change studies. This process is highlighted by the arrows on the right-hand side of Figure 5.1. In this case, the relevant large-scale predictor variables are extracted from GCM output and then used to drive the statistical models calculated earlier using observed data, with the result being station data for the time period corresponding to that of the GCM input data.

It must be remembered that the statistical models have been calibrated using observed data and are therefore valid only for the data range used in the model calibration process (Wilby *et al.*, 2004). For some variables, such as temperature, it is highly likely that the GCM-derived values will lie outside of the data range used for model calibration. If this is the case, then it is technically incorrect to use the statistical models for downscaling.

Wilby *et al.* (2004) indicate and discuss five major issues when statistical downscaling is considered as a viable approach.

- Choice of statistical method:
  - Statistical method should be determined by the nature of the predictand. Temperature and precipitation data distributions may require different types of linear or non-linear functions.
- Choice of predictors:
  - Predictors based on statistical analysis alone may sufficiently represent the local climate or may not be useful in linking the predictors and predictand.

- Handling of extremes:
  - Extreme events are difficult to replicate using statistical downscaling methods as they are generally devised for mean climate.
- Models for tropical regions:
  - Tropical climatic regions may require consideration of additional variables that accurately represent the ocean–atmosphere interactions.
- Feedback processes:
  - Climate subsystems and weak synoptic forcing should be considered as these may have profound effects on downscaling results.

The success of SD depends on the existence of a strong relationship between predictor and predictand. The assumption that the statistical relationship between small-scale and large-scale parameters is time invariant is essential for SD. Several variants of transfer function-based SD approaches are documented by Wilby and Fowler (2011). A few variants include multiple regression (Murphy, 1999), canonical correlation analysis (CCA) (von Storch *et al.*, 1993), and generalized linear modeling (Fealy and Sweeney, 2007). Canonical correlation analysis helps in identifying linear relationships between two sets of variables. In addition to CCA, PCA is also used in many SD studies to select predictor variables. Ghosh and Mujumdar (2006, 2008) discuss different statistical downscaling approaches using classification techniques, support vector machines, and fuzzy clustering schemes. The second book in this book series by Mujumdar and Nagesh Kumar provides an exhaustive description of these approaches.

5.5.2 Weather typing

This downscaling procedure using statistical/dynamic techniques is very similar to that described above for empirical/statistical methods. Weather typing requires the task of weather classification and the circulation-based schemes can be insensitive to future climate forcing (Wilby *et al.*, 2004). In this case, weather types, or atmospheric circulation patterns, are used rather than large-scale predictor variables and statistical relationships between the types, or patterns, and observed station data are calculated. Steps involved in weather typing are shown in Figure 5.2. The first step in this process is the identification of the weather types, or atmospheric circulation patterns, usually from atmospheric pressure information. These types or patterns may be based on existing subjectively derived weather classes, e.g., the Lamb weather types in the UK (Lamb, 1972) or the European Grosswetterlagen (Hess and Brezowsky, 1977), or they may be objectively derived using techniques such as PCA or ANN. Once the weather types have been identified, statistical models between the types and local station data are calibrated and then verified. If it is possible to

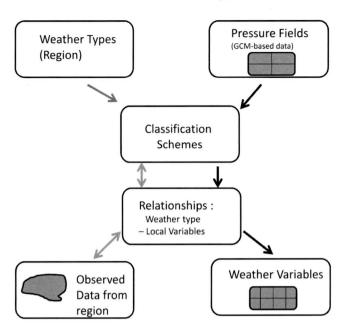

Figure 5.2 Spatial downscaling using the weather typing approach.

develop models that perform satisfactorily, then they can be used in climate change studies.

5.6 WEATHER GENERATORS

Although stochastic weather generators are more widely used in temporal downscaling, they may also be used for spatial downscaling (Wilks, 1999). The first step in the use of a stochastic weather generator is its calibration using observed daily weather data at a site. Parameter sets, which contain information about the statistical characteristics of the observed data, result from the calibration process and are subsequently used to simulate artificial weather data that have the same statistical characteristics as the observed data, but differ on a day-to-day basis. For SD, a weather generator can be calibrated using "average weather" data for a particular region, roughly corresponding to the size of an appropriate GCM grid size, with the resulting parameter set describing the statistical characteristics of that region's weather (Benestad et al., 2008). This average weather is calculated using a number of stations within the selected region. The weather generator can also be used at each of these individual stations, resulting in a suite of individual station parameter sets. Daily GCM data for the grid box corresponding to the area-average weather data are also used in the weather generator and parameter sets can be generated appropriately. The differences between the area-average and GCM grid box parameter sets are then applied to the individual station parameter sets, thus allowing the generation of synthetic data, corresponding to the time period of the GCM, for each station within the area. This process is described in more detail in

Wilks (1999) and Benestad et al. (2008). Spatial downscaling using stochastic weather generators requires a large amount of observed station data, which may or may not be readily available. Additional concepts related to weather generators are discussed later in this chapter.

5.7 REGIONAL CLIMATE MODEL: DYNAMIC DOWNSCALING

Dynamic downscaling is another approach used to produce high-resolution climate scenarios, where Earth-system processes are mathematically represented in a way similar to GCMs but at a much higher resolution than typical GCMs (Puma and Gold, 2011). Wilby and Fowler (2011) provide an extensive review of dynamic and statistical downscaling models. Regional climate models are numerical models that are similar to GCMs, but are generally available at higher (or finer) spatial resolution. It is generally agreed that these models can provide better representations of land surface features and other processes. Depending on the model resolution, they may also be able to resolve some of the atmospheric processes that are parameterized in a GCM. A nested approach is adopted generally so that an RCM is nested within the driving GCM so that the high-resolution model simulates the climate features and physical processes in much greater detail for a limited area of the globe, whilst drawing information about initial conditions, time-dependent lateral meteorological conditions, and surface boundary conditions from the GCM (CICS, 2011). Most nesting techniques are developed in such a way that only one-way linking is allowed. This linking suggests that no feedback is allowed from the RCM simulation to the driving GCM. The global model simulates the response of the global circulation to large-scale forcings, whilst the RCM accounts for sub-GCM grid-scale forcings, such as complex topographical features and land-cover inhomogeneity, in a physically based way and thus enhances the simulations of atmospheric and climatic variables at finer spatial scales. However, the RCM is susceptible to any systematic errors in the driving fields provided by the global models, and these may be exacerbated in the RCM, resulting in a poor simulation of the regional climate. Wilby and Fowler (2011) indicate that despite the clear benefits to precipitation modeling arising from improved vertical and horizontal resolution of terrain within RCMs, it is apparent that some landscape features are still poorly represented. They provide an example in which the topography is not accurately represented at 20-km resolution. Also, GCM fields at temporal resolutions at durations less than a day are essential to address the boundary conditions for the RCM; these are generally not made available by GCM simulations. Also, RCM simulations may be computationally intensive compared to statistical downscaling due to finer domain size and resolution.

5.8 OTHER APPROACHES

This section describes conceptually simple approaches for climate scenario development that do not rely on the GCM-based model simulations. There are two approaches that are referred to as the spatial and temporal analogue approach and the arbitrary approach. The spatial and temporal analogue approach involves the construction of spatial or temporal analogues using historic climate data (Puma and Gold, 2011). The data used as the analogues are either from the past or from another location (Wilby et al., 2004). Availability of data from another location is crucial for the success of this method. The analogues may not always represent the future conditions correctly and the method based on analogues may not be reliable for climate change to scenario development. However, this is one of the simplest approaches that uses no climate change model results.

The arbitrary change approach assigns random changes, at regular intervals, to various variables (e.g., temperature and precipitation) of the established baseline climate, bypassing the need for use of more complicated climate scenario techniques (Puma and Gold, 2011; Lu, 2006). This approach is essentially a method of sensitivity analysis to determine the responses of the system under consideration based on expected or predicted changes associated with future climate changes. Another approach referred to as the "unintelligent" downscaling technique involves simple spatial interpolation between the center points of GCM grid cells to obtain estimates at the desired intermediate points that correspond to those in a higher-resolution baseline climate data set (Wilby et al., 2004; Puma and Gold, 2011).

5.9 STATISTICALLY DOWNSCALED CLIMATE CHANGE PROJECTIONS: CONCEPT EXAMPLE

Statistically downscaled climate change projections that are corrected using specific bias-correction and spatial disaggregation (BCSD) (BCSD, 2011; Maurer, 2007) procedures are used in the current example to illustrate the skill of different climate change models. BCSD data have several advantages:

1. The data have been tested for several different applications in the USA.
2. Several web-based data retrieval mechanisms are available to efficiently obtain downscaled twenty-first-century climate projections to enable comprehensive assessment of uncertainty.
3. Downscaled data provide a statistical match with observed historical data.

4. Data set includes spatially continuous, fine-scale gridded output of precipitation and temperature useful for analysis of climate change impacts on water resources.

Data sets as comprehensive as the BCSD data set are difficult to produce using dynamic downscaling (USGS, 2009). The downscaled climate projections (DCP) archive from the BCSD data set available for different variables and years is computationally less expensive compared to RCM application. The list of BCSD data sets from different climate change models is provided in Table 5.1. The table also includes the model identification number and the modeling groups involved in the development of the models. Table 5.2 describes the available data for different SRES scenarios and the works in which these models are exhaustively discussed.

5.9.1 Special Report on Emissions Scenarios

The IPCC began the development of a set of emissions scenarios in 1996, effectively to update and replace the well-known IS92 scenarios. The new set of scenarios is described in the IPCC Special Report on Emissions Scenarios (SRES). Four different narrative storylines were developed to define the relationships between the forces driving emissions and their evolution. A set of 40 scenarios (35 of which contain data on the full range of gases required to force climate models) cover a wide range of the main demographic, economic, and technological driving forces of future greenhouse gas and sulfur emissions.

Any given scenario represents a specific quantification of one of the four storylines. All the scenarios based on the same storyline constitute a scenario "family." The main characteristics of the four SRES storylines and scenario families are explained below. These storylines are adopted from the IPCC SRES report (IPCC, 2000).

A1: The A1 storyline and scenario family mainly represents a future world with very rapid economic growth, global population that peaks in mid-century and declines thereafter, with rapid introduction of new and more efficient technologies. The A1 scenario family develops into three groups that emphasize alternative directions of technological change in the energy system. The three A1 groups are distinguished by their technological focus: fossil intensive (A1FI), non-fossil energy sources (A1T), or a balance across all sources (A1B) (with no reliance on one particular energy source).

A2: The A2 storyline and scenario family describes a very diverse world. Self-reliance and preservation of local identities are the main underlying influencing factors. Slow and continuously increasing population is expected. Economic development is primarily regionally oriented and per capita economic growth and technological changes are more fragmented and slower than other storylines.

Table 5.1 *List of climate change models (BCSD, 2011) used for projections in South Florida*

Model number	Modeling group and country of origin	WCRP CMIP3 i.d.
1	Bjerknes Centre for Climate Research	BCCR-BCM2.0
2	Canadian Centre for Climate Modeling and Analysis	CGCM3.1 (T47)
3	Meteo-France/Centre National de Recherches Meteorologiques, France	CNRM-CM3
4	CSIRO Atmospheric Research, Australia	CSIRO-Mk3.0
5	US Department of Commerce/NOAA/Geophysical Fluid Dynamics Laboratory, USA	GFDL-CM1.0
6	US Department of Commerce/NOAA/Geophysical Fluid Dynamics Laboratory, USA	GFDL-CM2.0
7	NASA Goddard Institute for Space Studies, USA	GISS-ER
8	Institute for Numerical Mathematics, Russia	INM-CM3.0
9	Institut Pierre Simon Laplace, France	IPSL-CM4
10	Center for Climate System Research (The University of Tokyo)	MIROC3.2 (medres)
	National Institute for Environmental Studies, and Frontier Research Center for Global Change (JAMSTEC), Japan	
11	Meteorological Institute of the University of Bonn, Meteorological Research Institute of KMA	ECHO-G
12	Max Planck Institute for Meteorology, Germany	ECHAM5/ MPI-OM
13	Meteorological Research Institute, Japan	MRI-CGCM2.3.2
14	National Center for Atmospheric Research, USA	CCSM3
15	National Center for Atmospheric Research, USA	PCM
16	Hadley Centre for Climate Prediction and Research/Met Office, UK	UKMO-HadCM3

Table 5.2 *SRES runs available for the models (BCSD, 2011) and associated references*

Model	SRES A2 runs	SRES A1B runs	SRES B1 runs	Main reference for model
1	1	1	1	Furevik *et al.*, 2003
2	1 . . . 5	1 . . . 5	1 . . . 5	Flato and Boer, 2001
3	1	1	1	Salas-Melia *et al.*, 2005
4	1	1	1	Gordon *et al.*, 2002
5	1	1	1	Delworth *et al.*, 2006
6	1	1	1	Delworth *et al.*, 2006
7	1	2, 4	1	Russell *et al.*, 2000
8	1	1	1	Diansky and Volodin, 2002
9	1	1	1	IPSL, 2005
10	1 . . . 3	1 . . . 3	1 . . . 3	K-1 model developers, 2004
11	1 . . . 3	1 . . . 3	1 . . . 3	Legutke and Voss, 1999
12	1 . . . 3	1 . . . 3	1 . . . 3	Jungclaus *et al.*, 2006
13	1 . . . 5	1 . . . 5	1 . . . 5	Yukimoto *et al.*, 2001
14	1 . . . 4	1 . . . 3, 5 . . . 7	1 . . . 7	Collins *et al.*, 2006
15	1 . . . 4	1 . . . 4	2 . . . 3	Washington *et al.*, 2000
16	1	1	1	Gordon *et al.*, 2000

B1: The B1 storyline and scenario family describes a convergent world with the same global population that peaks in mid-century and declines thereafter, as in the A1 storyline, but with rapid change in economic structures toward a service and information economy and introduction of clean and resource-efficient technologies. The emphasis is on global solutions to economic, social, and environmental sustainability, including improved equity, but without additional climate initiatives.

B2: The B2 storyline and scenario family describes a world in which the emphasis is on local solutions to economic, social, and environmental sustainability. Continuously increasing global population is expected, however, at a rate lower than A2, with intermediate levels of economic development and

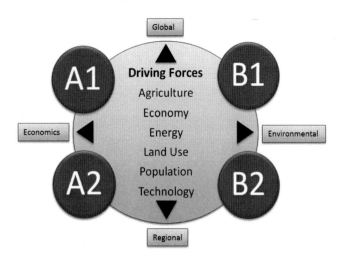

Figure 5.3 Schematic illustration of the four SRES storylines.

less rapid and more diverse technological change than in the A1 and B1 storylines. While the scenario is also oriented towards environmental protection and social equity, it focuses on local and regional levels.

The WCRP CMIP3 data sets are available for only three storylines, A1 (A1B), A2, and B1. Figure 5.3 represents the schematic illustration of the storylines as discussed above.

The SRES scenarios do not include climate initiatives that explicitly assume implementation of the United Nations Framework Convention on Climate Change or the emissions targets of the Kyoto Protocol. Policies adopted by different governments

may influence the greenhouse gas emission drivers, such as demographic change, social and economic development, technological changes, resource uses, and pollution management. Information about models and SRES available from BCSD (BCSD, 2011; Maurer, 2007) are provided in Table 5.2.

5.9.2 Bias-correction spatial disaggregation

The bias-correction part of the BCSD procedure used for generation of fine resolution temperature and precipitation data sets is illustrated in Figure 5.4. The bias-correction procedure is the first step and uses a quantile mapping technique (Wood *et al.*, 2002, 2004).

The observations are spatially aggregated to a specific scale from $1/8°$ to $2°$. The coarser scale for correction is chosen specifically for these data. However, there is no limitation on specific scale selection. The GCM simulations for the twentieth and twenty-first centuries are also spatially interpolated to conform to the aggregated scale. Common temporal data of observations and GCM simulations for the twentieth century are used for bias correction. Quantile maps are then generated for any specific variable of interest (e.g., temperature and precipitation). The quantile maps are then used to adjust the twentieth and twenty-first century GCM simulations. Before application of bias correction to twenty-first century projected values for temperature, the twenty-first century GCM trend is identified and is removed from the GCM twenty-first century data set and is finally then added back to the adjusted GCM data set (Maurer, 2007). The bias-correction

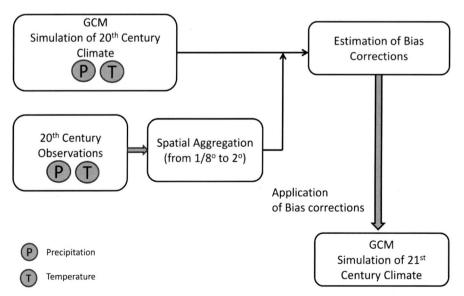

Figure 5.4 Bias-correction procedure (step 1) of BCSD climate change projections.

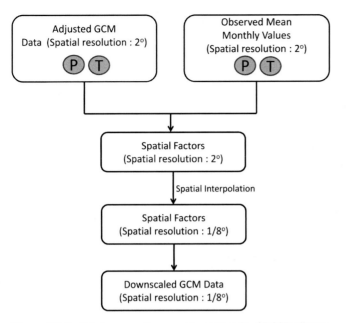

Figure 5.5 Spatial disaggregation procedures (step 2) of BCSD climate change projections.

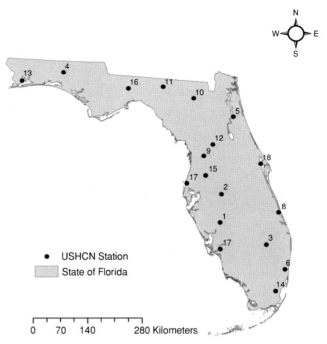

Figure 5.6 Location of USHCN stations in Florida.

methodology assumes that GCM biases have the same structure during the twentieth and twenty-first century simulations.

The second step in BCSD methodology is the spatial disaggregation procedure, which is explained as step 2 in Figure 5.5. The adjusted GCM data and observed values of same spatial resolution are used to obtain spatial correction factors. These factors at a coarser resolution are then interpolated to a resolution at which downscaled data are required. The interpolation algorithm used in BCSD methodology is SYMAP (Shepard, 1984). The downscaled data are available through an interactive website for downloading the data for a specific grid (of 1/8°) resolution over the entire USA.

5.9.3 Limitations

The BCSD procedure has several limitations, some of which are common to many statistical downscaling methods. The assumption of stationarity of climate is the main and weakest assumption in statistical downscaling. Assumptions are made in the case of BCSD:

1. Relationships between the main variables (e.g., precipitation and temperature) at large- and fine-scale will be the same in the future as they are in the past.
2. Any biases exhibited by a GCM are similar for the historical period and for future simulations.

Tests of these assumptions, using historic data, using the BCSD method suggest that the procedure is competitive with other downscaling methods (Wood *et al.*, 2004). The WCRP CMIP3 has produced data sets for only three storylines: A1 (A1B), A2, and B1.

Uses of BCSD data sets are extensive and have been reported in several published studies across the USA. Some of the studies are carried out for different regions: (1) Colorado River basin (Christensen *et al.*, 2004); (2) US Northeast (Hayhoe *et al.*, 2007); (3) California (Maurer and Duffy, 2005); and (4) Sacramento–San Joaquin River basin (Van Rheenen *et al.*, 2004). The BCSD method has been shown to provide downscaling capabilities comparable to other statistical and dynamic methods, especially in the context of hydrologic impacts (Wood *et al.*, 2004).

5.9.4 Evaluation of BCSD data for Florida

The BCSD GCM data are evaluated for several sites in Florida in the USA. The locations of these sites are shown in Figure 5.6.

5.9.5 Selection of downscaled climate change model

The availability of several GCM model outputs or simulations of twentieth and twenty-first century climate provides users with different SRES scenarios. Several questions need to be answered before any one or more models are used for climate change impact studies. These questions, adopted from USGS (2009), are: (1) How many projections should be retained, or should the analyst confine the number to a select few? (2) Should the projections be regarded equally or unequally based on perceived climate model skill? (3) What aspects of the projections should be related to study (e.g., impacts of climate change on regional or local water resource availability)? The good practice guidance paper on assessing and combining multi-model climate projections by Knutti *et al.* (2010) is an excellent resource that will help recommendations in using multi-model ensembles for detection and attribution, and model evaluations.

5.9.6   Performance measures

Several measures can be developed to assess the performance of climate change models in simulating the past historical climatology partially defined by variables of interest. The historical observations during the same time interval for which GCM simulations are available are used for comparison purposes. The temporal observations may need to be aggregated to the scale at which GCM simulations are available. Studies comparing statistical and dynamic downscaling methods use hydrologic indices for assessing the relative skill of the methods (Wilby *et al.*, 2004).

5.9.7   Error measures for model skill assessment

Error measures can be used to assess the relative performance of different GCM-based data. Summary statistics with the intention of checking to see if these are preserved should be evaluated first. Cumulative density function plots can be used to evaluate the distributions of the precipitation data based on historical and GCM-based data. The downscaled precipitation data are compared with available historical data at a site. The performance measures that can be used are:

1. Mean absolute error (MAE): absolute difference in precipitation totals based on GCM-based and historical data.
2. Correlation coefficient (ρ): correlation between historical and GCM-based precipitation data.
3. Absolute probabilistic error: absolute difference in values based on cumulative density plots based on historical and GCM-based data.
4. Probabilistic correlation: correlation based on magnitudes from historical and GCM-based data obtained from pre-specified non-exceedence probabilities.
5. Absolute deviations in annual values: absolute deviation in annual values as opposed to monthly deviations (GCM-based temporal resolutions).
6. Absolute deviation in month/year ratios: absolute deviation in ratios.
7. Calculation of absolute probabilistic error is shown in Figure 5.7. Probabilistic correlation can be calculated based on one of the schematics shown in Figures 5.8 and 5.9.

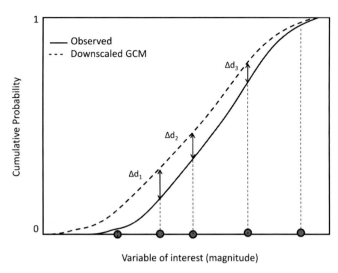

Figure 5.7 Cumulative probability plots of observed and downscaled GCM simulations.

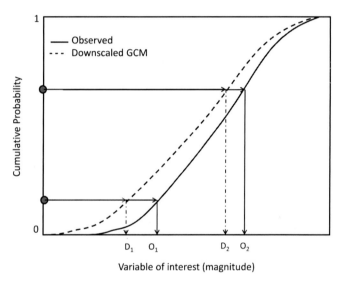

Figure 5.8 Cumulative probability plots of observed and downscaled GCM simulations.

5.9.8   Results and analysis

The GCM-downscaled precipitation and temperature data and observed data for the same variables at 18 United States Historical Climatology Network (USHCN) stations in Florida are compared in the exercise reported in this section. All the climate change model projections are for the SRES A1B scenario. The spatial resolution of the BCSD grid and the locations of USHCN stations are shown in Figure 5.10.

Observed and GCM-downscaled precipitation data are compared for four different stations using CDF plots as shown in Figure 5.11. Similar CDF plots for temperature are shown in Figure 5.12.

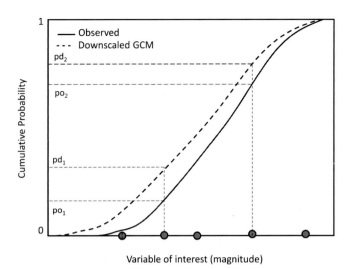

Figure 5.9 Cumulative probability plots of observed and downscaled GCM twentieth-century simulations used for probabilistic correlation measure.

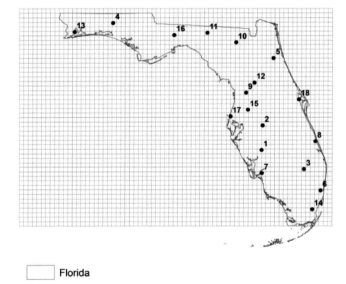

Figure 5.10 Location of USHCN stations used in the case study region with BCSD grid shown over the region of interest.

The CDF plots for all the stations for temperature and precipitation are shown in Figures 5.13 and 5.14 respectively. Visual assessment and evaluation of performance measures clearly indicate that temperature can be downscaled with better skill than precipitation. This observation was already confirmed by an exhaustive study by Fowler *et al.* (2007). They also indicate that winter climate can be downscaled with more skill than summer climate because of stronger relations with large-scale circulation, and wetter climates can be downscaled with more skill than drier climates. Considering the number of statistical performance measures and the variables of interest evaluated, it is difficult to clearly indicate superiority of one model over the other.

Correlation coefficients estimated from monthly observed and GCM projections of precipitation and temperature are provided in Tables 5.3 and 5.4 respectively. It is evident from the tables that the correlations are higher for temperature compared to those for precipitation.

The performance and error measures are weighted using observed and GCM-based precipitation depths at all the stations and the total weight is estimated to rank the models. Higher total weight indicates a better model based on the aggregated performance measure. The rankings of the models are provided in Table 5.5. It is evident from the rankings that the NCAR_PCM model is the best and it is therefore used for further evaluation at each station or observation site.

Cumulative density function plots of observed and specific GCM-based precipitation data for station 14 are shown in Figure 5.15. The selected GCM model is PCM and it is ranked as the best model based on different error measures and performance measures discussed earlier. A quantile–quantile map is also shown using the same data sets in Figure 5.16. It is evident from Figures 5.15 and 5.16 that there is general agreement between observed and GCM-based data. However, the skill of the best model is inferior for precipitation compared to temperature.

Monthly precipitation values at station 14 (from years 1950 to 2099) from GCM models are shown in Figure 5.17. Monthly extremes (from years 1950 to 2099) at station 14 for all GCM models are shown in Figure 5.18. Figure 5.19 shows the annual monthly extremes (from years 1950 to 2099) using the best model at all stations. The projected precipitation data from all the GCM models are summarized by box plots in Figure 5.20.

Two-sample Kolmogorov–Smirnov (KS) tests are conducted to check if the distributions of observed and GCM-based precipitation data are from the same continuous distribution of observed precipitation data. Tables 5.6 and Tables 5.7 provide results from KS tests for observed precipitation data and GCM-based data for the twentieth and twenty-first centuries. The KS tests allow comparison of the distributions of the values in the two data sets.

The null hypothesis is that observed and GCM-based precipitation data are from the same continuous parent distribution of observed precipitation data. The alternative hypothesis is that they

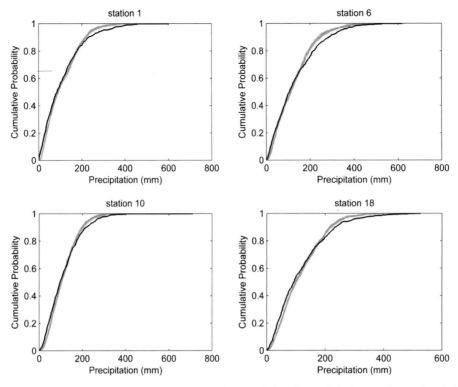

Figure 5.11 Cumulative probability plots of observed and downscaled GCM simulations for precipitation at select stations in Florida.

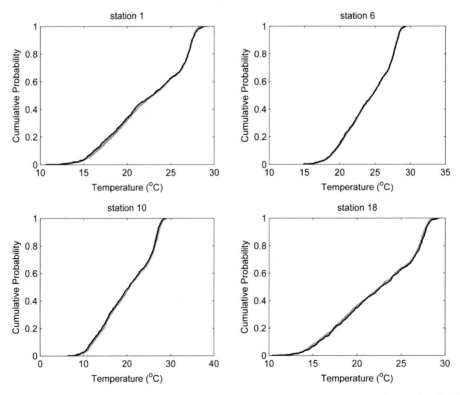

Figure 5.12 Cumulative probability plots of observed and downscaled GCM simulations for temperature at select stations in Florida.

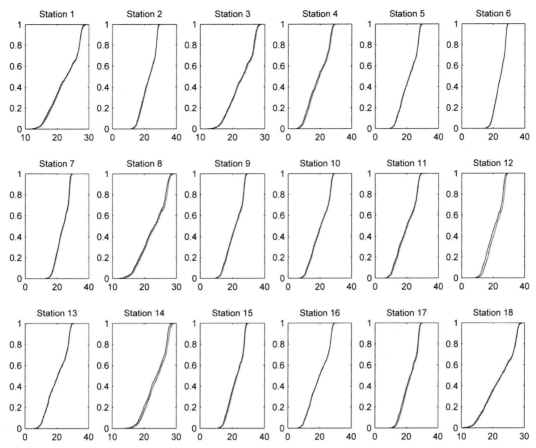

Figure 5.13 Cumulative density plots for temperature-based climate model data for USHCN stations (*x*-axis: temperature (°C), *y*-axis: cumulative probability).

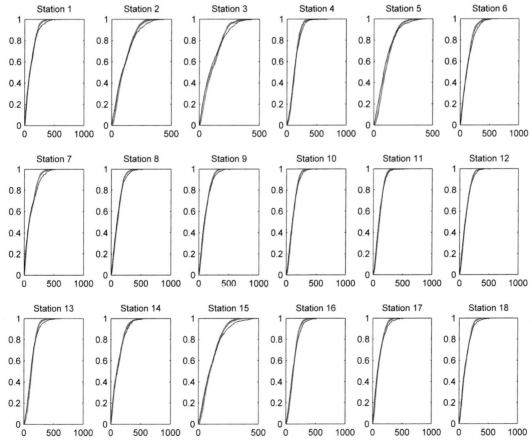

Figure 5.14 Cumulative density plots for precipitation-based climate model data for USHCN stations (*x*-axis: precipitation depth (mm), *y*-axis: cumulative probability).

Table 5.3 *Monthly correlations for precipitation data*

Station	BCCR-BCM2.0	CGCM	CNRM-CM	CSIRO-Mk	GFDL-CM	GISS-ER	INM-CM	IPSL-CM	MIROC	ECHO-G	ECHAM5	MRI-CGCM	CCSM3	PCM	UKMO-HadCM
1	0.58	0.49	0.55	0.55	0.51	0.57	0.52	0.49	0.51	0.51	0.56	0.52	0.58	0.57	0.56
2	0.54	0.49	0.49	0.51	0.49	0.55	0.51	0.46	0.49	0.48	0.52	0.51	0.50	0.54	0.51
3	0.53	0.49	0.52	0.52	0.51	0.54	0.53	0.51	0.51	0.51	0.52	0.49	0.57	0.55	0.54
4	0.19	0.16	0.12	0.13	0.17	0.14	0.20	0.20	0.23	0.16	0.19	0.12	0.23	0.21	0.16
5	0.41	0.37	0.37	0.34	0.36	0.37	0.44	0.33	0.40	0.37	0.39	0.39	0.39	0.44	0.39
6	0.45	0.40	0.43	0.45	0.44	0.38	0.44	0.38	0.42	0.45	0.41	0.44	0.46	0.43	0.44
7	0.66	0.63	0.66	0.63	0.63	0.59	0.63	0.60	0.65	0.62	0.63	0.60	0.65	0.64	0.66
8	0.38	0.34	0.38	0.37	0.37	0.39	0.38	0.39	0.39	0.35	0.35	0.36	0.42	0.38	0.36
9	0.47	0.45	0.46	0.48	0.48	0.52	0.48	0.44	0.48	0.45	0.48	0.51	0.46	0.52	0.51
10	0.29	0.34	0.30	0.32	0.35	0.37	0.36	0.32	0.35	0.34	0.34	0.34	0.39	0.39	0.33
11	0.20	0.22	0.15	0.16	0.20	0.26	0.20	0.24	0.24	0.22	0.22	0.19	0.30	0.28	0.21
12	0.38	0.39	0.34	0.40	0.42	0.43	0.43	0.37	0.42	0.43	0.40	0.42	0.41	0.46	0.44
13	0.14	0.15	0.10	0.10	0.10	0.08	0.18	0.15	0.14	0.10	0.10	0.08	0.17	0.15	0.11
14	0.55	0.53	0.55	0.55	0.58	0.52	0.55	0.51	0.56	0.56	0.55	0.53	0.60	0.56	0.59
15	0.43	0.40	0.42	0.46	0.45	0.49	0.46	0.39	0.45	0.43	0.48	0.48	0.45	0.48	0.46
16	0.23	0.25	0.16	0.18	0.22	0.21	0.27	0.27	0.26	0.22	0.25	0.19	0.32	0.29	0.24
17	0.47	0.45	0.45	0.47	0.50	0.46	0.47	0.43	0.46	0.45	0.46	0.50	0.48	0.51	0.49
18	0.44	0.42	0.38	0.41	0.42	0.45	0.44	0.39	0.43	0.38	0.42	0.43	0.43	0.45	0.44

Table 5.4 *Monthly correlations for temperature data*

Station	BCCR-BCM2.0	CGCM	CNRM-CM	CSIRO-Mk	GFDL-CM	GISS-ER	INM-CM	IPSL-CM	MIROC	ECHO-G	ECHAM5	MRI-CGCM	CCSM3	PCM	UKMO-HadCM
1	0.92	0.91	0.92	0.91	0.92	0.92	0.92	0.92	0.92	0.91	0.91	0.92	0.92	0.92	0.91
2	0.92	0.91	0.92	0.92	0.93	0.93	0.93	0.92	0.92	0.91	0.92	0.92	0.92	0.92	0.92
3	0.91	0.90	0.91	0.90	0.91	0.91	0.91	0.91	0.90	0.90	0.90	0.91	0.90	0.91	0.91
4	0.94	0.94	0.94	0.94	0.94	0.94	0.94	0.94	0.94	0.94	0.94	0.94	0.94	0.94	0.94
5	0.93	0.92	0.92	0.92	0.93	0.93	0.94	0.93	0.93	0.92	0.93	0.93	0.93	0.93	0.93
6	0.90	0.89	0.89	0.89	0.89	0.90	0.90	0.90	0.89	0.89	0.89	0.90	0.89	0.89	0.89
7	0.92	0.91	0.92	0.91	0.92	0.92	0.92	0.92	0.91	0.91	0.91	0.92	0.91	0.91	0.91
8	0.91	0.90	0.91	0.90	0.91	0.91	0.91	0.91	0.91	0.90	0.90	0.91	0.90	0.91	0.90
9	0.93	0.92	0.92	0.92	0.93	0.93	0.93	0.92	0.92	0.92	0.92	0.92	0.92	0.93	0.92
10	0.94	0.93	0.93	0.93	0.94	0.94	0.94	0.94	0.93	0.93	0.94	0.93	0.94	0.94	0.93
11	0.94	0.94	0.93	0.93	0.94	0.94	0.94	0.94	0.94	0.93	0.94	0.93	0.94	0.94	0.94
12	0.93	0.92	0.92	0.92	0.93	0.93	0.93	0.93	0.93	0.92	0.92	0.92	0.93	0.93	0.92
13	0.94	0.94	0.94	0.94	0.94	0.94	0.94	0.94	0.94	0.94	0.94	0.94	0.94	0.94	0.94
14	0.90	0.90	0.90	0.89	0.90	0.90	0.90	0.90	0.90	0.89	0.90	0.90	0.89	0.89	0.90
15	0.92	0.91	0.92	0.92	0.93	0.93	0.93	0.92	0.92	0.91	0.92	0.92	0.92	0.93	0.92
16	0.94	0.94	0.94	0.93	0.94	0.94	0.94	0.94	0.94	0.94	0.94	0.94	0.94	0.94	0.94
17	0.93	0.92	0.92	0.92	0.93	0.93	0.93	0.92	0.92	0.92	0.92	0.92	0.92	0.92	0.92
18	0.92	0.91	0.92	0.91	0.92	0.92	0.92	0.92	0.92	0.91	0.91	0.92	0.92	0.92	0.91

Climate model

Table 5.5 *Weighted performance measures for different GCM models*

	Weight for each performance measure								
Model (with scenario)	RMSE	MAE	MC	PE	DYM	PC	RMY	TW	Rank
bccr_bcm2_0.1.sresa1b	380	360	734	613	1306	1749	169	5310	5
cccma_cgcm3_1.1.sresa1b	360	362	697	613	1322	1748	168	5271	6
cnrm_cm3.1.sresa1b	359	346	681	542	1336	1744	170	5179	15
csiro_mk3_0.1.sresa1b	378	364	703	576	1314	1743	161	5239	12
gfdl_cm2_0.1.sresa1b	371	357	721	622	1287	1748	159	5264	7
giss_model_e_r.2.sresa1b	375	370	734	629	1300	1751	156	5314	4
inmcm3_0.1.sresa1b	395	373	750	540	1263	1744	187	5251	9
ipsl_cm4.1.sresa1b	348	332	685	658	1322	1751	151	5247	10
miroc3_2_medres.1.sresa1b	389	388	739	637	1259	1752	194	5356	3
miub_echo_g.1.sresa1b	362	344	702	656	1330	1749	119	5261	8
mpi_echam5.1.sresa1b	376	364	728	587	1269	1747	176	5246	11
mri_cgcm2_3_2a.1.sresa1b	370	369	708	541	1315	1745	166	5213	14
ncar_ccsm3_0.1.sresa1b	400	390	779	570	1136	1750	214	5238	13
ncar_pcm1.1.sresa1b	414	397	785	630	1319	1747	207	5500	1
ukmo_hadcm3.1.sresa1b	388	375	745	677	1320	1752	183	5440	2

RMSE: root mean squared error, MAE: mean absolute error, MC: monthly correlation, PE: probabilistic error, DYM: difference in yearly mean, PC: probabilistic correlation, RMY: ratio of month to year, TW: total weight.

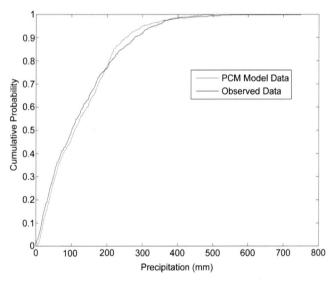

Figure 5.15 Cumulative distribution plot of the selected best GCM model and the observed data at station 14.

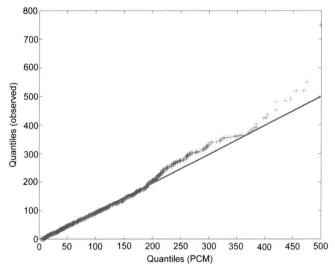

Figure 5.16 Quantile–quantile plot of the selected best GCM model and the observed data at station 14.

are from different continuous distributions. The results reported in Table 5.6 are for historical observations and GCM-based data for the twentieth century and those in Table 5.7 are for future projections. Binary values representative of hypothesis test results are reported in Table 5.6 and Table 5.7. A value of 1 represents that the null hypothesis is rejected at 5% significance level and a value of 0 suggests acceptance of the null hypothesis. Results from KS tests provided in Tables 5.6 and 5.7 indicate that in the majority of cases (different models and stations), comparative KS tests resulted in alternative hypotheses being true, with one exception for station

14. The results from the PCM model resulted in the lowest number of alternative hypotheses being true.

5.10 WEATHER GENERATOR: CONCEPTS

Weather generator (WG), as the name suggests, generates weather in essence at a particular location. Synthetic or artificial sequences of weather patterns for infinite time can be generated using

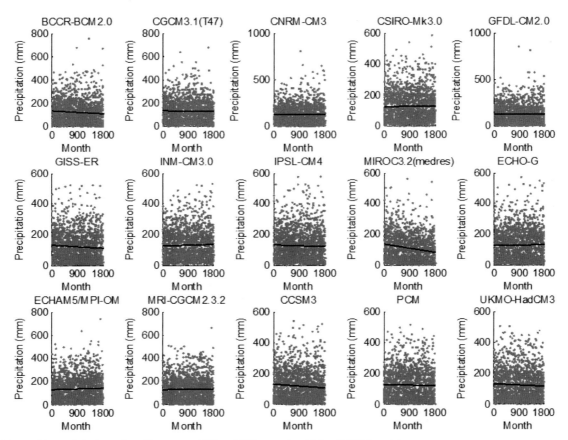

Figure 5.17 Monthly precipitation values (from years 1950 to 2099) at station 14 for all GCM models.

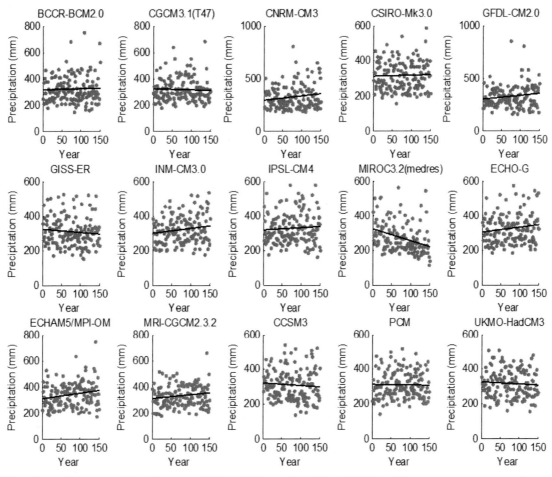

Figure 5.18 Annual monthly extreme values (from years 1950 to 2099) at station 14 for all GCM models.

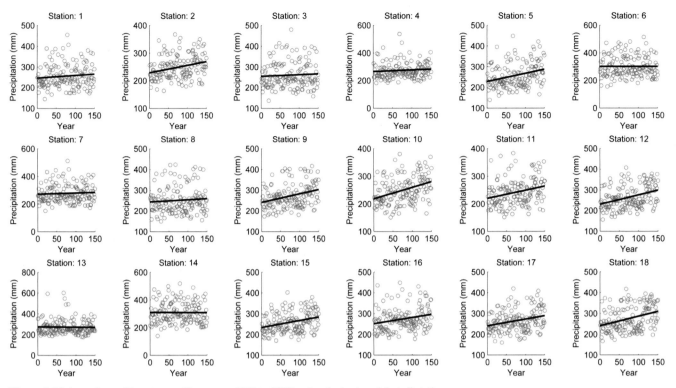

Figure 5.19 Annual monthly extremes (from years 1950 to 2099) using the best model at all stations.

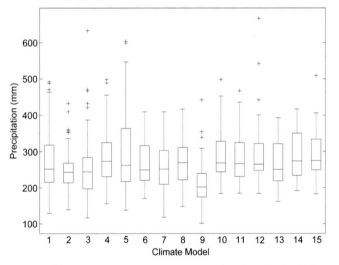

Figure 5.20 Five-number summary represented in a box plot for all the GCM models.

these generators based on statistical properties of observed data at a location. These generators are referred to as stochastic weather generators and they provide series of precipitation and also other meteorological variables (e.g., temperature, solar radiation, wind speed). Conditional dependencies are used in the generation of other variable values. Observed data used in this process of generation are used in validation of generators in preserving the statistical properties of the series, seasonal variations, and cross correlations. In general, the temporal scale of SD output is a month.

The first step of any stochastic weather generation process is to characterize wet and dry states using a two-stage Markov process. The transitional probabilities obtained from the Markov process are used in the procedure for generating wet and dry spells. An exhaustive review of the basic principles of WGs is provided by Wilby and Wigley (1997). The review also provides the limitations of several downscaling approaches along with WGs. One of the major limitations of the traditional WGs is that the parameters used to generate the weather time series are site-specific (limited to a location in space). The spatial statistics of the data based on several locations are not preserved. Rosenzweig and Hillel (2008) explain the process of downscaling using WGs at a site level from GCM simulations using the following three steps:

1. Daily precipitation is represented by a two-state, first-order Markov chain model.
2. Two transition probabilities are estimated: probability of wet day followed by wet day (P_{11}) and probability of wet day followed by a dry day (P_{01}).
3. Precipitation amounts are then simulated for rainy days using a gamma distribution. Usually a two-parameter gamma distribution is used for simulation of daily precipitation intensity in a Richardson-type WG (Benestad et al., 2008; Liao et al., 2004).

Table 5.6 *Two-sample Kolmogorov–Smirnov test results for historical observed data and GCM-based data for years 1950–2010*

Models/ stations	BCCR-BCM2.0	CGCM3.1 (T47)	CNRM-CM3	CSIRO-Mk3.0	GFDL-CM2.0	GISS-ER	INM-CM3.0	IPSL-CM4	MIROC3.2 (Medres)	ECHO-G	ECHAM5/ MPI-OM	MRI-CGCM2.3.2	CCSM3	PCM	UKMO-HadCM3
1	1	1	1	1	0	1	1	1	1	0	1	1	1	0	1
2	1	0	1	1	1	1	1	1	1	0	1	1	1	0	0
3	1	1	1	1	1	1	1	1	1	1	1	1	1	1	1
4	0	0	1	0	0	0	0	0	1	0	0	1	1	0	0
5	1	1	1	1	1	1	1	1	1	1	1	1	0	1	1
6	0	0	1	1	0	0	1	0	1	0	0	0	1	0	0
7	1	1	1	1	1	1	1	1	1	1	1	1	1	1	1
8	1	1	1	1	0	1	1	1	0	0	1	1	1	0	0
9	0	1	1	1	1	1	1	1	0	0	0	1	1	0	1
10	0	1	1	1	0	0	1	1	0	0	0	0	1	0	0
11	1	1	1	1	1	1	1	1	1	1	1	1	1	1	1
12	0	1	1	1	1	1	1	1	1	1	1	1	1	0	1
13	1	1	1	1	1	1	1	1	1	1	1	1	1	1	1
14	0	0	0	0	0	0	1	0	0	0	0	0	0	0	0
15	1	0	1	0	1	1	1	1	0	0	1	1	1	0	1
16	0	1	0	1	1	0	1	1	1	0	1	1	1	1	1
17	1	1	1	1	1	1	1	1	1	1	0	1	1	1	1
18	0	1	1	1	1	1	0	1	0	0	0	1	1	0	0

Table 5.7 *Two-sample Kolmogorov–Smirnov test results for historical observed data and future GCM-based data for years 2011–99*

Models/ stations	BCCR-BCM2.0	CGCM3.1 (T47)	CNRM-CM3	CSIRO-Mk3.0	GFDL-CM2.0	GISS-ER	INM-CM3.0	IPSL-CM4	MIROC3.2 (medres)	ECHO-G	ECHAM5/MPI-OM	MRI-CGCM2.3.2	CCSM3	PCM	UKMO-HadCM3
1	0	1	1	1	0	0	1	1	1	0	1	1	1	0	0
2	1	1	1	1	0	0	1	0	1	0	1	1	1	0	1
3	1	1	1	1	0	0	1	1	1	0	1	1	1	0	1
4	0	1	1	1	0	1	1	1	1	0	1	1	1	0	1
5	0	1	1	1	1	1	1	0	1	1	0	1	1	1	0
6	1	1	0	1	1	1	1	1	1	0	0	0	1	0	0
7	1	1	1	1	1	1	1	0	1	1	1	1	1	1	1
8	0	1	1	1	0	0	1	0	1	1	1	1	1	0	1
9	0	1	1	1	1	1	1	0	1	0	1	1	1	0	1
10	0	1	1	1	1	1	1	0	1	0	1	1	1	0	1
11	0	1	1	1	1	1	1	0	1	1	1	1	1	1	1
12	0	1	1	1	0	1	1	0	1	0	1	1	1	1	1
13	0	1	1	0	0	0	1	0	1	1	1	1	1	0	0
14	1	1	0	1	0	0	1	0	1	0	1	1	1	0	1
15	0	1	1	1	0	1	1	0	1	0	1	1	1	0	1
16	0	1	1	1	0	1	1	0	1	0	1	1	1	0	1
17	0	1	1	1	1	1	1	1	1	0	1	1	1	1	1
18	0	1	1	1	0	1	1	1	1	0	1	1	1	1	1

5.10.1 Types of weather generators

Two types of stochastic WGs are widely used and they are (1) the Richardson type (Richardson and Wright, 1984) and (2) the serial type (Semenov *et al.*, 1998). An exhaustive review of these WGs is provided by Wilks and Wilby (1999). A comprehensive comparison of WGs and downscaling techniques is carried out by Solaiman *et al.* (2010) and Dibike and Coulibaly (2005). Computationally efficient and robust WGs that are tested in several climatic regions of the world are now available. One of the generators that have been used for climate change impact assessment is the Long Ashton Research Station Weather Generator (LARS-WG). This stochastic WG is primarily designed and developed for climate change impact studies (Semenov and Barrow, 1997). It has been tested for diverse climates and found to be better than some other generators (Semenov *et al.*, 1998). A recent study by Semenov (2008) has tested LARS-WG for different sites across the world. Also, these weather generators are modified to preserve the spatial statistics among observation locations in space.

5.11 DOWNSCALING PRECIPITATION: MAJOR ISSUES

The available statistical and dynamic downscaling methods are still plagued by several limitations. The major issues related to precipitation downscaling are discussed by Wilby and Fowler (2011). The one-way nesting in RCMs limits their ability to provide feedback to GCMs and accurate precipitation estimates in a region. The spatial resolution at which RCMs are run needs to be improved in the future to resolve several issues related to precipitation downscaling. Improved representation of complex terrain (i.e., topography) and local land surface features in RCMs is another issue that requires attention. Selection of ideal predictors in statistical downscaling is a tedious exercise and a contentious issue. Some predictors that are useful for the current climate downscaling may not be appropriate for future climate conditions. Wilby and Fowler (2011) indicate that predictor variables that rely on humidity indices are now being used increasingly in downscaling studies. Their usage in the models has provided good success when compared to models using only predictors reflecting atmospheric circulation. Downscaling of extremes is yet another crucial issue that needs to be addressed by future downscaling models. Statistical downscaling methods that use regression-based transfer functions are known to underpredict variance (Burger, 1996). The STARDEX (STAtistical and Regional dynamical Downscaling of EXtremes) is one of the most recent recognized efforts for the European regions to compare and evaluate different statistical and dynamic downscaling techniques addressing extremes.

5.12 CONCLUSIONS AND SUMMARY

This chapter discusses the use of different downscaling methods for producing fine-resolution precipitation data from coarse GCM-based simulations. Spatial and temporal downscaling methods are also described. The methodology and utility of weather generators and their applications for climate impact studies are elaborated. The chapter also presents an illustration of a spatial downscaling procedure referred to as bias-correction spatial disaggregation (BCSD) for a case study in a subtropical climatic region of the USA. Several performance measures are used to evaluate the relative performance of different climate change models in representing the regional climatology. These measures are used to assess the skill of the BCSD models in spatial downscaling and ultimately selection of a model for use at a site.

EXERCISES

5.1   Use statistically downscaled and bias-corrected precipitation data for different regions in the USA and evaluate the future precipitation estimates for different climate change scenarios.

5.2   Identify the differences between statistical and dynamic downscaling procedures. Discuss the advantages or disadvantages one procedure has over the other.

5.3   Develop performance measures to identify the best model among competing models providing downscaled precipitation or temperature observations.

5.4   Discuss the key statistical tools that can be used to assess the performance of the downscaled model projections of precipitation in replicating past hydroclimatology and replication of patterns (magnitudes and frequency).

5.5   Observed monthly precipitation data at a USHCN rain gage site (Station 7) and downscaled GCM-based precipitation data from the PCM model from NCAR, using a BCSD procedure are provided in Table 5.8. The data sets are from year 2000 to 2010. Evaluate the model skill using the performance measures discussed in this chapter. The observed and GCM-based precipitation data are referred to as OBS and GCM respectively in the table.

5.6   Use the website provided below to download BCSD climate change projections from different models and SRES scenarios for different regions in the continental USA to assess the skill of each model. http://gdo-dcp.ucllnl. org/downscaled_cmip3_projections/dcpInterface.html# Welcome

Table 5.8 *Observed and downscaled monthly precipitation data in millimeters*

OBS	GCM	OBS	GCM	OBS	GCM	OBS	GCM	OBS	GCM	OBS	GCM
32.3	51.8	359.2	220.7	72.1	136.8	11.4	34.2	256.8	198.4	20.6	77.6
2.8	45.0	56.1	110.4	370.8	110.7	85.6	42.4	21.6	76.6	181.6	273.7
85.6	10.4	10.4	11.0	74.2	182.3	167.4	20.1	28.2	42.8	335.0	199.1
48.5	109.8	46.7	18.8	426.7	198.4	97.5	115.7	39.6	18.5	262.4	200.5
70.9	81.5	54.1	46.3	437.1	202.4	66.5	106.8	12.4	43.3	377.7	196.3
203.5	171.8	24.4	28.9	19.3	256.7	467.9	332.3	18.8	28.4	87.1	39.0
221.7	290.3	4.8	135.6	53.8	67.4	320.0	165.3	3.3	32.5	2.3	21.2
213.4	228.7	41.7	29.6	118.9	13.6	196.6	200.5	52.3	6.9	52.6	25.4
305.3	188.3	59.4	106.7	86.1	4.5	173.2	198.4	123.4	29.1	7.6	49.8
45.2	121.4	200.4	95.5	87.4	7.7	207.5	104.5	155.2	91.4	6.9	50.6
0.0	33.1	181.6	173.9	9.1	58.1	95.3	18.6	168.4	161.0	16.5	55.3
11.4	56.1	320.8	220.7	92.7	42.0	2.3	5.7	316.7	206.6	7.4	52.2
0.0	79.9	173.2	186.2	11.9	55.3	9.1	49.8	229.6	202.4	163.3	250.3
0.0	96.4	26.7	48.8	382.8	152.1	59.4	65.3	46.0	108.3	137.9	169.4
50.0	25.6	144.0	13.6	262.9	245.9	8.1	9.6	2.3	73.7	109.2	161.1
13.2	26.9	90.9	6.2	427.2	257.2	0.3	19.9	66.0	81.1	243.1	242.9
5.1	93.2	47.5	98.6	114.6	400.9	51.6	73.8	33.3	28.2	125.7	247.1
114.8	184.7	22.6	23.8	17.8	78.6	348.0	141.1	57.2	42.2	70.4	33.9
354.1	237.4	87.4	61.5	27.9	36.9	366.5	220.4	40.1	32.5	30.0	15.6
262.1	155.8	63.8	18.3	50.0	72.5	240.0	153.8	77.0	98.0	95.8	102.8

5.7 Explain the advantages and disadvantages of statistical downscaling with respect to precipitation data compared to dynamic downscaling.

5.8 Use the performance measures discussed in Chapter 3 for comparative assessment of spatial interpolation methods for comparison of downscaled models for precipitation and temperature.

USEFUL WEBSITES

http://gdo-dcp.ucllnl.org/downscaled_cmip3_projections/dcpInterface.html#Welcome (accessed December, 2011).

http://www-pcmdi.llnl.gov/ipcc/about_ipcc.php (accessed September, 2011).

https://webspace.utexas.edu/hs8238/www/surfacehydrology/surfacehydrology_Project.htm (accessed September, 2011).

http://www.cics.uvic.ca/scenarios/index.cgi?More_Info-Downscaling_Background (accessed January, 2012).

RESOURCES FOR STUDENTS

An original NetCDF code for extracting data from the BCSD-GCM simulations is available at: https://webspace.utexas.edu/hs8238/www/surfacehydrology/surfacehydrology_Project.htm.

A modified code is available upon request from the author.

6 Precipitation variability and teleconnections

6.1 INTRODUCTION

Variability of extreme precipitation events in several parts of the world has been investigated by many researchers. These studies have focused on a specific region of the world with or without teleconnections that are used to explain the variability. Internal modes of climate variability originating in one region can have short- or long-term global impacts referred to as teleconnections (Cronin, 2010). Long-term historical data available in several research studies have been used to assess these trends. Studies relevant to variability assessments in the USA were done by Karl *et al.* (1995) and Groisman *et al.* (1999). Studies by Goswami *et al.* (2006) and Manton *et al.* (2001) focused on trends in heavy precipitation in the Indian sub-continent and studies relevant to Southeast Asia and the South Pacific. Trends in extreme rainfall and dry events in Australia were investigated by Suppiah and Hennessy (1998). The reader is referred to studies by Haylock and Goodess (2004), Iwashima and Yamamoto (1993), and Hellstrom and Malmgren (2004) dealing with precipitation trends in Europe, Japan, and Sweden respectively. Recently, the spatio-temporal variability of dependence among precipitation extremes was investigated over the whole of South America for the period 1940–2004 using a new approach (Kuhn, 2006; Khan, 2007; Kuhn *et al.*, 2007).

Several research studies have reported changes in extreme precipitation and flood events due to possible reasons associated with global warming. Changnon and Kunkel (1995) examined trends in the heavy multi-day precipitation events associated with floods for the period from 1921 to 1985 and found upward trends in both floods and heavy precipitation events over portions of the central USA. Karl and Knight (1998) indicate an approximately 8% increase in precipitation across the contiguous USA since 1910, which was primarily due to increases in the intensity of heavy to extreme precipitation events. Kunkel *et al.* (1999) found statistically significant upward trends in 1-year return period, 7-day duration events of about 3% per decade and in 5-year, 7-day events of about 4% per decade for the period from 1931 to 1996. Groisman *et al.* (2001) reported a 50% increase in the frequency of extreme precipitation days (with precipitation exceeding 101.6 mm) in the US Upper Midwest during the twentieth century.

6.1.1 Internal modes of climate variability: teleconnections

Internal climate variability refers to what is frequently called "unforced" climate changes or "modes" of variability and these modes represent climate anomalies that originate in a particular region and have significant impacts elsewhere around the globe (Cronin, 2009). Teleconnections are defined as linkages between climate anomalies at some distances from each other (Glantz, 2001). The term teleconnection was first adopted and used by Angstrom (1935).

6.1.2 Oscillatory behavior

Variability of climate along with climate change is generally attributed to natural fluctuations about a mean climate. Rosenzweig and Hillel (2008) explain the oscillatory behavior as modes of climate variability in relation to climate change. They provide five cases of oscillatory behavior and these are: case 1, constant mean climate with regular variations about the mean with constant frequency and amplitude; case 2, gradually changing climate mean with variations about the mean with constant frequency and amplitude; case 3, variations with constant frequency and varying amplitude with gradually increasing mean climate; case 4, similar to case 3, with constant frequency and varying amplitudes of variations; and case 5, varying mean, gradually increasing frequency and changing amplitude. Rosenzweig and Hillel (2008) and Cronin (2009) list several climate variability systems that lead to quasi-regular, low- and high-frequency oscillations. Major oscillations are shown in Figure 6.1 along with their general location of occurrence. These oscillations are Arctic Oscillation (AO), North Atlantic Oscillation (NAO), Pacific Decadal Oscillation (PDO), El Niño Southern Oscillation (ENSO), and Madden–Julian Oscillation (MJO). The data and details for this figure are adopted from Rosenzweig and Hillel (2008). The locations and

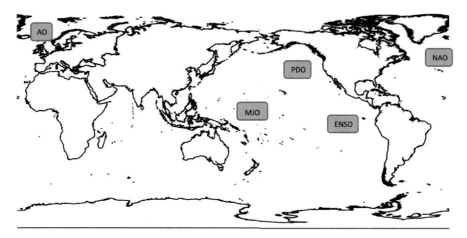

Figure 6.1 Approximate locations of major climate variability systems.

their calculation are detailed in textbooks by Cronin (2009) and Rosenzweig and Hillel (2008).

The location of the southern oscillation index (SOI) is the tropical Pacific and it has a frequency of two to eight years. It is calculated based on MSLP observations at Tahiti or Easter Island and Darwin, Australia. The NAO is manifest in the North Atlantic Ocean and is based on monthly or seasonal sea level pressure at Iceland and the Azores. It is a decadal and multi-decadal oscillation. The PDO is based on the North Pacific Index (NPI) and index of North Pacific SST north of $20°$ N. It is a multi-decadal oscillation. The AO is a decadal oscillation with its location in Arctic high Northern Hemisphere latitudes. It is based on winter monthly Northern Hemisphere MSLP. The AMO is based on SST in the Atlantic Ocean based on data available from 1856. ENSO has the same location as SOI within the regions NINO 1–4. There are some other oscillations with lesser global impacts, as discussed by Cronin (2009), and they are the Pacific North American (PNA) pattern and the Inter-decadal Pacific Oscillation (IPO). The former is located in the North Pacific and is based on mean normalized 500 hPa height and the latter is located in mid-latitude Pacific and South Pacific Convergence zone and is based on SST.

6.2 SOUTHERN OSCILLATION

The Southern Oscillation (SO) is quantified by the difference in surface air pressure between Darwin, Australia, and Tahiti. The index is referred to as the Southern Oscillation Index (SOI). The oscillation was discovered by Walker (1908, 1918), who indicated that there was a large-scale pattern in surface air pressure that extended over the entire tropical Pacific region. The index is based on pressure differences between two points and therefore the observations (i.e., pressures) are prone to local disturbances (IRI, 2011) and therefore are not smooth when evaluated at time scales

less than or equal to a month. Therefore, the index is generally averaged over several months.

6.3 EL NIÑO SOUTHERN OSCILLATION

The atmosphere is also capable of translating signals between different ocean basins, through what are known as atmospheric bridges. Warming of the Atlantic corresponds to a weakening in the anomalous easterly winds that blow over the tropical Pacific (Earthgauge, 2011). This leads to less instability between the surface and deep layers of the ocean and a weakening of ENSO. Observations and models of past climates demonstrate an Atlantic–Pacific "see-saw," where periods of Atlantic meridional overturning circulation (MOC) shut-down are accompanied by the development of a North Pacific MOC and warming of the waters there (Earthgauge, 2011). Analysis of twentieth century climate records indicates that there are interactions between ENSO and the PDO as well. La Niña phases are more pronounced during cool (negative) PDO phases, while the connection between El Niño events and the different PDO phases is less clear.

ENSO is a slow oscillation in which the atmosphere and ocean in the tropical Pacific region interact to produce a slow, irregular variation between two extremes. ENSO has two phases: the warm phase and the cool phase. These variations are more commonly known as El Niño (the warm phase) and La Niña (the cool phase). Although ENSO is centered in the tropics, the changes associated with El Niño and La Niña events affect climate around the world. ENSO events tend to form between April and June and typically reach full strength in December. During El Niño events, for example, the southern part of Florida tends to experience above-normal rainfall during the winter (dry season) months, whereas dry-season rainfall tends to be below normal during La Niña. Generally, the SOI is negative during El Niño, and positive during La Niña.

6.3.1 El Niño and La Niña

NINO regions are established to obtain the long-term indices related to El Niño and La Niña. These indices are based on SST and are generally obtained by using average values over some specified region of the ocean. Four regions of interest in the tropical Pacific Ocean are essential for monitoring and identifying El Niño and La Niña. The data provided in Table 6.1 are adopted from NCEP and the positive numbers and negative numbers refer to warm and cold episodes based on a threshold of $\pm0.5\ ^\circ$C for the Oceanic Niño Index (ONI) (3 month running mean of extended reconstructed sea surface temperature (ERSST) anomalies in the NINO 3.4 region (5° N–5° S, 120–170° W)) (NCEP, 2011), based on the 1971–2000 base period. The International Comprehensive Ocean–Atmosphere Data Set (ICOADS) is used for ERSST. The ERSST analysis is updated with the available ship and buoy data for that month. Details of this data set and analysis are reported by Xue *et al.* (2003) and Smith *et al.* (2008). The cold and warm phases are defined when the threshold is met for a minimum of five consecutive overlapping seasons. The values for ONI are provided in Table 6.1. The spatial locations of different NINO regions are shown in Figure 6.2. The geographical coordinates of four NINO regions are given in Table 6.2.

The sea surface indices are based on SST measurements using a network of buoys in the equatorial Pacific. A link between NINO 3.4 and Southern Oscillation indices is evident from Figure 6.3.

An optimal characterization of both the distinct character and the evolution of each El Niño or La Niña event requires at least two indices: (1) SST anomalies in the NINO 3.4 region (referred to as N3.4), and (2) a new index referred to as the Trans-Niño Index (TNI) suggested by Trenberth and Stepaniak (2001). The new index is given by the difference in normalized anomalies of SST between NINO 1+2 and NINO 4 regions. The first index is the mean SST throughout the equatorial Pacific east of the dateline and the second index is the gradient in SST across the same region.

6.3.2 Global conditions for La Niña and El Niño

Global conditions for different months are shown in Figures 6.4 and 6.5 for La Niña and in Figures 6.6 and 6.7 for El Niño. For La Niña during the months from December to February the precipitation trends are generally opposite to those conditions for El Niño during the same months. Conditions during December to February are considered to be more active for El Niño. The maps shown in Figures 6.4–6.7 are drawn based on data and information from NOAA's Climate Prediction Center (NCEP).

6.3.3 ENSO influences in the USA

The effects of ENSO on the southeastern USA, especially Florida (Florida, 2011), are more prominent in the months from December to February. The NOAA Climate Diagnostics Center confirms that ENSO is a major source of inter-annual climatic variability in South Florida. The following are the El Niño and La Niña influences on Florida according to Florida (2011).

El Niño:

- Rainfall: Above average rainfall.
- Severe weather: During El Niño the jet stream is oriented from west to east over the northern Gulf of Mexico and northern Florida. Thus this region is most susceptible to severe weather.

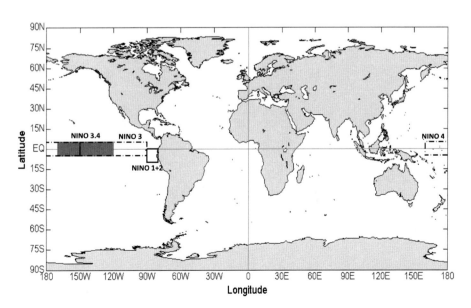

Figure 6.2 Locations of NINO regions.

Table 6.1 *Oceanic Niño index (source NWCPC)*

Year	DJF	JFM	FMA	MAM	AMJ	MJJ	JJA	JAS	ASO	SON	OND	NDJ
1950	−1.70	−1.50	−1.30	−1.40	−1.30	−1.10	−0.80	−0.80	−0.80	−0.90	−0.90	−1.00
1951	−1.00	−0.90	−0.60	−0.30	−0.20	0.20	0.40	0.70	0.70	0.80	0.70	0.60
1952	0.30	0.10	0.10	0.20	0.10	−0.10	−0.30	−0.30	−0.20	−0.20	−0.10	0.00
1953	0.20	0.40	0.50	0.50	0.50	0.50	0.40	0.40	0.40	0.40	0.40	0.40
1954	0.50	0.30	−0.10	−0.50	−0.70	−0.70	−0.80	−1.00	−1.20	−1.10	−1.10	−1.10
1955	−1.00	−0.90	−0.90	−1.00	−1.00	−1.00	−1.00	−1.00	−1.40	−1.80	−2.00	−1.90
1956	−1.30	−0.90	−0.70	−0.60	−0.60	−0.60	−0.70	−0.80	−0.80	−0.90	−0.90	−0.80
1957	−0.50	−0.10	0.30	0.60	0.70	0.90	0.90	0.90	0.90	1.00	1.20	1.50
1958	1.70	1.50	1.20	0.80	0.60	0.50	0.30	0.10	0.00	0.00	0.20	0.40
1959	0.40	0.50	0.40	0.20	0.00	−0.20	−0.40	−0.50	−0.40	−0.30	−0.20	−0.20
1960	−0.30	−0.30	−0.30	−0.20	−0.20	−0.20	−0.10	0.00	−0.10	−0.20	−0.20	−0.20
1961	−0.20	−0.20	−0.20	−0.10	0.10	0.20	0.00	−0.30	−0.60	−0.60	−0.50	−0.40
1962	−0.40	−0.40	−0.40	−0.50	−0.40	−0.40	−0.30	−0.30	−0.50	−0.60	−0.70	−0.70
1963	−0.60	−0.30	0.00	0.10	0.10	0.30	0.60	0.80	0.90	0.90	1.00	1.00
1964	0.80	0.40	−0.10	−0.50	−0.80	−0.80	−0.90	−1.00	−1.10	−1.20	−1.20	−1.00
1965	−0.80	−0.40	−0.20	0.00	0.30	0.60	1.00	1.20	1.40	1.50	1.60	1.50
1966	1.20	1.00	0.80	0.50	0.20	0.20	0.20	0.00	−0.20	−0.20	−0.30	−0.30
1967	−0.40	−0.40	−0.60	−0.50	−0.30	0.00	0.00	−0.20	−0.40	−0.50	−0.40	−0.50
1968	−0.70	−0.90	−0.80	−0.70	−0.30	0.00	0.30	0.40	0.30	0.40	0.70	0.90
1969	1.00	1.00	0.90	0.70	0.60	0.50	0.40	0.40	0.60	0.70	0.80	0.70
1970	0.50	0.30	0.20	0.10	0.00	−0.30	−0.60	−0.80	−0.90	−0.80	−0.90	−1.10
1971	−1.30	−1.30	−1.10	−0.90	−0.80	−0.80	−0.80	−0.80	−0.80	−0.90	−1.00	−0.90
1972	−0.70	−0.40	0.00	0.20	0.50	0.80	1.00	1.30	1.50	1.80	2.00	2.10
1973	1.80	1.20	0.50	−0.10	−0.60	−0.90	−1.10	−1.30	−1.40	−1.70	−2.00	−2.10
1974	−1.90	−1.70	−1.30	−1.10	−0.90	−0.80	−0.60	−0.50	−0.50	−0.70	−0.90	−0.70
1975	−0.60	−0.60	−0.70	−0.80	−0.90	−1.10	−1.20	−1.30	−1.50	−1.60	−1.70	−1.70
1976	−1.60	−1.20	−0.80	−0.60	−0.50	−0.20	0.10	0.30	0.50	0.70	0.80	0.70
1977	0.60	0.50	0.20	0.20	0.20	0.40	0.40	0.40	0.50	0.60	0.70	0.70
1978	0.70	0.40	0.00	−0.30	−0.40	−0.40	−0.40	−0.40	−0.40	−0.30	−0.20	−0.10
1979	−0.10	0.00	0.10	0.10	0.10	−0.10	0.00	0.10	0.30	0.40	0.50	0.50
1980	0.50	0.30	0.20	0.20	0.30	0.30	0.20	0.00	−0.10	−0.10	0.00	−0.10
1981	−0.30	−0.50	−0.50	−0.40	−0.30	−0.30	−0.40	−0.40	−0.30	−0.20	−0.10	−0.10
1982	0.00	0.10	0.10	0.30	0.60	0.70	0.70	1.00	1.50	1.90	2.20	2.30
1983	2.30	2.00	1.50	1.20	1.00	0.60	0.20	−0.20	−0.60	−0.80	−0.90	−0.70
1984	−0.40	−0.20	−0.20	−0.30	−0.50	−0.40	−0.30	−0.20	−0.30	−0.60	−0.90	−1.10
1985	−0.90	−0.80	−0.70	−0.70	−0.70	−0.60	−0.50	−0.50	−0.50	−0.40	−0.30	−0.40
1986	−0.50	−0.40	−0.20	−0.20	−0.10	0.00	0.30	0.50	0.70	0.90	1.10	1.20
1987	1.20	1.30	1.20	1.10	1.00	1.20	1.40	1.60	1.60	1.50	1.30	1.10
1988	0.70	0.50	0.10	−0.20	−0.70	−1.20	−1.30	−1.20	−1.30	−1.60	−1.90	−1.90
1989	−1.70	−1.50	−1.10	−0.80	−0.60	−0.40	−0.30	−0.30	−0.30	−0.30	−0.20	−0.10
1990	0.10	0.20	0.20	0.20	0.20	0.20	0.30	0.30	0.30	0.30	0.30	0.40
1991	0.40	0.30	0.30	0.40	0.60	0.80	1.00	0.90	0.90	1.00	1.40	1.60
1992	1.80	1.60	1.50	1.40	1.20	0.80	0.50	0.20	0.00	−0.10	0.00	0.20
1993	0.30	0.40	0.60	0.70	0.80	0.70	0.40	0.40	0.40	0.40	0.30	0.20
1994	0.20	0.20	0.30	0.40	0.50	0.50	0.60	0.60	0.70	0.90	1.20	1.30
1995	1.20	0.90	0.70	0.40	0.30	0.20	0.00	−0.20	−0.50	−0.60	−0.70	−0.70
1996	−0.70	−0.70	−0.50	−0.30	−0.10	−0.10	0.00	−0.10	−0.10	−0.20	−0.30	−0.40
1997	−0.40	−0.30	0.00	0.40	0.80	1.30	1.70	2.00	2.20	2.40	2.50	2.50
1998	2.30	1.90	1.50	1.00	0.50	0.00	−0.50	−0.80	−1.00	−1.10	−1.30	−1.40
1999	−1.40	−1.20	−0.90	−0.80	−0.80	−0.80	−0.90	−0.90	−1.00	−1.10	−1.30	−1.60
2000	−1.60	−1.40	−1.00	−0.80	−0.60	−0.50	−0.40	−0.40	−0.40	−0.50	−0.60	−0.70
2001	−0.60	−0.50	−0.40	−0.20	−0.10	0.10	0.20	0.20	0.10	0.00	−0.10	−0.10
2002	−0.10	0.10	0.20	0.40	0.70	0.80	0.90	1.00	1.10	1.30	1.50	1.40
2003	1.20	0.90	0.50	0.10	−0.10	0.10	0.40	0.50	0.60	0.50	0.60	0.40
2004	0.40	0.30	0.20	0.20	0.30	0.50	0.70	0.80	0.90	0.80	0.80	0.80
2005	0.70	0.50	0.40	0.40	0.40	0.40	0.40	0.30	0.20	−0.10	−0.40	−0.70
2006	−0.70	−0.60	−0.40	−0.10	0.10	0.20	0.30	0.50	0.60	0.90	1.10	1.10
2007	0.80	0.40	0.10	−0.10	−0.10	−0.10	−0.10	−0.40	−0.70	−1.00	−1.10	−1.30
2008	−1.40	−1.40	−1.10	−0.80	−0.60	−0.40	−0.10	0.00	0.00	0.00	−0.30	−0.60
2009	−0.80	−0.70	−0.50	−0.10	0.20	0.60	0.70	0.80	0.90	1.20	1.50	1.80
2010	1.70	1.50	1.20	0.80	0.30	−0.20	−0.60	−1.00	−1.30	−1.40	−1.40	−1.40
2011	−1.30	−1.20	−0.90	−0.60	−0.20	0.00	0.00					

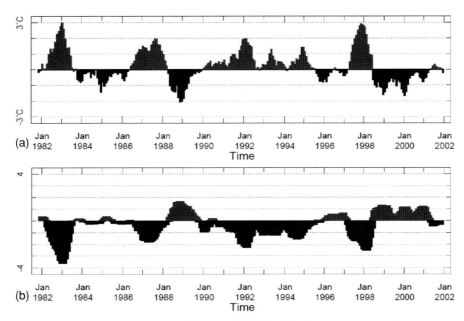

Figure 6.3 (a) NINO 3.4 sea surface temperature index and (b) Southern Oscillation index (image courtesy IRI).

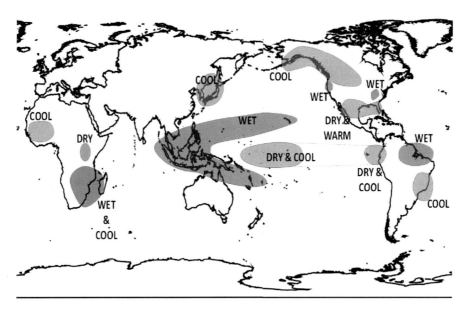

Figure 6.4 Global climate conditions for La Niña (December–February).

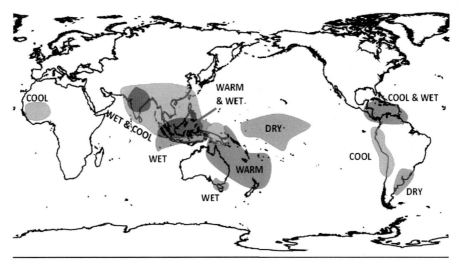

Figure 6.5 Global climate conditions for La Niña (June–August).

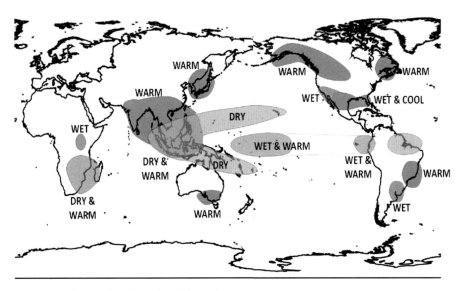

Figure 6.6 Global climate conditions for El Niño (December–February).

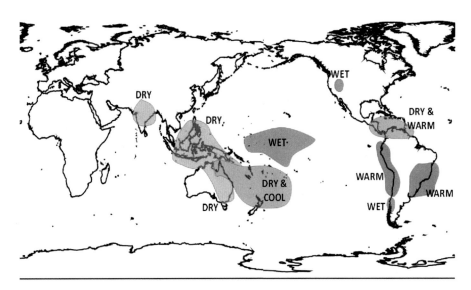

Figure 6.7 Global climate conditions for El Niño (June–August).

Table 6.2 *Geographic locations of NINO regions*

NINO region	Longitude (range)	Latitude (range)
1 + 2	90° W−80° W	10° S−0°
3	150° W−90° W	5° S−5° N
3.4	170° W−120° W	5° S−5° N
4	160° E−150° W	5° S−5° N

- Temperatures: Below normal temperatures.
- Winter storms: Increased cyclogenesis (low-pressure systems) in the Gulf of Mexico.
- Hurricanes: El Niño almost always reduces the frequency of storms.

La Niña:

- Rainfall: Generally dry conditions prevail during La Niña events in late fall, winter, and early spring.
- Wildfires: Increased risk of wildfires in spring/summer months.
- Temperatures: Temperatures average slightly above normal during La Niña events.
- Hurricanes: The chances for the continental USA and the Caribbean Islands to experience hurricane activity increases substantially during La Niña.

The rainfall patterns for a strong El Niño (1997–8 year) for the continental USA are shown in Figure 6.8.

Table 6.3 *Years of El Niño and La Niña*

El Niño	La Niña
1951	1950
1953	1954
1957	1955
1958	1956
1963	1964
1965	1970
1968	1971
1969	1973
1972	1974
1976	1975
1977	1983
1979	1984
1982	1988
1986	1995
1987	1998
1990	1999
1991	2000
1992	
1993	
1994	
1997	
2002	

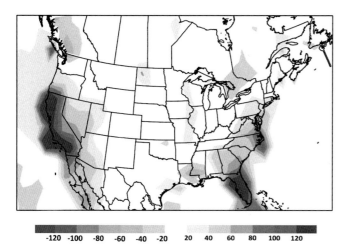

Figure 6.8 Rainfall patterns for strong El Niño for year 1998 (source: http://www.srh.noaa.gov/mlb/?n=enso_florida_rainfall). See also color plates.

6.3.4 El Niño and La Niña cycles: occurrences

The frequency of El Niño and La Niña cycles is about two to seven years. The frequency was not constant over the last century, with some of the cycles being absent for a prolonged period. The list of El Niño and La Niña years based on oceanic index from NINO 3.4 region are provided in Table 6.3. The strength of these El Niño

cycles is evaluated based on SST anomalies and warm pools in the Pacific Ocean (Rosenzweig and Hillel, 2008). Accordingly, three years 1972–3, 1982–3, and 1997–8 are classified as the strongest El Niño years.

6.4 DECADAL OSCILLATIONS

Decadal oscillations such as AMO, PDO, and NAO are considered to be prominent sources of inter-annual climate variability (Rosenzweig and Hillel, 2008).

6.4.1 Atlantic Multi-decadal Oscillation

The AMO is defined as a pattern of Atlantic climate variability, detected as SST variations over the Atlantic Ocean, between the equator and Greenland. It is a long-range climatic oscillation that causes periodic changes in the surface temperature of the Atlantic Ocean, which may persist for several years or decades, usually 20 to 40 years. AMO as a multi-decadal pattern of climate variability was introduced by Kerr (2000) and Schlesinger *et al.* (2000). The works of Schlesinger and Ramankutty (1994) and Andronova and Schlesinger (2000) discuss the pattern of global mean temperature with pre-defined multi-decadal frequency. AMO describes temperature deviations in the ocean surface that appear to be driving shifts ("warm" and "cool" phases) in South Florida's climate (Enfield *et al.*, 2001). There are also significant year-to-year fluctuations in the instrumental records of ocean temperature (AOML, 2011). The AMO index for different months based on smoothened and unsmoothened data is shown in Figures 6.9 and 6.10 respectively.

Kaplan *et al.* (1998) and Enfield *et al.* (2001), using long-term instrumental records of SST, suggest that AMO is a 65–80 year quasi-cycle variability of SST with 0.4 °C range. There is a strong statistical correlation between AMO-related SST variability and rainfall in the southeastern USA and Mississippi river flow (Cronin, 2009). Proxy temperature records developed from tree-ring sequences from the southeastern USA and northern Europe show that AMO-like cycles in the surface-air temperature adjacent to the North Atlantic Ocean have been present in the climate system for at least the last 400 years (Gray *et al.*, 2004). Variability in the AMO-like cycles present in these proxy records is similar to the variability in the instrumental record of the AMO. The 1900–25 period seems to have been dominated by a cool phase of the AMO, followed by a warm phase (1926–69) and another cool phase (1970–94). The warm phase that started in the year 1995 continues. It is difficult to predict the end of this phase because the exact years of warm and cool phases of AMO are difficult to identify and vary among researchers. These changes are natural and have been occurring for at least the last 1,000 years. Accordingly, the following four periods of AMO in the twentieth century can

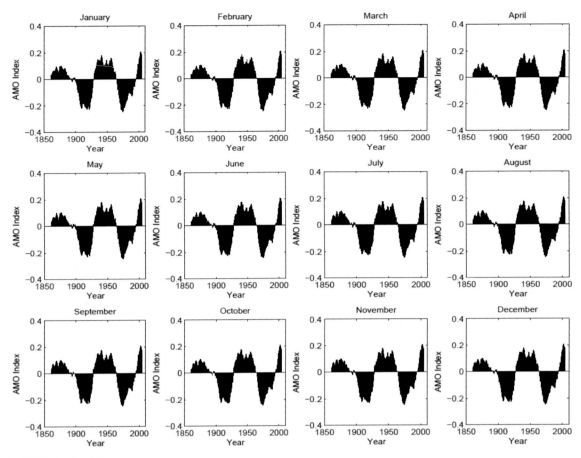

Figure 6.9 AMO index for different months.

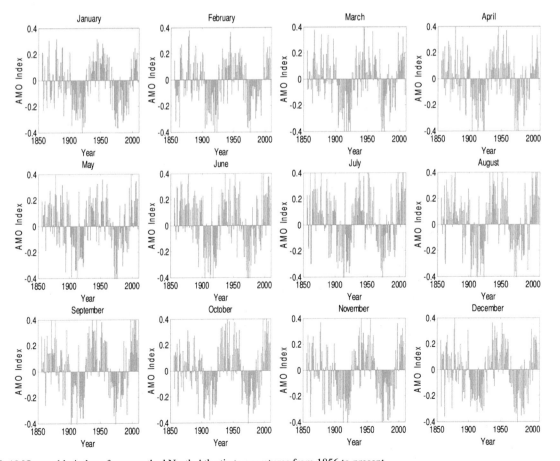

Figure 6.10 AMO monthly index of unsmoothed North Atlantic temperatures from 1856 to present.

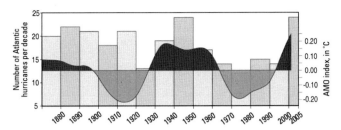

Figure 6.11 Number of Atlantic hurricanes over the last 100 years and link to AMO phases (source: Poore *et al.*, 2005).

be identified: AMO (cool), 1895–1925; AMO (warm), 1926–69; AMO (cool), 1970–94; AMO (warm), 1995–present.

6.4.2 AMO and the southeastern USA

Increased hurricane activity was noted in AMO warm phase years as shown in Figure 6.11.

It is evident from Figure 6.12 that the eastern USA, especially Florida, has seen increased hurricane activity in the AMO warm phases. The list of major hurricane landfalls is provided in Table 6.4.

6.5 TELECONNECTIONS AND EXTREME PRECIPITATION

Internal modes of climate variability defined by oscillations are referred to as teleconnections when the impacts of these

oscillations are felt somewhere other than the main locations of their origin. Extreme precipitation events influenced by these oscillations can be evaluated considering the magnitude and the duration of the events. The following are some of the key issues to be considered when the links are evaluated.

- Temporal variation of extreme precipitation events considering magnitudes combined with durations under different oscillation modes defined by yearly, decadal, or multi-decadal frequencies.
- Spatial variation of extreme precipitation events within a region influenced by oscillations.
- Spatial and temporal variation of events influenced by one or more oscillations over a specific time domain.
- Seasonal variation of extreme precipitation events considering magnitude, duration, and frequency.

6.5.1 Assessment of AMO: example from southeastern USA

Kerneldensity estimation (KDE) as a non-parametric smoothing technique can be used to analyze historical precipitation extremes. KDEs serving as alternatives to histograms do not require arbitrary rounding to bin centers, and provide smooth curves by applying kernel density smoothing (Wilks, 2006). The application of kernel smoothing to the frequency distribution of a data set produces the KDE, which is a non-parametric alternative to the fitting of a parametric probability density function. Different types of kernels

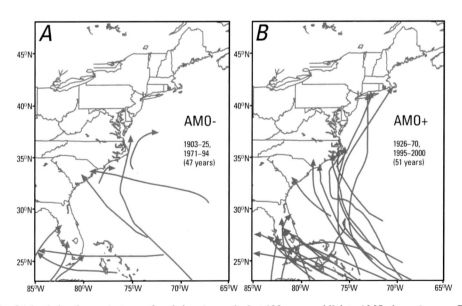

Figure 6.12 Landfalls of Atlantic hurricanes (category 3 and above) over the last 100 years and link to AMO phases (source: Goldenberg *et al.*, 2001; Sutton and Hodson, 2005).

Table 6.4 *List of hurricane landfalls on the eastern side of the USA*

| Decade | Saffir–Simpson category | | | | | All 1, 2, 3, 4, 5 | Major: 3, 4, 5 |
	1	2	3	4	5		
1851–1860	8	5	5	1	0	19	6
1861–1870	8	6	1	0	0	15	1
1871–1880	7	6	7	0	0	20	7
1881–1890	8	9	4	1	0	22	5
1891–1900	8	5	5	3	0	21	8
1901–1910	10	4	4	0	0	18	4
1911–1920	10	4	4	3	0	21	7
1921–1930	5	3	3	2	0	13	5
1931–1940	4	7	6	1	1	19	8
1941–1950	8	6	9	1	0	24	10
1951–1960	8	1	5	3	0	17	8
1961–1970	3	5	4	1	1	14	6
1971–1980	6	2	4	0	0	12	4
1981–1990	9	1	4	1	0	15	5
1991–2000	3	6	4	0	1	14	5
2001–2004	4	2	2	1	0	9	3
1851–2004	109	72	71	18	3	273	92
Average per decade	7.1	4.7	4.6	1.2	0.2	17.7	6

can be used in developing the smoothing function and they are uniform, triangular, biweight, triweight, Epanechnikov, normal, and others. Kernel density estimates using a Gaussian smoothing function and optimal bandwidth are common. The bandwidth plays a major role in representing the data and can be selected by cross-validation or optimization. The choice of kernel bandwidth controls the smoothness of the probability density curve. Different bandwidths should be experimented with before an optimal one can be selected for representation of KDEs. The histograms of the temporal occurrences of extreme precipitation events of different durations are superimposed on scaled KDEs to confirm the match between the two. Figure 6.13 shows the KDEs and histograms for occurrences of extreme events in the warm phase of AMO for different durations. Similar information is shown in Figure 6.14 for the cool phase of AMO.

The principal modes of the distributions are seen to occur at two virtually identical positions for the durations during the warm phase of AMO. The occurrences of extreme precipitation events were confined to the months of June and September. The probability density functions show clear bi-modal characteristics as the durations change from 6 hours to 72 hours.

The probability density functions show no clear principal modes as the durations change from 1 hour to 72 hours. The majority of the extreme events during these intervals are occurring in the month of June. The bi-modal pattern does not

emerge until the duration of 120 hours, with the majority of the extreme precipitation events occurring in months of June and September. The principal modes of the distributions are seen to occur at two positions for the intervals during the cool phase of AMO, suggesting that the occurrences of extreme precipitation events are confined to June and September for durations of 96 and 120 hours.

6.5.2 Temporal windows

Temporal windows can be defined based on fixed time intervals or to coincide with the cycles of internal modes of climate variability (e.g., AMO warm and cool phases). Windows of constant, increasing, or varying temporal lengths can be selected for evaluation of precipitation patterns and trends in different phases (Figure 6.15).

Cumulative density functions (CDF) are developed based on raw extreme precipitation data for different durations in two different phases of AMO. Temporal window lengths are identified so that they coincide with AMO phases and they conform to the most recent data. The CDF plots for two phases: warm phase (1942–69) and cool phase (1970–94) are shown in Figure 6.16. For shorter durations (1–12 hours), the probability of exceedence of annual extreme precipitation depth above any threshold value is the same

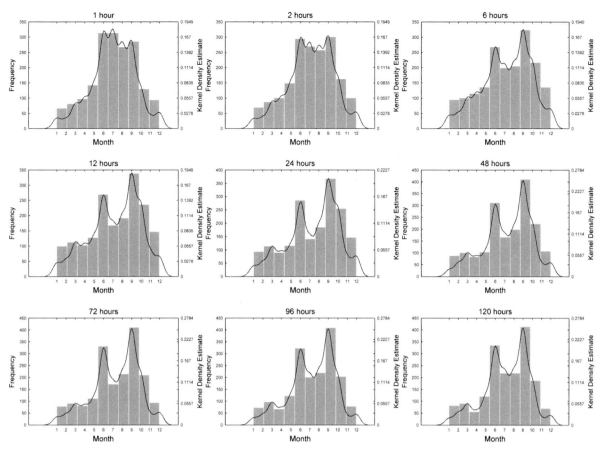

Figure 6.13 Kernel density estimate plots and histograms of occurrence of extreme events (months) for different durations for warm phase of AMO.

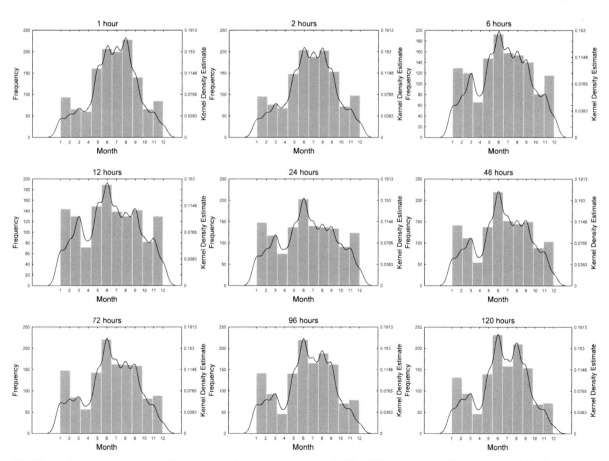

Figure 6.14 Kernel density estimate plots of occurrence of extreme events (months) for different durations for cool phase of AMO.

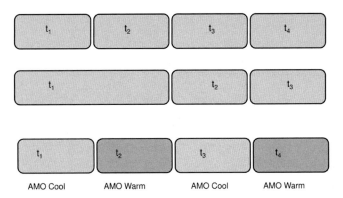

Figure 6.15 Definition of temporal windows for trend analysis of precipitation extremes.

for the cool and warm phases. However, as the duration increases from 24 hours to 120 hours this probability increases, suggesting higher precipitation depths are possible in the warm phase compared to the cool phase. Figure 6.17 shows the CDF plots for the current warm phase (1995–2010) and cool phase (1970–94).

For shorter durations (1–6 hours), the probability of exceedence of annual extreme precipitation depth above any threshold value is higher for the cool phase compared to that for the warm phase. However, the trend is reversed for durations longer than 24 hours.

Extreme precipitation depths for different durations are compared for two warm phases (1942–69 and 1995–2010) combined and those from the cool phase and are shown in Figure 6.18. For durations less than 12 hours, the combined warm phase and cool phase years provide similar CDF plots. However, when the duration is higher than 12 hours, the warm phase suggests higher precipitation totals are expected compared to the cool phase. Three CDF plots are shown in Figure 6.19, with plots associated with warm (1942–69), cool (1970–94), and warm (1995–2010) phases. The probability of exceedence of a precipitation depth above a specific value is higher for the earlier warm phase compared to the most recent warm phase. In general, the precipitation totals are higher in the warm phase compared to the cool phase. In Figure 6.20 comparisons are made for two warm phases of AMO.

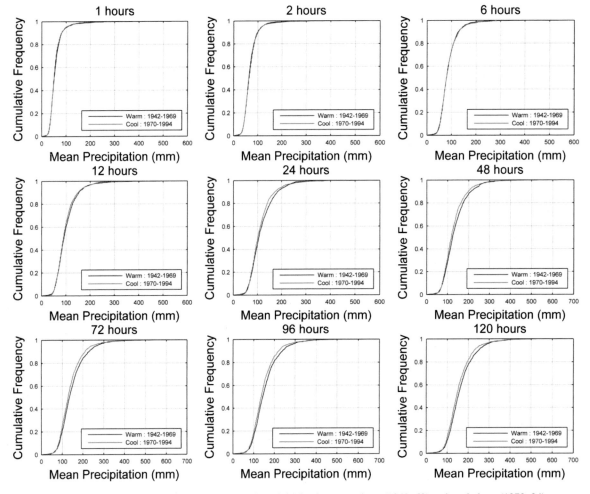

Figure 6.16 Cumulative density plots of the mean extreme annual precipitation in warm phase (1942–69) and cool phase (1970–94).

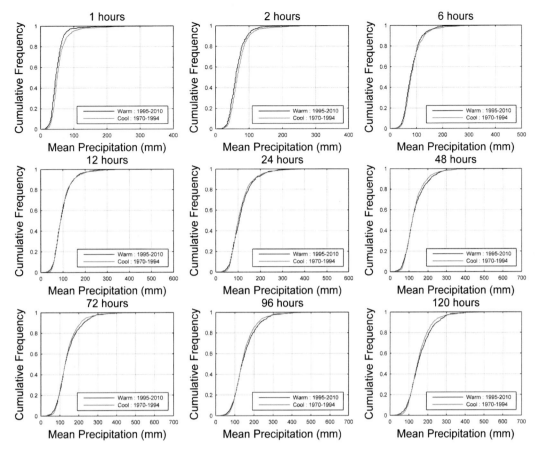

Figure 6.17 Cumulative density plots of the mean extreme annual precipitation in warm phase (1995–2010) and cool phase (1970–94).

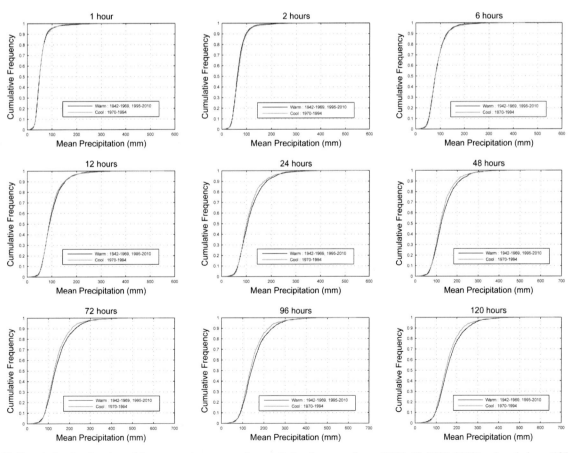

Figure 6.18 Cumulative density plots of the mean extreme annual precipitation in warm phases (1942–69; 1995–2010) and cool phase (1970–94).

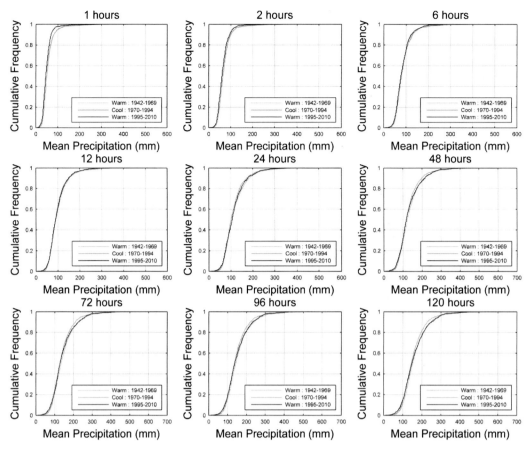

Figure 6.19 Cumulative density plots of the mean extreme annual precipitation in warm phase (1942–69), cool phase (1970–94), and warm phase (1995–2010).

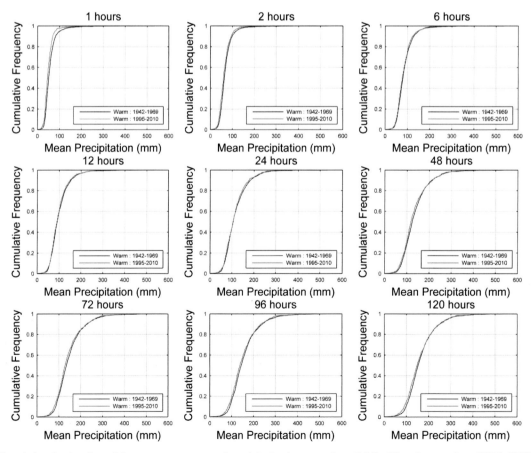

Figure 6.20 Cumulative density plots of the mean extreme annual precipitation in warm phase (1942–69) and warm phase (1995–2010).

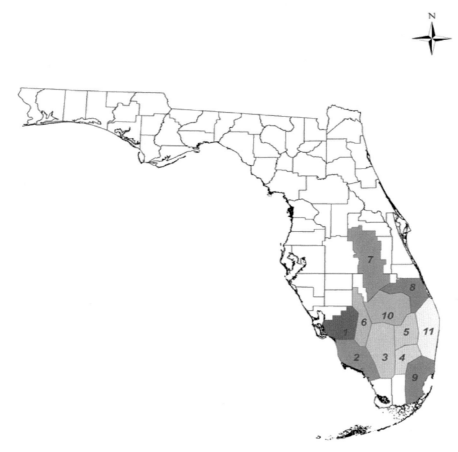

Figure 6.21 Meteorologically homogeneous rainfall areas identified in South Florida.

6.5.3 Homogeneous rainfall areas

Homogeneous rainfall areas (Figure 6.21) in South Florida are used to evaluate spatial variability of extreme precipitation magnitudes in two phases of AMO. KDEs are used along with bootstrap sampling whenever the length of data available in one phase is different from the data set in another phase. KDE plots shown in Figures 6.22 and 6.23 suggest that the magnitudes of extreme precipitation for different durations are generally higher in the AMO warm phase. However, these conclusions are not applicable and valid universally across the region. Region 10 on the southeastern side of Florida experiences increases in precipitation extremes for AMO warm phases for longer duration events.

6.5.4 Spatial variation of extreme events and AMO phases

Spatial variation of extreme precipitation in different phases of AMO can be evaluated for different durations. The interpolated values of extreme precipitation depths for different durations for two different phases of AMO are shown in Figure 6.24 for duration of 96 hours. The interpolation is carried out using IDWM using four nearest neighbors. In general, the higher precipitation magnitudes are realized throughout Florida in the warm phase compared to the cool phase of AMO. At durations of 1 hour both phases of AMO have similar variation of extreme values throughout Florida. However, as the duration increases the southeastern and northwest panhandle regions experience higher precipitation magnitudes. Higher precipitation magnitudes in the AMO warm phase can be attributed to increased landfall of hurricanes in the most recent warm phase.

6.5.5 Data windows: extreme precipitation events

Temporal windows that coincide with teleconnections can provide insights into the probability of occurrence of extreme events generally useful for hydrologic design. Figure 6.25 shows an example of CDF plots of annual daily maxima plotted for different AMO phases based on data from Florida.

It is evident from Figure 6.25 that probability of exceedence of an extreme precipitation event is higher for the AMO 2 (warm)

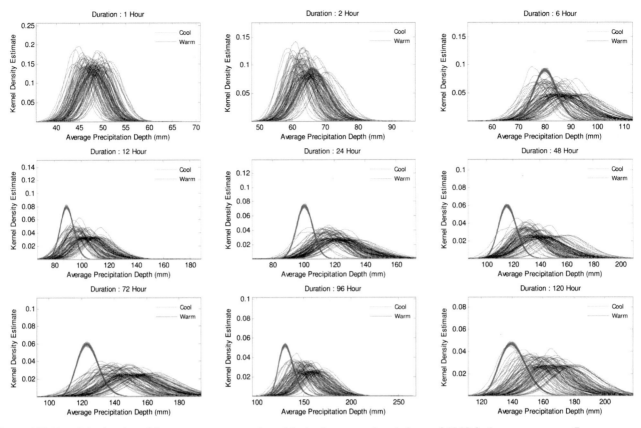

Figure 6.22 Kernel density plots of the mean extreme annual precipitation in warm and cool phases of AMO for homogeneous area 7.

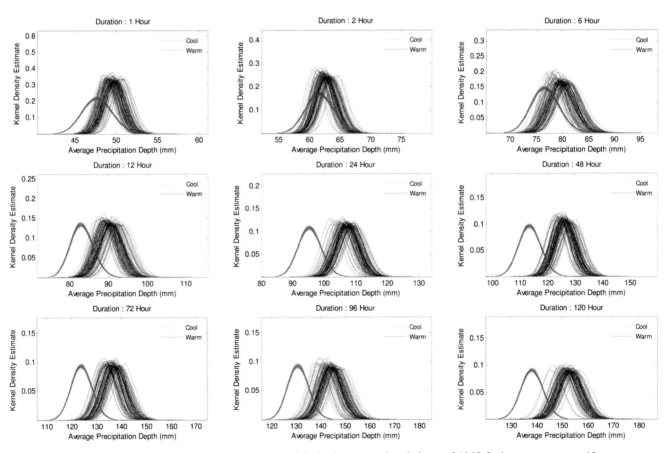

Figure 6.23 Kernel density plots of the mean extreme annual precipitation in warm and cool phases of AMO for homogeneous area 10.

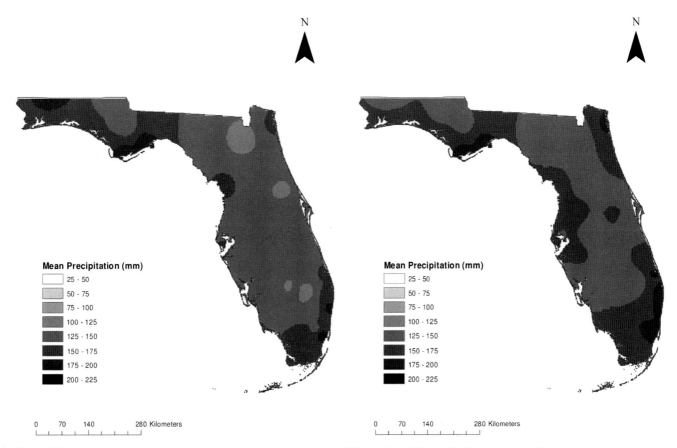

Figure 6.24 Spatial variation of extreme precipitation depth for duration of 96 hours for AMO cool (left) and warm (right) phases.

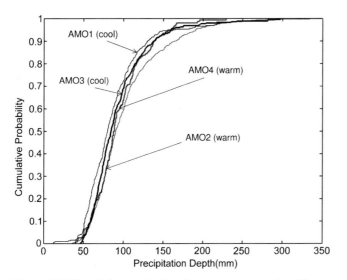

Figure 6.25 Cumulative density plot of annual extremes under different AMO phases.

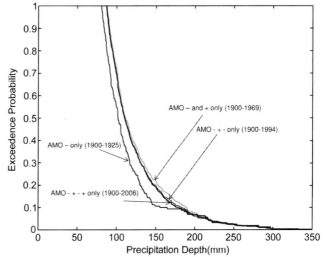

Figure 6.26 Exceedence plot of annual extremes under different AMO phases.

phase. Experiments with different combinations of temporal windows that coincide with AMO cycles are shown in Figure 6.26. Exceedence probability is highest for events above a specific threshold value for data that use combined AMO cycles from 1900 to 1969.

6.6 ENSO AND PRECIPITATION

Thevariability of monthly rainfall during ENSO warm and cool phases in Florida is assessed. In general, higher precipitation amounts are associated with El Niño conditions in the months

Table 6.5 *Variation in total precipitation values (mm) for each month in dry season*

	November	December	January	February	March	April	Total
El Niño	75	80	82	89	116	71	513
La Niña	45	52	44	60	73	69	343
Percentage change (%)	39.94	34.67	46.49	32.70	36.91	3.19	32.46

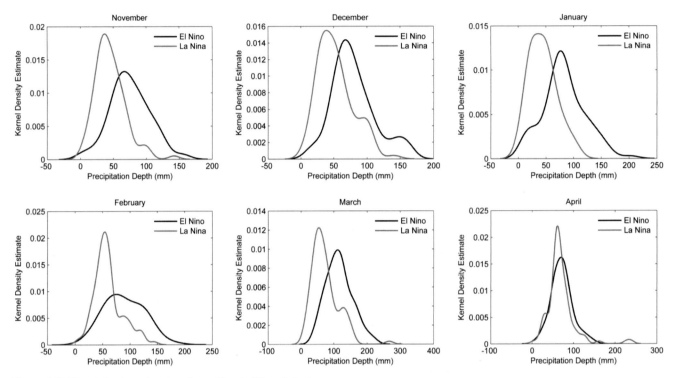

Figure 6.27 Kernel density estimation of monthly rainfall totals in dry season.

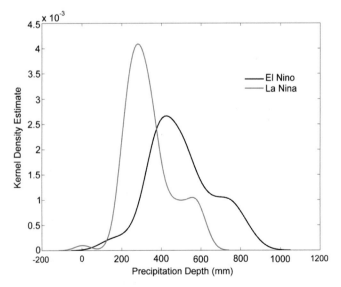

Figure 6.28 Kernel density estimates of total precipitation in dry season during different phases of ENSO.

of December, January, and February. In Florida, November to April (of the following year) is classified as the dry season. Precipitation totals in different months of the dry season are evaluated for El Niño and La Niña phases. Figure 6.27 shows the KDE plots of precipitation totals for each month of the dry season. The KDE plots suggest an increase in the precipitation amounts. For the month of April the distributions of precipitation totals for both La Niña and El Niño overlap. Figure 6.28 shows the distributions for the complete dry season. Precipitation totals for the entire dry season in the El Niño phase are higher compared to La Niña. The percentage increase in precipitation totals from La Niña to El Niño is given in Table 6.5. As seen in the table there is a significant increase (~30–45%) in total precipitation in the months November through March in El Niño with maximum increase in the month of January.

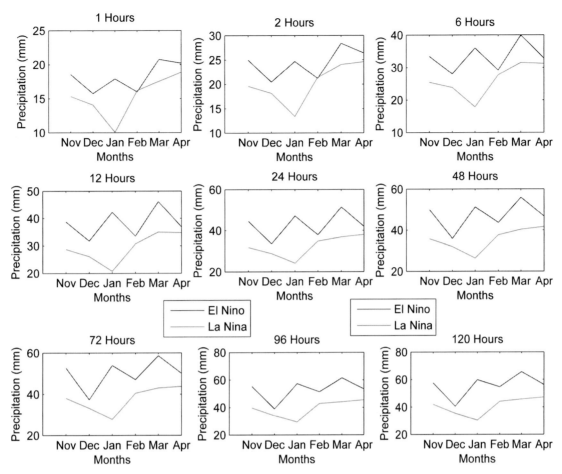

Figure 6.29 Mean extreme precipitation in dry season for AMO cool phase.

6.7 COMBINED INFLUENCE OF AMO–ENSO PHASES

This section discusses the combined influence of AMO–ENSO phases on precipitation extremes for different durations. AMO is a decadal oscillation and ENSO cycles occur with frequency of approximately 4 years. For the purpose of the analysis, available data can be evaluated by using eight categories:

1. AMO – Warm phase in El Niño;
2. AMO – Warm phase in La Niña;
3. AMO – Cool phase in El Niño;
4. AMO – Cool phase in La Niña;
5. El Niño in AMO warm phase;
6. El Niño in AMO cool phase;
7. La Niña in AMO warm phase;
8. La Niña in AMO cool phase.

Precipitation extremes during an AMO cool phase in El Niño and La Niña are shown in Figure 6.29. Variations in mean values of precipitation during the dry season of El Niño during AMO cool and warm phases are shown in Figure 6.30. Again, variations in

mean values of precipitation during the dry season of La Niña during AMO cool and warm phases are shown in Figure 6.31.

6.8 PACIFIC DECADAL OSCILLATION

The PDO is similar to ENSO in many respects. However, the location and the frequency of the PDO are different from ENSO. In general the PDO occurs over 20–30 years and has similar footprints of warming ocean waters (Rosenzweig and Hillel, 2008; Trenberth and Hurrell, 1994). The causes of PDO are not very clear; however, the understanding of this oscillation is particularly helpful for North America.

6.9 NORTH ATLANTIC OSCILLATION

A major oscillation that influences the climate of Europe, North America, and Asia is the NAO. The NAO index is based on the difference of normalized pressures between Lisbon, Portugal, and Stykkisholmur/Reykjavik, Iceland.

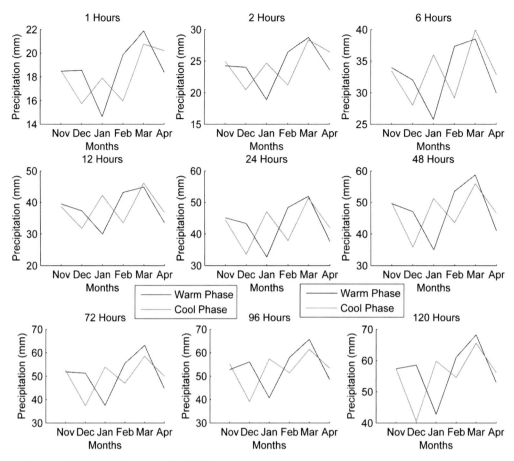

Figure 6.30 Mean extreme precipitation in dry season for El Niño.

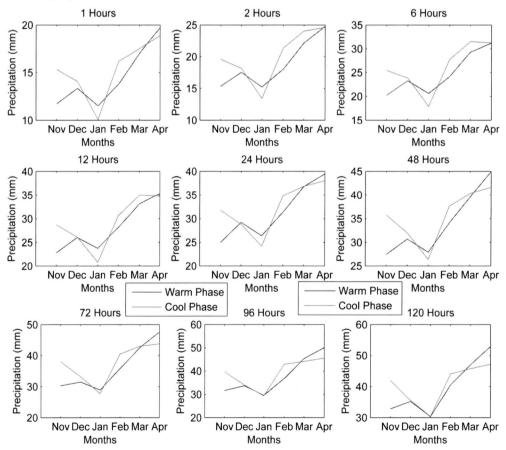

Figure 6.31 Mean extreme precipitation in dry season for La Niña.

6.10 FORECASTS BASED ON TELECONNECTIONS

Forecasts made based on information about the climatic variables and the link to teleconnections can be beneficial to water management agencies (Garbrecht and Piechota, 2006). Strong relationships between rainfall occurrence and streamflows have been observed in several parts of the world with variables that relate to the manifestation of teleconnections. One example of a strong correlation between SSTs and streamflows was provided by Dawod and El-Rafy (2002) and discussed in detail by Beek (2010). In their work, Dawod and El-Rafy related the annual Nile flows to SSTs at different locations in the Indian Ocean and the Pacific Ocean. Prediction is possible using information available at the end of June for the flow in the next hydrologic year (Beek, 2010). A high correlation between the predicted set of flows and observed flows suggests the utility of multiple linear regression equations linking flow and SSTs.

Seasonal forecasts using climate change information are linked to analogue years by considering historical climate records (Ludwig, 2009). ENSO is strongly correlated with climate in several countries. This strong relationship can be used for seasonal rainfall and streamflow forecasts. Souza and Lall (2003) report the use of NINO3.4 (an indicator used to define the ENSO state) and the North Atlantic Dipole to forecast streamflows in northeast Brazil. Australian rainfall amounts have been linked to El Niño (Chiew *et al.*, 1998) and SOI (Chiew *et al.*, 2003). These links help in futuristic seasonal to yearly forecasts of rainfall, helping water resources management professionals.

In general, teleconnections can be used for seasonal climate forecasts with benefits to several water management sectors. Water allocation can be improved if seasonal forecasts are available (Stone *et al.*, 1996) and future rainfall forecasts will be extremely beneficial to agrarian communities especially in arid and semi-arid regions. Seasonal forecasts can be used for water-pricing and also water-use restrictions (Ludwig, 2009; Chiew *et al.*, 2003). Once the forecasts are made, their utility may be evaluated with the help of a Taylor–Russell diagram (Figure 6.32; Sarewitz *et al.*, 2000; Hagemeyer, 2006) or contingency tables generally used in weather forecast evaluation. The diagram is a scatter plot of observed and predicted values of variables of interest. The main chart can help show the uncertainty in the predictions and assist in the decision-making process (Hagemeyer, 2006), with the horizontal line representing a boundary between situations requiring action, which are above the line, and situations requiring no action, which are below the line. The line is referred to as the threshold criterion line. If the observed event exceeds the criterion, then some sort of preventative or protection action is required. The vertical line shown in the chart is the

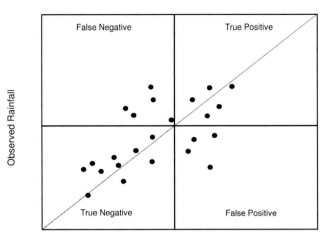

Figure 6.32 Taylor–Russell diagram to evaluate prediction uncertainty.

decision cutoff line. Generally decisions are made by decision makers using predictions.

The two lines on the Taylor–Russell diagram divide the scatter plot into four regions (Hagemeyer, 2006). The regions are labeled according to standard decision research terminology (Sarewitz *et al.*, 2000; Hagemeyer, 2006) as follows:

- **True positive:** Appropriate action is taken. Prior warning of extreme weather (extreme conditions) is provided and the observed event occurs.
- **False positive:** Inappropriate action is taken that leads to a false alarm. Prior warning provided with the actual extreme event not occurring.
- **False negative:** No prior warning was provided and action was not taken when it was actually needed.
- **True negative:** No prior warning or action was taken and there was no realization of a severe event.

6.11 PRECIPITATION AND TELECONNECTIONS: GLOBAL IMPACTS

ENSO is by far the most studied teleconnection and probably the most publicized due to the strong El Niño events of 1982–3 and 1997–8, which led to the greatest damage to agricultural and other sectors. ENSO is considered as the single largest cause of extremes in precipitation (as well as the cause of inter-annual variability) accounting for 15–20% of the global variance of precipitation (Dai *et al.*, 1997; Strangeways, 2007). The contribution is higher than 20% in ENSO regions. These conclusions were

Table 6.6 *Selected works dealing with teleconnections with primary focus on precipitation*

Teleconnection(s)	Focus of the work	Reference
AMO	Rainfall and river flows in the continental USA	Enfield *et al.*, 2001
AMO	Climate impacts	Knight *et al.*, 2006
AMO	Hurricane activity in Atlantic	Goldenberg *et al.*, 2001
AMO	Variability and climate impacts	Ting *et al.*, 2011
PDO, NAO, and SOI	Summer rainfall in Beijing in relation to inter-decadal variability	Wei *et al.*, 2008
PDO and AMO	Temperature variability in Upper Colorado River flow	McCabe *et al.*, 2004
ENSO, PDO, and AMO	Streamflow based upon the interactions of the teleconnections	Rogers and Coleman, 2003
ENSO	Extreme rainfall and temperature frequencies	Gershunov and Barnett, 1998
ENSO	Linear relation between precipitation and SST anomalies	Montroy, 1997
ENSO and NAO	Seasonal rainfall	Rodo *et al.*, 1997
ENSO	Children of the tropics: El Niño and La Niña	Henson and Trenberth, 1998
ENSO	Seasonal rainfall and river discharges in Florida	Schmidt *et al.*, 1999
ENSO	Rainfall anomalies in East Asia	Wu *et al.*, 2003
ENSO	Evolution, global atmospheric surface temperatures	Trenberth *et al.*, 2002a
ENSO	Extreme weather events of 1997 and 1998	Trenberth, 1999
ENSO	Indices of El Niño evolution	Trenberth and Stepaniak, 2001
ENSO	Inter-annual variations in the atmospheric heat budget	Trenberth *et al.*, 2002b
ENSO	Sea level pressures, surface temperatures, and precipitation	Treberth and Caron, 2000
ENSO	Future projections	Stevenson *et al.*, 2011

based on monthly gridded data sets from 1900 to 1988. Studies performed by Teegavarapu and Goly (2011) indicated that El Niño is responsible for higher precipitation totals compared to La Niña in the months of December to February in the southern part of the USA, especially in Florida. The conclusions were based on gridded precipitation data sets from CRU and GPCC and also historical precipitation data sets from USHCN rain gage stations. The NAO is associated with inter-annual and decadal effects on temperature and precipitation in several regions of the Middle East (Rosenzweig and Hillel, 2008; Cullen and deMoncal, 2000).

Large multi-decadal SST anomalies have had significant impacts on regional and global climate. Linkages of SSTs with the variability in regional hydrologic phenomena, such as temperature, precipitation, streamflow, and flood and drought, have been well documented. McCabe *et al.* (2004) reported that the frequencies of drought in the conterminous USA are highly correlated with the PDO and AMO. They concluded that these two components explain about 52% of the variance in droughts over the conterminous USA. Rogers and Coleman (2003) showed

that the high (low) streamflows in the upper Mississippi River basin are consistent with cold (warm) phases of the AMO winter signal. The strong relationship between SSTs and variability in precipitation over the USA and other regions of the world has been discussed in several research studies (Enfield *et al.*, 2001; Lachniet *et al.*, 2004; Goodrich and Ellis, 2008). Enfield *et al.* (2001) demonstrated that the cool (warm) AMO phases are directly related to above (below) normal rainfall over most of the USA. They also showed that the inflow to the Mississippi River basin, in the eastern USA, varies by about 10% and the inflow to Lake Okeechobee, in South Florida, varies by about 40% in response to the changes in AMO phases. Strong correlation of El Niño (La Niña) Southern Oscillation phases with the increase (decrease) in precipitation and snowfall over the USA and Canada has been documented in several research findings (e.g., Groisman and Easterling, 1994; Groisman *et al.*, 1994; Latif and Barnett, 1994).

Many studies have demonstrated a strong relationship between streamflow variability and the atmospheric–oceanic circulations (Kahya and Dracup, 1993; McCabe, 1996; Rogers and Coleman,

Table 6.7 *Extreme precipitation depths (in millimeters) for two phases of AMO*

Cool phase				Warm phase			
Hours							
48	72	96	120	48	72	96	120
175	176	196	196	119	121	121	121
216	216	218	219	121	140	162	162
109	121	168	194	113	146	148	152
69	70	97	100	75	95	105	114
44	46	46	47	123	135	154	196
104	105	106	119	145	156	194	225
114	120	124	125	208	208	208	208
152	152	176	205	125	131	136	142
129	129	129	129	219	229	236	236
73	73	73	73	138	144	144	144
105	109	123	129	110	110	110	110
161	165	165	171	59	59	59	59
308	342	411	411	158	233	249	251
136	143	149	149	136	141	145	147
183	215	250	250	146	176	176	176
178	178	178	178	90	90	100	104
156	161	181	209	128	128	139	146
78	82	101	102	141	160	160	166
152	161	161	161	88	88	112	116
105	105	105	111	135	135	135	135
148	154	155	167	90	90	93	98
196	199	199	199	187	187	187	187
78	78	97	106	152	157	160	184

under the warmer climate scenarios. Therefore, long-term trend analyses of extreme events, along with SST-related variability should be investigated. The availability and quality of long-term precipitation data required to perform such analyses are often limited. Most of the long-term precipitation data contain gaps due to instrument malfunctioning, inability of operators to collect data, discontinuity, etc. Continuous time series data are a preliminary requirement before analyses of long-term data can be performed.

A few selected works dealing with teleconnections with primary focus on precipitation are provided in Table 6.6.

6.12 CONCLUSIONS AND SUMMARY

Teleconnections that are defined by internal modes of climate variability, if properly understood, can be evaluated for their links to extreme precipitation events and other climate variables in a region influenced by such modes. Generally these links are established using statistical tools and historical climate analogues to aid seasonal forecasts for water resources planning and management. This chapter discusses several teleconnections and provides an in-depth review of two teleconnections that influence the regional climate of Florida, USA, as examples. Spatial and temporal variability of extreme precipitation events for different durations and increased activity of cyclonic precipitation under the warm phase of the AMO, and variability of precipitation extremes and totals during different phases of ENSO are discussed in detail.

2003). A strong influence of ENSO on streamflow patterns over the western USA has been documented in numerous research findings (e.g., Kahya and Dracup, 1993; Dettinger and Diaz, 2000; Clark *et al.*, 2001; Tootle *et al.*, 2005). Tootle *et al.* (2005) found a strong influence of ENSO on streamflow variability over the western USA, while a weaker influence due to other atmospheric–oceanic circulations such as PDO, NAO, and AMO was also reported. McCabe *et al.* (2004) reported the influence of AMO and PDO on the droughts over the USA. Relationships between SSTs and streamflow variability have been utilized by several researchers to forecast streamflow 3 to 12 months in advance (Gutierrez and Dracup, 2001; Grantz *et al.*, 2005; Tootle *et al.*, 2005, 2007, 2008; Soukup *et al.*, 2009). Analyses of precipitation data are vital for various water resources design and management applications. However, variability due to SSTs should be considered while performing analyses of long-term data. Long-term analyses of precipitation data are also required to study the changes due to global warming. Increased frequency of extreme precipitation events over the eastern USA has been predicted

EXERCISES

6.1 Using Figures 6.4–6.7, identify the regions in which teleconnections influence the local climate and evaluate the influences based on available historical climatological data.

6.2 Identify the variables such as precipitation or streamflows in your region that have a strong relation with SSTs or other teleconnection-related variables for developing multiple regression models to aid in seasonal forecasts.

6.3 Data for extreme precipitation depths for different durations from years coinciding with cool and warm phases of AMO at one rain gage station in Florida, USA, are provided in Table 6.7. Evaluate the influence of the warm phase on extremes compared to that of the cool phase.

6.4 Use the GPCC or CRU observed gridded data sets (discussed in Chapter 2) for assessment of variability of precipitation trends in temporal windows that coincide with internal modes (oscillations) of climate variability.

6.5 Discuss the utility of using non-parametric probability density functions such as kernel density functions for evaluation of climatic variables over time. Investigate the use of bootstrap sampling for assessment.

USEFUL WEBSITES

http://www.cesm.ucar.edu/events/ws.2011/Presentations/Ocean/stevenson.pdf (accessed September, 2011).

http://www.ucar.edu/communications/factsheets/elnino/ (accessed September, 2011).

7 Precipitation trends and variability

7.1 HISTORICAL AND FUTURE TRENDS

Long-term data on precipitation occurrence, intensity, amount, and spatial and temporal distribution are vital for water resources management. These data provide important information for the design of hydrologic structures (such as dams, culverts, and detention basins), water supply and water quality modeling, and other hydrologic and water quality issues. Hydrologic structures are normally designed to safely handle most extreme precipitation and flood events that are possible in the design life of the structure. Therefore, analyses of extreme precipitation events are often performed for hydrologic structure design, flood control planning and assessment, and various other water management issues. Several research studies have reported changes in extreme precipitation and flood events due to global warming. Changnon and Kunkel (1995) examined trends in the heavy multi-day precipitation events associated with floods for the period from 1921 to 1985 and found upward trends in both floods and heavy precipitation events over portions of the central USA. Karl and Knight (1998) pointed to an approximate 8% increase in precipitation across the contiguous USA since 1910, which was primarily due to increases in the intensity of heavy to extreme precipitation events. Kunkel et al. (1999) found statistically significant upward trends in 1-year return period, 7-day duration events of about 3% per decade and in 5-year, 7-day events of about 4% per decade for the period from 1931 to 1996. Groisman et al. (2001) reported a 50% increase in the frequency of extreme precipitation days (with precipitation exceeding 101.6 mm) in the Upper Midwest USA during the twentieth century.

7.1.1 Historical data analysis from existing studies

Increasing temperatures tend to increase evaporation, which leads to more precipitation (IPCC, 2007d). As average global temperatures have risen, average global precipitation has also increased. According to the IPCC (2007), the following precipitation trends have been observed:

- Precipitation generally increased over land north of 30°N from 1900 to 2005, but has mostly declined over the tropics since the 1970s. Globally there has been no statistically significant overall trend in precipitation over the past century, although trends have varied widely by region and over time.
- It has become significantly wetter in eastern parts of North and South America, northern Europe, and northern and central Asia, but drier in the Sahel, the Mediterranean, southern Africa, and parts of southern Asia.
- Changes in precipitation and evaporation over the oceans are suggested by freshening of mid- and high-latitude waters (implying more precipitation), along with increased salinity in low-latitude waters (implying less precipitation and/or more evaporation).
- There has been an increase in the number of heavy precipitation events over many areas during the past century, as well as an increase since the 1970s in the prevalence of droughts – especially in the tropics and subtropics.

In the Northern Hemisphere's mid and high latitudes, the precipitation trends are consistent with climate model simulations that predict an increase in precipitation due to human-induced warming. By contrast, the degree to which human influences have been responsible for any variations in tropical precipitation patterns is not well understood or agreed upon, as climate models often differ in their regional projections (IPCC, 2007d).

7.2 GLOBAL PRECIPITATION TRENDS

The global precipitation anomalies shown in Figure 7.1 are based on historical data from the Australian Government Bureau of Meteorology. A linear trend of 2.08 mm/decade was observed. The anomaly calculations were based on 30-year climatology (1961–90). Global precipitation trends by UNEP are shown in Figure 7.2.

The UNEP/GRID provides the following summary of precipitation trends. Precipitation very likely increased during the twentieth century by 5 to 10% over most mid and high latitudes of the

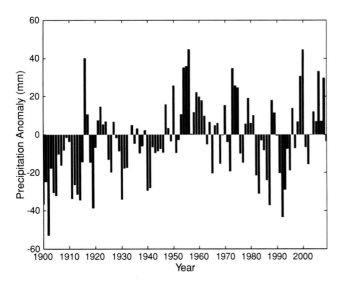

Figure 7.1 Global precipitation anomaly (data source: Australian Government Bureau of Meteorology).

Northern Hemisphere continents, but in contrast rainfall likely decreased by 3% on average over much of the subtropical land areas. There has likely been a 2 to 4% increase in the frequency of heavy precipitation events in the mid and high latitudes of the Northern Hemisphere over the latter half of the twentieth century. There were relatively small long-term increases over the twentieth century in land areas experiencing severe drought or severe wetness, but in many regions these changes are dominated by inter-decadal and multi-decadal climate variability with no significant trends evident over the twentieth century.

7.3 USA PRECIPITATION CHANGES

Precipitation changes over the USA are evaluated and discussed by the USEPA (United States Environmental Protection Agency) based on observations compiled by NOAA's National Climatic Data Center. These changes are shown in Figure 7.3. In general, for the contiguous USA, the total annual precipitation has increased at an average rate of 6.1% per century since 1900, although there was considerable regional variability. The greatest increases came in the East North Central climate region (11.6% per century) and the South (11.1%). Hawaii was the only region to show a decrease (−9.25%).

7.4 ASSESSMENT OF EXTREME PRECIPITATION TRENDS: TECHNIQUES

Assessments of extreme precipitation data require several exploratory data analysis techniques for preliminary analysis and statistical techniques for trend analysis. The following sections discuss exploratory data analysis techniques for assessment of precipitation time series data.

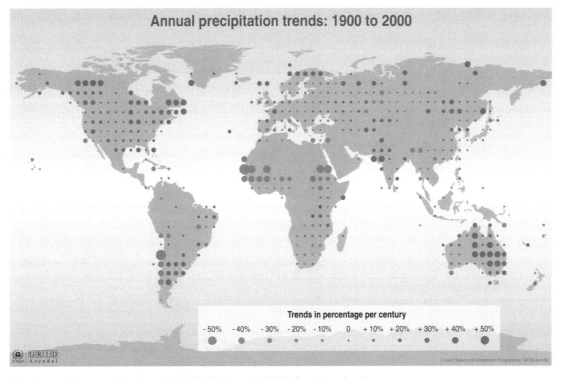

Figure 7.2 Global precipitation trends (source: UNEP/GRID-Arendal, 2005). See also color plates.

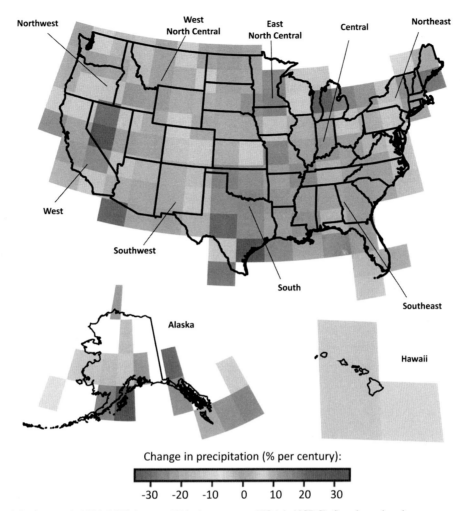

Figure 7.3 Annual precipitation trends 1901–2005 (source: EPA; data courtesy NOAA, NCDC). See also color plates.

7.4.1 Exploratory data analysis

Exploratory data analysis (EDA) is the preliminary step in understanding data and is a critical component in statistical analysis that will be carried out in future. EDA involves graphical description of data using time series plots, bivariate plots, histograms, autocorrelation functions, spatial plots of data (trends and magnitudes), box plots, and probability plots and smoothing curves based on scatter plots or time series plots and kernel density estimates for non-parametric characterization of data. Grubb and Robson (2000) provide an excellent treatise on exploratory and visual data analysis. They indicate the utility of EDA for understanding temporal patterns, seasonal variations, regional and spatial variations, data problems, correlations among variables, independent and autocorrelated variables, and details of seasonal structure.

Many data analysis and data mining books (e.g., Han and Kamber, 2006; Tan *et al.*, 2006; Myatt and Johnson, 2009) provide extensive details of exploratory data analysis and techniques. A list of EDA tools and their uses is provided in Table 7.1. The list

is not comprehensive in describing all the available tools and their utilities. The list also provides tools and utilities based on data type.

EDA techniques can also be used for four aspects of preliminary data analysis. These four aspects (Chakrabarti *et al.*, 2008) are cleaning, integration, transformation, and reduction. According to Chakrabarti *et al.* (2008) cleaning mainly refers to filling missing values, smoothing out noise with identification of outliers, and correcting inconsistencies in data. Data integration is a process of combining data from multiple sources to obtain coherent data sets. Transformation is mainly carried out to convert data to appropriate forms for statistical analysis.

7.5 FITTING PROBABILITY DISTRIBUTIONS FOR EXTREME RAINFALL DATA

The following steps are used to fit and evaluate probability distributions to rainfall data. Before the steps are executed, the

Table 7.1 *Examples of tools and their uses for EDA*

Data type	Tool	Use
Univariate	Summary statistics	Assess central tendency, spread (variability), symmetry, peakedness
	Time series and run sequence plots	Variations at different temporal scales, trend
	Histogram	Distribution of data, frequency of observations
	Kernel density function	Smoothed distribution of data with no assumptions of distribution
	Cumulative density function	Empirical density function
	Box plots	Visual depiction of entire data (summary statistics, outliers, etc.)
	Probability plots	Assess the distribution of the sample
	Quantile–quantile plots	Compare empirical and fitted cumulative density functions
Bivariate	Scatter plots	Understand linear and non-linear associations, outliers
	Pearson correlation	Quantitative evaluation of strength of correlation
	Spearman rank correlation (ρ)	Evaluation of strength of correlation using rank-ordered data
	Kendall's τ	Alternative to Pearson correlation
	Autocorrelation function	Serial dependence at different temporal lags, persistence
Multivariate	Scatter plots	Visualize data, associations between variables
	Star plot	Multi-dimensional plot explaining variability
	Glyph scatter plot	Glyphs are used to provide additional information in scatter plots
	Correlation matrix	Quantitative evaluation of correlations in matrix form
	Scatter plot matrix	Plots in matrix format to visualize associations
	Correlation maps	Geographical variation of a process

historical precipitation data at a site are evaluated for any problems associated with missing data and homogeneity issues. A computational code is required to sequentially search the historical precipitation data to obtain the extreme values data for different durations.

1. Obtain annual extreme rainfall depth data for different durations.
2. Obtain the empirical cumulative density functions for rainfall depths for different durations.
3. Develop a hypothesis for the underlying cumulative density function for data.
4. Estimate the parameters for the hypothesized distribution function.
5. Select a goodness-of-fit test along with a level of significance.
6. Evaluate the validity of the hypothesized distribution to data using a goodness-of-fit test.
7. Repeat steps 3–6 if the hypothesized distribution does not fit the data.

The annual extreme value or peak over threshold (POT) series should satisfy four conditions (Ashkar, 1996) and they are (1) randomness, (2) independence, (3) homogeneity, and (4) stationarity. Generally, annual extreme rainfall events are considered random and independent. Homogeneity and stationarity are two main issues that need to be thoroughly addressed when dealing with extreme precipitation data.

Table 7.2 *Values of constant a for different plotting position formulae*

Plotting position formula	a
Weibull	0.000
Blom	0.375
Cunnane	0.400
Gingorten	0.440
Hazen	0.500

7.5.1 Development of cumulative distribution functions

A quantile plot (i.e., an empirical cumulative distribution function plot or empirical cumulative density plot) provides a graph of the ordered data and the corresponding estimated cumulative probabilities (Millard and Neerchel, 2001). The estimated probabilities are defined as the plotting positions and several formulae are available for obtaining these probabilities. A general expression described by Blom (1958) and Cullen and Frey (1999) is provided in Equations 7.1 and 7.2 for development of empirical cumulative density functions.

$$P(X < x_i) = \frac{i - a}{n - 2a + 1} \ \forall i \tag{7.1}$$

$$x_1 < x_2 < x_3 < \cdots < x_n \tag{7.2}$$

Table 7.2 provides different values of parameter a for obtaining different plotting position formulae based on Equation 7.1.

7.6 STATISTICAL DISTRIBUTIONS

Several statistical distributions are appropriate for characterizing the extreme precipitation values. The following sections briefly describe the distributions that are commonly used.

7.6.1 Normal distribution

Normal distributions are not generally applicable to characterize hydrologic variables as the normally distributed random variables vary from $-\alpha$ to $+\alpha$. Assigning probabilities to negative variable values, however small the values, is theoretically possible but not logical. However, several studies have used normal distributions to characterize extreme annual rainfall values.

$$f(x) = \frac{1}{\sigma_x \sqrt{2\pi}} e^{\left[-\frac{1}{2}\left(\frac{x-\mu_x}{\sigma_x}\right)^2\right]} \qquad (7.3)$$

7.6.2 Log-normal distribution

The log-normal distribution can be used to characterize the distributions of extreme values using logarithmic ($\log_e$ or ln) transformations (i.e., $y = \log_e x$).

$$f(x) = \frac{1}{x\sigma_y \sqrt{2\pi}} e^{\left[-\frac{(\ln x - \mu_y)^2}{2\sigma_y^2}\right]}, x > 0 \qquad (7.4)$$

7.6.3 Three-parameter log-normal distribution

A three-parameter log-normal distribution function is similar to the density function provided by Equation 7.4 with an additional parameter referred to as the threshold parameter. The transformation $y = \log_e(x - \gamma)$ with the threshold parameter γ is used for the three-parameter log-normal distribution. When the threshold parameter is associated with location and when it is equal to zero, the distribution reduces to a two-parameter log-normal distribution (Millard and Neerchal, 2006).

7.6.4 Extreme value distribution

The general form of extreme value distribution is given by:

$$f(x) = \frac{1}{ao} e^{\left\{\pm \frac{x-bo}{ao} - \exp\left[\pm \frac{x-bo}{ao}\right]\right\}}, \qquad (7.5)$$

$$-\infty < x < \infty, \infty < bo < \infty, ao > 0$$

where ao and bo are scale and location parameters, bo is the mode of the distribution. The minus sign in $\pm$ is used for maximum values. The distribution is also referred to as a Type I or Gumbel distribution.

7.6.5 Extreme value Type III

The probability density function for extreme value Type III is given by Equation 7.6 and is generally used for characterizing low streamflows bounded on the left by zero (Chin, 2006). The distribution is also referred to as the Weibull distribution.

$$f(x) = aox^{ao-1} bo^{-ao} e^{\left[-\left(\frac{x}{bo}\right)^{ao}\right]}, x \geq 0, ao, bo > 0 \quad (7.6)$$

7.6.6 Generalized extreme value

The generalized extreme value (GEV) distribution is based on extreme-value Type I, II, and III distributions for maxima (Chin, 2006). The CDF for GEV is given by Equation 7.7. The variables ao, bo, and co in Equation 7.7 are the location, scale, and shape parameters of the distribution.

$$F(x) = \begin{cases} e^{\left\{-\left[1 - \frac{co(x-ao)}{bo}\right]^{\frac{1}{co}}\right\}} & co \neq 0 \\ e^{\left\{-\exp\left[-\frac{(x-ao)}{bo}\right]\right\}} & co = 0 \end{cases} \qquad (7.7)$$

7.6.7 Gamma-type distribution

The general form of the one-parameter gamma distribution (Chin, 2006; Millard and Neerchal, 2006) is given by:

$$f(x) = \frac{1}{\Gamma(\alpha)} x^{\alpha-1} e^{-x}, x \geq 0 \qquad (7.8)$$

7.6.8 Pearson Type III distribution

The probability density function for a three-parameter gamma distribution, which is also referred to as a Pearson Type III distribution is given by:

$$f(x) = \frac{1}{B^{\alpha}\Gamma(\alpha)}(x-\gamma)^{\alpha-1} e^{-(x-\gamma)/\beta}, x \geq \gamma \qquad (7.9)$$

The log-Pearson Type III is essentially the same distribution function with the data being logarithmically transformed.

7.6.9 Other distributions

A few other distributions, generalized logistic, generalized normal, generalized Pareto, and five-parameter Wakeby, are also appropriate for characterizing precipitation extremes. These distributions should be evaluated along with the others discussed previously.

7.7 PARAMETER ESTIMATION

Parameter estimation involves estimation of population parameters from the sample data. Six characteristics of the estimation methods are described by Small (1990) and they are (1) consistency, (2) lack of bias, (3) efficiency, (4) sufficiency, (5) robustness, and (6) practicality.

7.7.1 Method of moments

The method of moments (MOM), as the name suggests, indicates that the parameters of a distribution can be estimated using the moments of the distribution. This method is also referred to as the method of matching moments. These estimates are unbiased estimates of mean, standard deviation, and skewness. The estimates are obtained from the sample data and they are assumed to be equal to the population parameters.

$$\hat{\mu}_x = \frac{1}{no}\sum_{j=1}^{no} x_j \tag{7.10}$$

$$\hat{\sigma}_{x^2} = \frac{1}{no-1}\sum_{j=1}^{no}(x_j - \hat{\mu})^2 \tag{7.11}$$

$$\hat{g}_x = \frac{no}{(no-1)(no-2)}\frac{\sum_{j=1}^{no}(x_j - \hat{\mu})^3}{\hat{\sigma}_{x^3}} \tag{7.12}$$

7.7.2 Maximum likelihood estimation method

The maximum likelihood estimation (MLE) method is the most common method used for estimation of parameters. A likelihood function (L°) given by Equation 7.13 is maximized. The variable np is the number of parameters, no is the number of observations, and j is the index for the observation number. Partial derivatives of the likelihood function (Equation 7.14) with respect to different parameters are obtained to solve a set of equations. The solutions of these equations provide the estimated parameter values.

$$L^\circ\left(\theta_1, \theta_2, \ldots, \theta_{np}\right) = \prod_{j=1}^{no} p_X\left(x_j | \theta_1, \theta_2, \ldots, \theta_{np}\right) \tag{7.13}$$

$$\frac{\partial L^\circ}{\partial \theta_{jp}} = 0, \quad jp = 1, \ldots, np \tag{7.14}$$

7.7.3 L-moments approach

The concept of L-moments as linear combinations of probability weighted moments (PWM) was proposed by Hosking (1990). The theory of PWM was described by Greenwood *et al.* (1979). The PWM is described by Equation 7.15. The variable ω_r is the rth order PWM and $F_X(x)$ is the cumulative distribution function of x. The unbiased sample estimators of PWMs are defined by Hosking and Wallis (1998) as ϕ_1, ϕ_2, ϕ_3, and ϕ_4 four L-moments. Unbiased sample estimates of the PWM for any distribution can be obtained using Equations 7.15–7.19.

$$\omega_r = \int_{-\infty}^{+\infty} x\left[F_X(x)\right]^r f_X(x)\,dx \tag{7.15}$$

$$\phi_1 = \omega_0 \tag{7.16}$$

$$\phi_2 = 2\omega_1 - \omega_0 \tag{7.17}$$

$$\phi_3 = 6\omega_2 - 6\omega_1 + \omega_0 \tag{7.18}$$

$$\phi_4 = 20\omega_3 - 30\omega_2 + 12\omega_1 - \omega_0 \tag{7.19}$$

The values of ω_0, ω_1, ω_2, and ω_3 are obtained by equating them to derived constants (bc_0, bc_1, bc_2, and bc_3) from the sample using Equations 7.20–7.23.

$$bc_0 = \frac{1}{no}\sum_{j=1}^{no} x_j \tag{7.20}$$

$$bc_1 = \frac{1}{no(no-1)}\sum_{j=2}^{no}(j-1)x_j \tag{7.21}$$

$$bc_2 = \frac{1}{no(no-1)(no-2)}\sum_{j=3}^{no}(j-1)(j-2)x_j \tag{7.22}$$

$$bc_3 = \frac{1}{no(no-1)(no-2)(no-3)}\sum_{j=4}^{no}(j-1)(j-2)(j-3)x_j \tag{7.23}$$

The ratios of ϕ_2 and ϕ_1, ϕ_3 and ϕ_2, and ϕ_4 and ϕ_2 are defined as L-coefficient of variation, L-skewness, and L-kurtosis, respectively. These ratios are referred to as L-moment ratios. Parameters for different statistical distributions and associated L-moments (Hosking and Wallis, 1998; Chin, 2006) to obtain these parameters are given in Table 7.3. The coefficients of L-variation, L-skewness, and L-kurtosis provide measures of dispersion, symmetry, and peakedness, respectively.

7.8 FREQUENCY FACTORS

Frequency factors (Chin, 2006; Rakhecha and Singh, 2009) are available for several distributions and they can be used to obtain the magnitude of an event for a specific return period, T. Frequency factors are used in the method proposed by Chow (1951) referred to as the frequency method or general frequency equation for hydrologic frequency analysis. The general equation is given as:

$$x_T = \bar{x} + K_T s \tag{7.24}$$

where x_T is the magnitude of an event for a return period T, K_T, $\bar{x}$ and s are the frequency factor, mean, and standard deviation values calculated based on the sample, respectively. The frequency factors depend on the type of distribution used. In general for two-parameter distributions, the factors depend on return period. The factors for normal and gamma (Pearson Type III) and extreme

Table 7.3 *Moments of the distributions and corresponding L-moments (Chin, 2006)*

Distribution type	Parameters	Moments	L-moments
Normal	μ_X, σ_X	$\mu_X = \mu_X$ $\sigma_X = \sigma_X$	$\phi_1 = \mu_X$ $\phi_2 = \dfrac{\sigma_X}{\pi^{1/2}}$
Log-normal ($Y = \ln X$)	μ_Y, σ_Y	$\mu_Y = \mu_Y$ $\sigma_Y = \sigma_Y$	$\phi_1 = \exp\left(\mu_Y + \dfrac{\sigma_Y^2}{2}\right)$ $\phi_2 = \exp\left(\mu_Y + \dfrac{\sigma_Y^2}{2}\right)$ $\operatorname{erf}\left(\dfrac{\sigma_Y}{2}\right)$
Exponential (two-parameter)	ξ, η	$\mu_X = \xi + \dfrac{1}{\eta}$ $\sigma_X = \xi + \dfrac{1}{\eta^2}$	$\phi_1 = \xi + \dfrac{1}{\eta}$ $\phi_2 = \dfrac{1}{2\eta}$
Gumbel	ξ, α	$\mu_X = \xi + 0.5772\alpha$ $\sigma_X^2 = 1.645\alpha^2$	$\phi_1 = \xi + 0.5772\alpha$ $\phi_2 = 0.6931\alpha$
Generalized extreme value	ξ, α, κ	$\mu_X = \xi + \dfrac{\alpha}{\kappa}[1 - \Gamma(1+\kappa)]$ $\sigma_X^2 = \left(\dfrac{\alpha}{\kappa}\right)^2 \left\{\Gamma(1+2\kappa) - [\Gamma(1+\kappa)]^2\right\}$	$\phi_1 = \xi + \dfrac{\alpha}{\kappa}[1 - \Gamma(1+\kappa)]$ $\phi_2 = \dfrac{\alpha}{\kappa}(1 - 2^{-\kappa})\Gamma(1+\kappa)$

value Type I distributions are provided by Chin (2006) and are given in Equations 7.25 and 7.26 respectively.

$$K_T = w - \frac{2.515517 + 0.802853w + 0.010328w^2}{1 + 1.432788w + 0.189269w^2 + 0.001308w^3} \quad (7.25)$$

$$w = \left[\ln\left(\frac{1}{p^2}\right)\right]^{\frac{1}{2}}, \quad 0 < p \le 0.5 \quad (7.26)$$

In Equation 7.26, the value of p is the exceedence probability, and when the value of p is greater than 0.5, then p is substituted with $1 - p$ with modification of sign for the value of K_T.

$$K_T = \frac{1}{3k}\left\{\left[(x'_T - k)k + 1\right]^3 - 1\right\} \quad (7.27)$$

$$k = \frac{g_x}{6} \quad (7.28)$$

$$K_T = -\frac{\sqrt{6}}{\pi}\left\{0.5772 + \ln\left[\ln\left(\frac{T}{T-1}\right)\right]\right\} \quad (7.29)$$

7.9 PARAMETRIC AND NON-PARAMETRIC TESTS

A list of parametric and non-parametric tests used in several phases of statistical analysis of trends is provided in the Table 7.4. The list is by no means exhaustive; however, it includes some of the most commonly used tests. Corder and Foreman (2009) provide a comprehensive review of non-parametric statistical tests.

7.9.1 Goodness-of-fit tests for normal distributions

The extreme precipitation data sets may sometimes be characterized by a normal distribution. Transformations can be carried out for precipitation data using traditional transformation methods and also Box–Cox transformation if the data need to be normally distributed. The transformed or untransformed values of extreme rainfall using optimal Box–Cox transformed parameters can be tested using several goodness-of-fit tests.

An example of a Box–Cox transformation is provided using the data in Table 7.5.

The data are not normal, as evident from the probability plot on normal paper shown in Figure 7.4. Box–Cox transformation of data is carried out and the optimal Box–Cox parameter is estimated to be 0.2661. The transformed values are provided in Table 7.6 and the probability plot using these values is shown in Figure 7.5. Using the plot based on transformed values, normality can be confirmed. However, one or more statistical hypothesis tests can be carried out to check the normality of data.

Visual assessments of probability plots provide a preliminary indication of the normality of the sample data. However, several other tests such as Chi-square, Kolmogorov–Smirnov, Anderson–Darling, Lilliefor, Shapiro–Wilk, and D'Agostino–Pearson goodness-of-fit tests can be carried out to assess the normality of the sample. One major limitation of the Kolmogorov–Smirnov (KS) test, which is not often indicated in the literature, is that population parameters are required for the test as opposed to sample parameters. Also, the KS test is a supremum test.

Table 7.4 *List of parametric and non-parametric tests*

Type of statistical test	Test	Use
Parametric	Chi-square	Goodness-of-fit test
	Kolmogorov–Smirnov	Goodness-of-fit test for continuous distributions
	Anderson–Darling	Goodness-of-fit test for specific distributions
	Lilliefor	Goodness-of-fit test for normality
	t-test	One- and two-sample tests (requires normality assumption)
	F-test	Two-sample test for equality of variances (requires normality assumption)
	Jarque–Bera	Test for normality
Non-parametric	Wilcoxon rank sum	Test for equality of medians (independent samples)
	Wilcoxon signed rank	Test for zero median for symmetrical distributions
	Ansari–Bradley	Test for variances (of similar distributions), independent samples
	Runs (Wald–Wolfowitz)	Randomness of samples
	Kruskal–Wallis	Comparison of three or more unrelated samples
	Friedman	Comparison for related samples
	Mann–Whitney	Comparison of two unrelated samples

Table 7.5 *Annual precipitation depths (daily extremes)*

Precipitation depth (mm)		
106	26	122
90	69	81
105	109	112
122	108	69
75	135	160
88	81	81
124	84	91
87	141	71
123	124	173
115	79	79
81	112	165
109	66	178
85	58	91
77	91	99
199	208	130
85	130	137
53	71	
142	104	
113	99	
238	91	

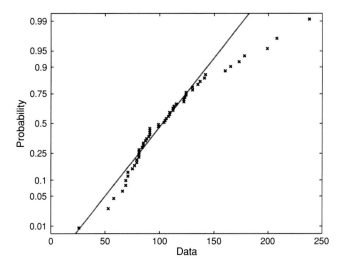

Figure 7.4 Normal probability plot of precipitation data.

7.9.2 Goodness-of-fit tests for other distributions

Two common goodness-of-fit tests used for evaluation of other distributions include the KS test and the Chi-square test. The KS test is recommended when the number of samples is less than 30. The Chi-square test is more sensitive to the number of bins (i.e., intervals) used. The probability-plot correlation coefficient (PPCC) approach (Dingman, 2008) involves calculation of the correlation coefficient between the sample value and values that would exist at the corresponding exceedence probabilities if the data were from a specific distribution. A distribution quantile function (Stedinger *et al.*, 1992) can be used to calculate values associated with specific exceedence probability for the proposed distribution. The hypothesis that the data are not from a specific distribution is not rejected if the correlation coefficient is closer

Several modified versions of the KS test are now available and these modified tests overcome this limitation. The most preferred tests are D'Agostino–Pearson, Anderson–Darling, and Lilliefor.

The Lilliefor test for transformed data provided in Table 7.6 confirms normality, while the same test carried out on original data provided in Table 7.5 returns an alternative hypothesis at 0.05 significance level.

Table 7.6 *Box–Cox transformed precipitation depths*

Precipitation depth (mm)		
9.24	5.18	9.73
8.69	7.84	8.34
9.21	9.34	9.43
9.73	9.30	7.84
8.10	10.10	10.74
8.61	8.34	8.34
9.79	8.46	8.72
8.57	10.26	7.92
9.76	9.79	11.05
9.52	8.26	8.26
8.34	9.43	10.86
9.34	7.70	11.16
8.50	7.31	8.72
8.18	8.72	9.01
11.61	11.79	9.96
8.50	9.96	10.16
7.05	7.92	
10.29	9.17	
9.46	9.01	
12.36	8.72	

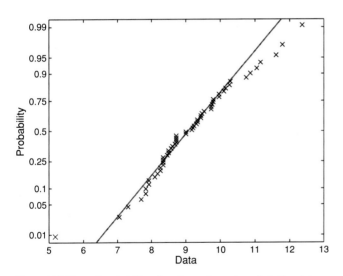

Figure 7.5 Normal probability plot of transformed precipitation data.

to 1. Critical values of the correlation coefficient for the PPCC approach as a function of sample size are available for rejecting or accepting the hypothesis that data are from a specific distribution.

7.9.3 Quantitative measures

The two indices defined by Equations 7.30 and 7.31 – mean absolute deviation index (MADI) and mean square deviation index (MSDI) proposed by Jain and Singh (1987) – can be used to measure the relative goodness-of-fits when several distributions are evaluated for the data. The MADI and MSDI are given by:

$$MADI = \frac{1}{N} \sum_{i=1}^{N} \left| \frac{x_i - z_i}{x_i} \right| \tag{7.30}$$

$$MSDI = \frac{1}{N} \sum_{i=1}^{N} \left(\frac{x_i - z_i}{x_i} \right)^2 \tag{7.31}$$

where x_i is the observed value (i.e., extreme precipitation) and z_i is the estimated value of extreme precipitation obtained for a specific exceedence probability. The empirical probability of exceedence is obtained from the Gringorten formula. The non-exceedence cumulative probability using the Gringorten formula is given by:

$$P(X < x_i) = \frac{i - 0.44}{n + 0.12}$$

The distribution providing the smallest values of indices can be selected as the best distribution characterizing the extreme precipitation data.

7.10 REGIONAL FREQUENCY ANALYSIS

In many situations, data available at a site (i.e., rain gage) may not be adequate for frequency analysis of extreme precipitation. Therefore, data from a number of rain gages in a region can be used for the analysis in a procedure referred to as regional or pooled frequency analysis. The analysis based on the "index storm" approach is beneficial for quantile estimations of extreme precipitation events with the help of data augmentation for data-scarce sites. Regional frequency analysis (Hosking and Wallis, 1997) requires delineation of a homogeneous region that is classified according to a homogeneous pooling group using a classification method and a homogeneity criterion. Once the homogeneous group is established, the augmented data are used for fitting a statistical distribution. Parameter estimation for a specific distribution can be carried out using the maximum likelihood or L-moments estimation method discussed previously in this chapter.

7.11 ILLUSTRATIVE EXAMPLES

7.11.1 Daily precipitation time series

Precipitation data at different temporal resolutions other than annual extremes of peaks over a specific threshold can be characterized by different probability distributions. Exponential and gamma distributions are generally found to be the best fits for daily precipitation depths and inter-event times (Adams and Howard, 1986; Adams and Papa, 2000). Daily precipitation data series are

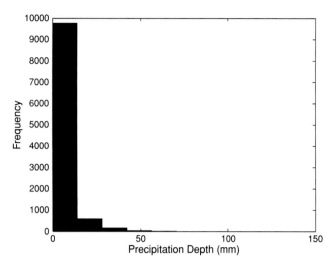

Figure 7.6 Histogram of daily precipitation time series at a rain gage.

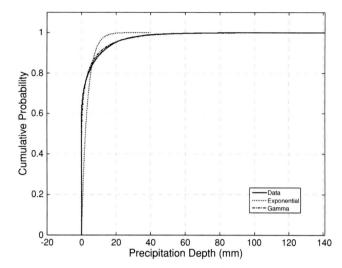

Figure 7.7 Cumulative density function plots of daily precipitation data using exponential and gamma distributions.

used to illustrate this point. Figures 7.6 and 7.7 show the histogram and cumulative density plots for daily precipitation data from a site in Lexington, Kentucky. The time series consists of 60% zero values.

It is evident from Figure 7.7 that the gamma distribution with two parameters provides the best fit for the data. Maximum likelihood estimates and method of moments are used for parameter estimations for exponential and gamma distributions. The likelihood function becomes unbound due to the presence of zero values in the daily precipitation data series. Probability distributions for non-zero precipitation data can be developed. Figure 7.8 shows the scaled exponential and gamma distributions superimposed over histograms. It can be seen from Figure 7.8 that the gamma distribution fits well for the non-zero daily precipitation data.

Wilks (2006) indicates that the Gaussian distribution is a poor choice for representing the positively skewed precipitation data, since too little probability is assigned to the largest precipitation amounts and non-negligible probability is assigned to impossible negative precipitation amounts. As can be seen from Figure 7.6, the distribution of daily precipitation data with a high percentage of zero values is positively skewed.

7.11.2   Annual extremes for different durations

Annual precipitation extreme values available at a rain gage in Florida for different durations are used to illustrate the fitting of several distributions using the MLE parameter estimation procedure. Details of the rainfall extremes are provided in Table 7.7.

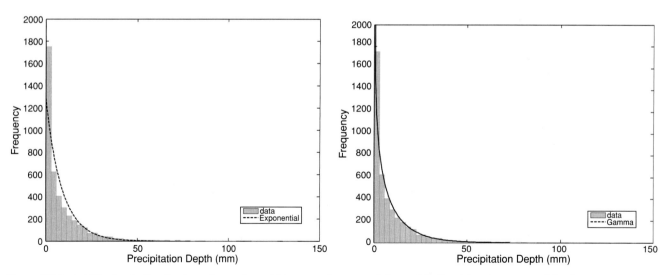

Figure 7.8 Probability density functions superimposed over histograms of non-zero daily precipitation data using exponential and gamma distributions.

Table 7.7 *Annual precipitation extreme (mm) at rain gage in Florida*

Duration (hours)								
1	2	6	12	24	48	72	96	120
148	148	151	151	208	213	223	262	262
42	52	56	57	66	81	103	103	104
54	80	97	135	137	194	223	241	252
34	52	97	137	150	158	198	201	208
64	99	154	161	161	161	161	161	161
41	58	129	157	187	189	193	193	200
38	53	93	113	154	198	234	234	255
74	100	198	221	226	227	231	231	243
37	50	93	107	124	164	215	231	242
40	62	94	107	130	145	146	165	184
53	67	68	76	91	91	91	99	108
84	84	105	108	116	131	137	178	191
151	151	151	151	151	151	158	182	185
48	58	65	74	100	129	129	129	132
52	63	65	65	85	92	103	104	116
90	93	142	174	181	181	193	200	227
36	39	48	71	71	85	89	100	106
74	96	109	135	243	274	277	356	360
195	210	229	239	252	253	253	253	282
56	62	87	102	104	104	104	104	106
52	53	53	53	56	61	108	108	108
36	44	88	112	144	147	147	147	147
72	75	90	107	109	119	152	182	186
107	107	107	120	174	191	191	191	191
107	109	120	126	126	126	126	126	150
67	117	119	119	125	173	216	246	247
59	69	82	86	97	129	150	150	167
43	55	82	83	86	120	131	141	174
45	60	70	91	143	166	166	172	172
71	87	106	106	107	176	187	197	199
56	67	79	122	174	186	188	189	189
42	61	78	78	78	87	87	87	88
45	59	66	87	93	105	106	108	111
60	62	65	65	84	105	112	114	137
45	46	62	65	67	105	119	120	124
81	91	100	100	107	117	123	131	131

Data source: National Climatic Data Center

Figure 7.9 shows the cumulative probabilities calculated using three plotting position methods. Figures 7.10 and 7.11 show the probability plots based on normal and Weibull plotting positions.

Time series plots of precipitation data for 2-hour duration are shown in Figures 7.12 and 7.13. Figure 7.12 provides information about the Sen slope and a hypothesis test based on the Mann–Kendall tau. The null hypothesis (H_0) suggests that the slope of the line is not significant. Figure 7.13 shows the plot with the linear trend removed from the data. Similarly, Figure 7.14 shows

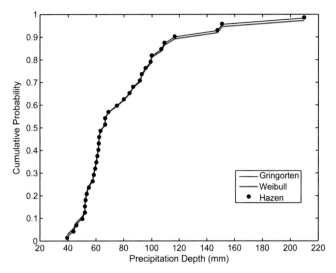

Figure 7.9 Cumulative probability plots using different plotting position formulae for annual extreme rainfall depths for duration of 2 hours.

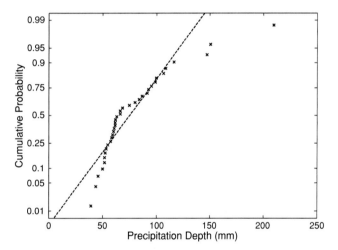

Figure 7.10 Normal probability plot of extreme precipitation data.

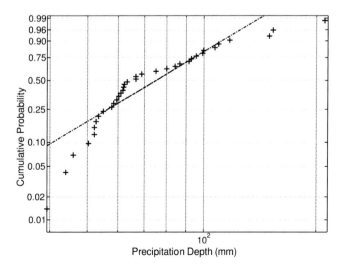

Figure 7.11 Weibull probability plot of extreme precipitation data.

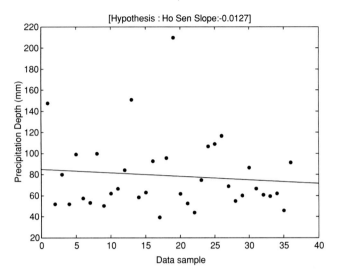

Figure 7.12 Time series plot of extreme daily precipitation data for 2-hour duration.

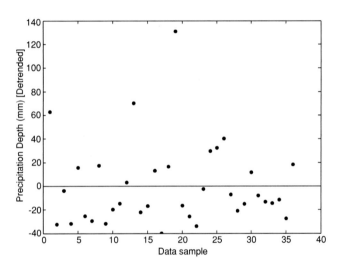

Figure 7.13 Time series plot of de-trended extreme daily precipitation data for 2-hour duration.

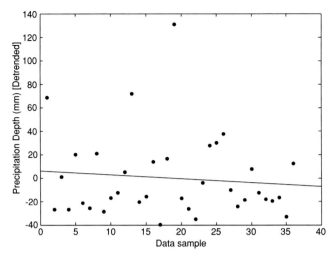

Figure 7.14 Time series plot of extreme daily precipitation data with mean removed for 2-hour duration.

the plot with the mean value removed from the data series. A decreasing trend is apparent in the plots shown in Figures 7.13 and 7.14.

Figure 7.15 shows the fitted and empirical statistical distributions. The GEV distribution was found to be appropriate for characterizing the distributions of extreme precipitation depths of all durations. This can be confirmed by visual evaluation of CDF plots based on data and fitted GEV parameters, and goodness-of-fit tests. Figure 7.16 shows the fitted and empirical cumulative density plots of extreme precipitation values using L-moments (Hosking *et al.*, 1985).

A recent survey of rainfall frequency estimation methods by Svensson and Jones (2010) indicated that GEV is the most used distribution for rainfall extrema among nine different countries (Canada, Sweden, France, Germany, USA, South Africa, New Zealand, Australia, and UK). Svensson and Jones (2010) indicate that in large parts of New Zealand the regional shape parameter in GEV is negative and there is a general increase in parameter value with the increase in duration. In some countries, the Gumbel distribution is used, partly because the estimation of a shape parameter from short records is not justified for individual sites. In contrast to it, L-moment estimators also compare favorably with those from MLE methods, particularly for small sample sizes (15–25 samples) in a GEV distribution (Hosking *et al.*, 1985). Also many researchers attribute the shorter records (around 15–25 years) for the negative shape parameters for GEV distributions. Conclusions based on a detailed study of the use of GEV distributions and parameters for extreme rainfall characterization (Teegavarapu and Goly, 2011) using 80 rain gages in Florida are given below.

- The GEV distribution is found to be the best distribution for characterizing extreme precipitation events for all durations based on several goodness-of-fit tests and visual comparison of cumulative density plots.
- The shape parameter of GEV distributions shows spatial and temporal variations. However, no specific pattern in space is identified.
- Evaluation of the shape parameter for two sets of data from two different sources suggests that the most positive shape parameter values occurred for the duration of 24 hours.
- Assessment of spatial variability of GEV distribution parameters suggests that no distinct patterns or clusters of negative or positive shape parameters are evident.

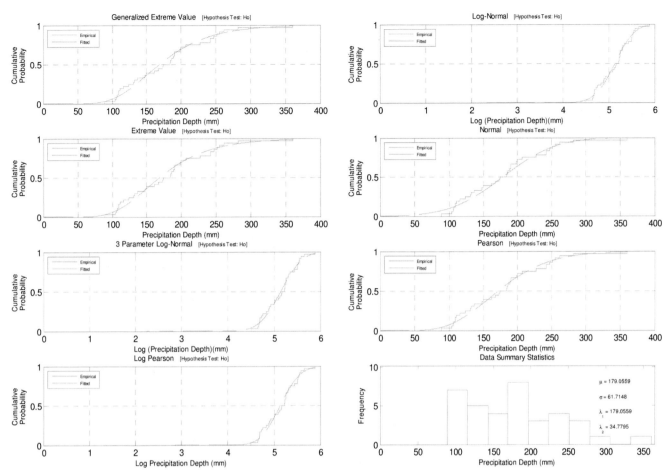

Figure 7.15 Fitted and empirical cumulative density plots of extreme precipitation values for duration of 96 hours.

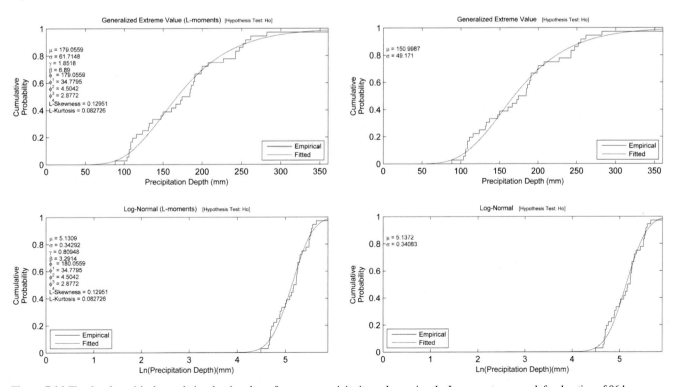

Figure 7.16 Fitted and empirical cumulative density plots of extreme precipitation values using the L-moments approach for duration of 96 hours.

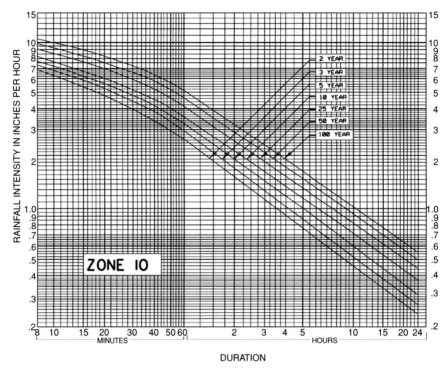

Figure 7.17 Typical IDF curve for a region in South Florida, USA (source: Florida Department of Transportation).

7.12 VALUE OF FITTING A PARAMETRIC FREQUENCY CURVE

Observed rainfall data are rarely used directly in hydrologic design for a variety of reasons. The main reasons include (1) lack of data at the region of interest and at the temporal resolution required; (2) existence of data gaps in chronological records; and (3) lack of homogeneous observations due to changes in the observation network over time and space. Design rainfall intensity defined for a given duration and frequency is commonly used in single event hydrologic simulation. Intensity–duration–frequency (IDF) curves are generally developed by local or regional water management agencies to aid the hydrologic and water resources design professionals to use these readily available curves. The use of design precipitation to estimate floods is particularly valuable in those situations where flood records are not available or not long enough at the site of interest, or they are not homogeneous due to changes of watershed characteristics such as urbanization and channelization (WMO, 2010). A typical IDF curve is shown in Figure 7.17 for a region in Florida.

Regional frequency analysis for spatially available rainfall observations are characterized using statistical distributions. The distributions are then used for development of IDF relationships. Generally, the distributions used are those that are asymmetric and skewed to the right (i.e., negatively skewed). The skewness is mainly due to the range of values observed for the hydrologic process variable. Positively skewed values for hydrologic variables

(i.e., precipitation) are bounded on the left by the lower boundary of the ranges. Distributions (e.g., Gaussian) that can characterize variables on the entire real-line of variables can be mathematically fit to precipitation data. Selection of statistical distributions is crucial for those variables that are physically constrained to be non-negative. Fitting a Gaussian distribution to precipitation data is practically possible, yet the probabilities obtained from the use of this distribution are not feasible. Parameters for distributions are generally estimated using MOM, maximum likelihood (ML), and L-moment methods. MOM is the easiest and the L-moment and ML methods are computationally intensive and require iterative approaches. For example, the gamma distribution often used in the fields of hydrology and operational climatology to characterize precipitation data is one of the statistical distributions that require MLE of parameters.

IDF curve development and characterization of regional rainfall extremes are areas of continuing research, as evidenced by many studies (Madsen *et al.*, 2002). IDF curves can be developed from historical annual maxima for different durations using the Weibull formula (Chin, 2006). Alternatively, IDF curves can be developed by assuming that the historical extremes can be best characterized by an extreme value Type I (Gumbel) distribution and using a frequency factor-based analysis.

IDF curves can be represented by one of the generalized functional forms (Dingman, 2008; WMO, 2010) provided below:

$$I = \frac{A}{(D+C)^B} \tag{7.32}$$

$$I = \frac{AT}{D^e + C} \qquad (7.33)$$

$$I = \frac{A}{(D - C)^e} \qquad (7.34)$$

$$I = \frac{A + B\log(T)}{(1 + D)^e} \qquad (7.35)$$

In the equations, I is the average intensity, D is the duration, T is the return period, and B, e, and C are constants that are region-specific and can be derived by optimal function approximation or regression. Site-specific data, after being characterized by a particular probability distribution, can be used to generate IDF curves and functional forms (Equations 7.32 and 7.33) can be developed. Generalized relationships similar to those in Equations 7.32 and 7.33 are developed for meteorologically homogeneous areas. Examples of these relationships can be found in studies by Bell (1969), Chen (1983), Kothyari and Garde (1992), Pagliara and Vitti (1993), and Alia (2000). Polynomial equations of order greater than three can be used to approximate the IDF curves. For example, the Australian Bureau of Meteorology uses a sixth-order polynomial for functional approximations of IDF curves.

The point rainfall depths for a given duration and return period obtained from IDF curves should be adjusted using areal reduction factors to obtain the average depth over a drainage area. The reduction factor is a function of area and duration (Dingman, 2008). The US National Weather Service (NWS) provide depth–area reduction curves applicable for the USA, which were approximated by an empirical functional form by Eagleson (1972) and described by Dingman (2008). The functional form is given by:

$$K(D, A) = 1 - \exp(-1.1D^{0.25}) + \exp(-1.1D^{0.25} - 0.01A) \qquad (7.36)$$

where $K(D, A)$ is a reduction factor that is a function of duration (D) in hours and area (A) in square miles. Equation 7.36 for reduction factors provides slightly higher values for durations less than 6 hours compared to the original curves provided by US NWS. When the probable maximum precipitation is to be applied to an area larger than about 25 km^2, it should be reduced (WMO, 2010).

7.12.1 Estimation of parameters for IDF functional forms

The parameters A, B, C, and e can be obtained by optimization formulation as given by Equations 7.37 and 7.38.

Minimize

$$\sqrt{\frac{1}{nd} \sum_{D=1}^{nd} \left(\hat{I}_D - I_D\right)^2} \qquad (7.37)$$

Table 7.8 *Duration and precipitation intensity values*

(*D*) Time (minutes)	(*I*) Intensity (mm/hr)
8	239
10	229
20	183
30	150
40	130
50	114
60	104
120	68
180	51

subject to:

$$I_D = \frac{A}{(D + C)^B} \quad \forall D \qquad (7.38)$$

The variable nd is the number of durations for which intensity values are available, and I_D and $\hat{I}_D$ are the estimated and observed values of intensities for a specific duration, D. A non-linear solver available under the Excel environment can be used for the solution of the optimization formulation 7.37–7.38. Conditions of non-zero constants can be included in the formulation using constraints. The above formulation is shown for one of the functional forms provided in Equations 7.32 –7.35.

7.12.2 Estimation example

Duration and precipitation intensity values are provided in Table 7.8 and these values are used in the optimization formulation to estimate the constants.

The formulation (7.37 and 7.38) is solved using the Excel solver and the values obtained for the constants A, B, and C are 3047.62, 18.64, and 0.77 respectively. The estimated values of intensities using these constants in the function form given by Equation 7.32 are provided in Table 7.9. It is evident from the values of intensities that the optimal values of the curves help to replicate the original data set.

7.13 EXTREME RAINFALL FREQUENCY ANALYSIS IN THE USA

This section provides a brief review of rainfall frequency analysis efforts in the USA. Use of observed rainfall data in continuous simulation models to derive peak discharges for hydrologic and hydraulic infrastructure elements is not common. Regional frequency analysis (Madsen *et al.*, 2002) for spatially available

Table 7.9 *Durations and estimated intensities*

(D) Time (minutes)	(I_t) Intensity (mm/hr)
8	240
10	227
20	180
30	151
40	131
50	116
60	104
120	67
180	51

rainfall observations is characterized using appropriate statistical distributions. The distributions are then used for development of IDF relationships. This section provides a brief review of frequency distributions applicable for the USA. Frequency distributions for precipitation data for the USA for different durations are discussed in different technical documents referred to as HYDRO-35 (Frederick *et al.*, 1977) and TP-40 (Hershfield, 1961) developed by the US NWS. The frequency distributions evaluated for use in HYDRO-35 are the Pearson Type III, the log-Pearson Type III, and the Gumbel Type I (referred to as the Fisher–Tippett Type I). The predictions from the 1-, 6-, and 24-hour durations were compared to determine the percentage of observations that equaled or exceeded calculated values. The analysis showed no significant differences in results obtained. The Gumbel Type I distribution was then chosen for use in HYDRO-35 since it was also the method that was adopted in the previous studies.

IDF maps were developed using both the 60-minute and 24-hour durations for TP-40. In the case of HYDRO-35, the 60-minute duration was used for the development of IDF maps. The studies (TP-40 and HYDRO-35) provided similar results

for precipitation data for a 2-year return period. For example, for the Florida peninsula, the intensities presented in TP-40 and HYDRO-35 were in general similar. However, the HYDRO-35 study considered the intensity of thunderstorms prevalent in a tropical climate and higher values for rainfall intensity for the interior portion of the peninsula. Bonnin *et al.* (2006) document the development of NOAA Atlas 14, Volume 2, which contains precipitation frequency estimates with associated confidence limits for the USA. The atlas also provides information related to temporal distributions and seasonality and is divided into volumes based on geographic sections of the country. The atlas is intended to be the official documentation of precipitation frequency estimates and associated information for the USA. The current status of development of a precipitation frequency atlas for the USA is shown in Figure 7.18. The schematic is accurate as of October 2011.

The atlas supersedes precipitation frequency estimates contained in Technical Paper No. 40 (Hershfield, 1961), Technical Paper No. 49 (Miller, 1964), and HYDRO-35 (Frederick *et al.*, 1977). Many studies point to the use and applicability of either extreme value Type I distribution (Gumbel distribution) or GEV distributions. However, the validity of a specific probability distribution needs to be tested before it can be applied in a particular situation.

7.14 UNCERTAINTY AND VARIABILITY IN RAINFALL FREQUENCY ANALYSIS

Variability is introduced into extreme rainfall frequency analysis due to several factors and these include (1) data measurements, (2) lack of reasonable length of data, (3) existence and length of missing data, (4) lack of stationarity, (5) non-homogeneous nature of storm events, (6) temporal window from which extreme values are considered, and finally (7) partial duration series (PDS). The following are a few initial preliminary data analysis steps that

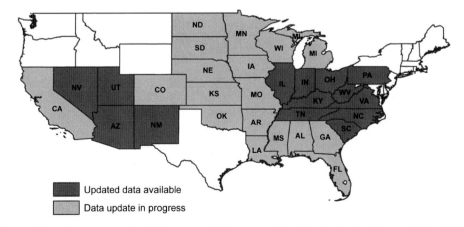

Figure 7.18 Status of precipitation frequency atlas for USA (source: Hydrometeorological Design Studies Center, NOAA, 2011).

can evaluate the influence of some of the factors on precipitation frequency analysis.

1. Check and confirm the validity of all the assumptions relevant to homogeneity, independence, randomness, and stationarity of time series.
2. Collect all the metadata related to the rain gage observations that are useful for assessing some of the assumptions indicated in the item 1.
3. Evaluate the completeness of precipitation data series. If data are missing for short periods in annual maximum duration series, data available before and after the missing periods can be used. However, the conditions related to what is an acceptable missing period need to be established.
4. Use statistical tests to assess existence or non-existence of linear trends in the annual duration series and variance of precipitation time series. Shift in mean and variance values can be tested by using data from two segments of the time series. In general each data segment should contain at least 30 years of data.
5. Assess the spatial variation of trends (increasing or decreasing) at different rain gage locations. This is an essential exercise if a regional frequency analysis needs to be conducted for development of iso-pluvial curves of rainfall intensity.

7.14.1 Data measurements

Precipitation data that are generally archived or compiled based on rain gage observations often are confined to a specific duration (e.g., day). However, when daily extremes are used for EVA, these values are based on fixed 24-hour period measurements as opposed to a moving window of 24 hours. Empirically derived correction factors (Storch and Zwiers, 1999) are often used to address this issue. The accumulations of precipitation based on fixed durations should be multiplied by these factors. Therefore, a recommendation is to deal with hourly data and to use a 24-hour moving temporal window to calculate annual extremes of 24-hour duration.

Adjustment factors for rainfall recorded at fixed intervals are recommended. Data collected at pre-defined fixed time intervals may not include the true maximum accumulations for those periods equal or close to the sampling period. Adjustment factors (Weiss, 1964; Young and McEnroe, 2003) are often based on average values of the ratios of true maximum and maximum value of accumulation in a fixed time interval. These factors are referred to as sampling adjustment factors.

7.14.2 Length of historical data

Guidelines on the requirements set for the minimum number of sample data (minimum number of annual extreme rainfall depths)

used for analysis are not available in literature. Data length is always a contentious issue in statistical analysis. A threshold number of years larger than ten is typically used. Justification for the use of data below this specific threshold is not advisable for any analysis. A minimum of 25 years of data is generally the accepted length of the record for statistical analysis of extreme values (Gupta, 2008). Methods are available (e.g., Sokolov *et al.*, 1976) to confirm the adequacy of data length. Comparison of summary statistics of observations from a gage with an incomplete data set can be compared with a nearby gage with complete data. The size of the sample (data length) should be large enough to warrant the estimation of parameters of the underlying probability distribution with the required reliability (Adams and Papa, 2000). Frequency analysis for precipitation extremes requires at least 25 years of data (Sevruk and Geiger, 1981) in humid regions. Sample record lengths have a substantial effect on the power of statistical tests (Yue *et al.*, 2002; Yue and Pilon, 2004) and therefore reasonable lengths exceeding 40 years are found to be satisfactory (Sevruk and Geiger, 1981; WMO, 2010). Availability of long-term precipitation data is critical for rainfall frequency analysis for determining statistically based rainfall estimates of reasonable reliability, especially for extreme rainfalls with high return periods, such as those greater than 100 years (WMO, 2010).

7.14.3 Missing data

Continuous precipitation data without any gaps are needed for frequency analysis, modeling, and design. Often precipitation data gaps of different length are unavoidable due to random and systematic errors. Incorrect recording and transcription of precipitation data creates gaps in the data to be filled and casts doubt on the reliability of data for statistical and trend analysis (Hosking and Wallis, 1998; Wallis *et al.*, 1991). Infilling methods may introduce significant biases when the gaps constitute more than 20% of the data. A recent study by Teegavarapu *et al.* (2010) indicated that bias introduced by infilling methods can alter the statistics of the data. Estimation of missing extreme precipitation data is difficult and should not be carried out by global interpolation methods that use all the available observation points in a region. In many instances, gages that are selected as single best estimators based on Euclidean distance or other statistical distances should be used for infilling extreme values. Schuenemeyer and Drew (2010) indicate that any statistical inference made from data with more than 15% of missing observations should be interpreted with caution.

Infilling of precipitation data at a site should focus on preservation of site-specific and regional statistics. Regional statistics are mainly site-to-site relationships, spatial autocorrelations, and variability across the region. Figure 7.19 shows the cumulative density plots based on precipitation data and filled precipitation data using interpolation methods. The missing data in this example are only 25% of the total data available. It is evident from

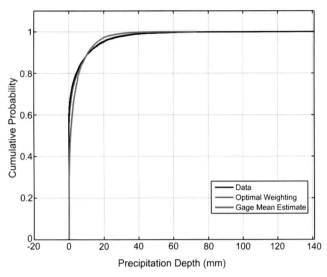

Figure 7.19 Cumulative density plots of precipitation data and filled data by using two interpolation methods.

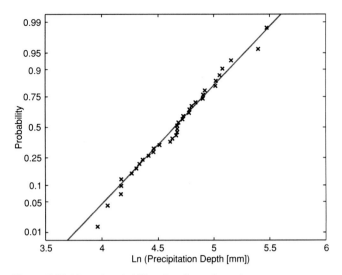

Figure 7.20 Normal probability plot of transformed extreme precipitation data for duration of 12 hours.

the Figure 7.19 that the data filled using a naïve method utilizing the mean of all the observations in the region do not preserve the site-specific statistics.

7.14.4 Stationarity

According to WMO (2010), stationarity means that, excluding random fluctuations, the data series is invariant with respect to time. Precipitation extremes are evaluated as block maxima or a single extreme value per year. Annual precipitation extremes time series should be checked for stationarity. This is an important element of the statistical analysis conducted after the initial phase of data collection. Trend analyses should be conducted for all the available annual extreme data for all durations and tests for statistically significant trends should be based on the Mann–Kendall or other tests. According to Ashkar (1996) two important forms of non-stationarity in a time series (e.g., streamflow time series) are jumps and trends. Another impact form is the existence of cycles that are associated with long-term climatic oscillations. Statistical tests for detecting stationarity include the Mann–Whitney test for jumps and the Wald–Wolfowitz test for trend (Bobee and Ashkar, 1991).

Betancourt (2009) argues that systems for management of water throughout the developed world have been designed and operated under the paradigm of hydrologic stationarity. The stationarity assumption suggests that hydrologic variables have time-invariant probability density functions whose properties can be estimated from the instrumental record (Betancourt, 2009). Given the magnitude and time lags of climate change associated with the buildup of greenhouse gases, stationarity may indeed be dead (Milly et al., 2008).

A viable successor to stationarity must encompass principles and methods for identifying non-stationary probabilistic models of relevant environmental variables and for using such models to optimize water systems (Betancourt, 2009; Milly et al., 2008). Non-stationary hydrologic variables can be handled stochastically to describe the temporal evolution of their means and variances, with estimates of uncertainty.

7.14.5 Detection of change in moments

Statistical analyses can be conducted to assess the changes in moments based on historical data. The data are split into two data sets (sample populations) and compared. Initially the sample populations are checked to see if they fit a specific probability distribution (e.g., normal or log-normal), and then the sample variances (S_1^2, S_2^2) are also checked to see if they are equal or unequal. Probability plots on normal or log-normal paper are generally used to visually check the hypothesized distributions. An example of a normal probability plot of log-transformed values of precipitation depths based on Table 7.5 is shown in Figure 7.20.

Annual precipitation totals can be characterized using normal and gamma distributions. Thompson (1999) indicates that annual precipitation tends to follow a normal distribution, particularly in humid climates, and existence of low values in arid and semi-arid climates can be characterized by positively skewed distributions. However, time dependence of these values may violate the assumption of independence due to persistence at inter-annual time scales. Droughts and wet years are examples of multi-year persistence (Thompson, 1999). Naoum and Tsanis (2003) report that annual rainfall amounts can be characterized using normal distributions for the island of Crete in Greece. When data seem to follow normal distributions, a few goodness-of-fit tests are

Table 7.10 *Statistical hypothesis tests used in assessment of changes in trend for normally distributed data*

Statistical tests	Purpose of the test	Null hypothesis (H_0)	Alternative hypothesis (H_a)
Kolmogorov–Smirnov Lilliefor Chi-square Anderson–Darling	Goodness-of-fit test	Data follow the hypothesized distribution (i.e., log-normal)	Data do not follow the hypothesized distribution
F-test	Equality of variances of sample data sets	$S_1^2 = S_2^2$	$S_1^2 \neq S_2^2$
T-tests (Equal variance – Student's two-sample t-test) (Unequal variance – Satterthwaite's two-sample t-test)	Tests for equality of means of sample data sets	$\bar{x}_1 = \bar{x}_2$	$\bar{x}_1 \neq \bar{x}_2$

applicable. One such test, the Kolmogorov–Smirnov test, can be used to ensure the validity of hypothesized data distributions. The Anderson–Darling test, which is sensitive to the tails of the distribution, can also be used. For data sets with equal variances, a comparison can be made using a t-test. For data sets with unequal variances, a comparison can be made using Satterthwaite's modified t-test (Satterthwaite, 1946; Mcbean and Rovers, 1998). Sample means ($\bar{x}_1, \bar{x}_2$) can then be calculated for the data sets and checked to see if they were statistically different at a specified significance level using hypothesis tests. A significance level of 0.05 can be used. F-tests are required to ascertain if the variances were equal or not equal. A t-test can then be used to make an inference about differences in sample mean values. The procedures for these tests are discussed in detail by Gilbert (1987) and Ott (1995). A summary of statistical hypothesis tests used in assessment of changes in trend for normally distributed data are provided in Table 7.10.

7.15 ASSESSMENT OF SAMPLE VARIANCES

The use of t-tests requires several conditions to be met before any inference can be made about sample means. The assumptions for the two-sample t-test are normality, independence of observations, and equal variances. Two different types of t-tests are used based on the knowledge of sample variances. The t-test is identified by Equations 7.39 and 7.40 when the sample variances based on two different sampling periods are equal.

$$t = \frac{|\mu_1 - \mu_2|}{\sqrt{\frac{n_1 + n_2}{n_1 n_2} \left(\frac{(n_1 - 1) S_1^2 + (n_2 - 1) S_2^2}{n_1 + n_2 - 2} \right)}} \quad (7.39)$$

The degrees of freedom (*df*) are defined by:

$$df = n_1 + n_2 - 2 \quad (7.40)$$

where n_1 is the number of samples in data set 1, n_2 is number of samples in data set 2. S_1^2 and S_2^2 are sample variances, and μ_1 and μ_2 are mean values of data sets 1 and 2 respectively. The t-test used for unequal sample variances is defined by Equations 7.41 and 7.42. The test is referred to as Satterthwaite's modified t-test (1946). The degrees of freedom in the case of this t-test are given in Equation 7.42.

$$t = \frac{|\mu_1 - \mu_2|}{\sqrt{\frac{S_1^2}{n_1} + \frac{S_2^2}{n_2}}} \quad (7.41)$$

$$df = \frac{\left[\frac{S_1^2}{n_1} + \frac{S_2^2}{n_2} \right]^2}{\frac{S_1^4}{n_1^2(n_1 - 1)} + \frac{S_2^4}{n_2^2(n_2 - 1)}} \quad (7.42)$$

7.16 NON-PARAMETRIC METHODS

Several non-parametric tests were listed previously (in Table 7.4) for use in statistical analysis of time series data. The following section discusses a technique that can be used to evaluate data characteristics with no assumptions made about the distribution of the data.

7.16.1 Kernel density estimation

Kernel density estimation (KDE) is a non-parametric technique to characterize data. In comparison to parametric estimators the non-parametric estimators have no fixed structure and depend on all the data points to reach an estimate.

The main reasons for using a KDE in place of a histogram are:

1. Histograms do not provide a smooth representation of the data.

2. The shape of the histograms depends on the end points of bins. The selection of the number of bins is subjective even though rules of thumb are available.

3. The visual representation of the data distribution also depends on the width of the bins. There are no clear guidelines for selection of the widths.

Recent studies by Shimazaki and Shinomoto (2007, 2010) provide approaches that provide optimum bin width in the case of histograms and bandwidth for kernel density functions. The use of optimum bin width eliminates the limitations associated with subjective bin selection. Kernel estimators center a kernel function (K) at each data point. Adoption of a smooth kernel function can overcome the limitations of the histograms. The contribution of data point x_i to the estimate at some point x depends on how far apart x_i and x are. The extent of this contribution is dependent upon the shape of the kernel function adopted and the width (bandwidth, h) assigned.

The estimated density at any point x is

$$f_h(x) = \frac{1}{nh} \sum_{i=1}^{n} K\left(\frac{x - x_i}{h}\right) \qquad (7.43)$$

A number of kernel functions are available for use with KDE. These are uniform, triangular, biweight, triweight, Epanechnikov, normal, and others.

7.16.2 Non-parametric test for independence: runs test

The runs test (Wald–Wolfowitz test) is a non-parametric test that is used to examine the randomness of samples. There is no parametric equivalent of this test. Nott (2006) used the runs test to confirm the evidence of non-randomness in a time series. The runs test can be used to decide if a data set is derived from a random process.

A run is generally defined as a series of increasing values or a series of decreasing values. The initial step in the runs test is to list the values in sequential order and count the number of runs. The number of increasing (or decreasing) values is the length of the run. In a random data set, the probability that the $(i + 1)$th value is larger or smaller than the ith value follows a binomial distribution, which forms the basis of the runs test (NIST, 2011). The first step in the runs test is to compute the sequential differences ($Y_i - -Y_{i-1}$). Positive values indicate an increasing value, whereas negative values indicate a decreasing value. In other terms, if $Y_i > Y_{i-1}$ a 1 (one) is assigned for an observation and a 0 (zero) otherwise. The series is then transformed to a series of 1s and 0s. To determine if the number of runs is the correct number for a series that is random, let n be the number of observations, n_1 be the number above the mean, n_2 be the number below the mean, and R be the observed number of runs. Then, using combinatorial methods, the probability $P(R)$ can be established and the mean

and variance of R can be derived (Cromwell *et al.*, 1994; Gibbons, 1997). When n is relatively large (>20) the distribution of R is approximately normal.

$$\overline{x_R} = \frac{2n_1 n_2}{n_1 + n_2} + 1 \qquad (7.44)$$

$$S_R = \sqrt{\frac{2n_1 n_2 (2n_1 n_2 - n_1 - n_2)}{(n_1 + n_2)^2 (n_1 + n_2 + 1)}} \qquad (7.45)$$

$$Z^* = \frac{R + h - \overline{x_R}}{S_R} \qquad (7.46)$$

The variable Z^* is the z-score for a normal approximation of the data, R (Corder and Foreman, 2009) is the number of runs, and h is the correction for continuity, ± 0.5, where

$$h = +0.5 \quad \text{if } R < \frac{2n_1 n_2}{(n_1 + n_2 + 1)} \qquad (7.47)$$

$$h = -0.5 \quad \text{if } R > \frac{2n_1 n_2}{(n_1 + n_2 + 1)} \qquad (7.48)$$

The null hypothesis is rejected if the calculated Z^* value is greater than the selected critical value obtained from the standard normal distribution table at 0.05 significance level. In other words, the $x(t)$ series is decided to be non-random.

7.17 HOMOGENEITY

Precipitation time series, like any other climatic time series, often exhibit spurious (non-climatic) jumps and/or gradual shifts due to changes in station location, environment (exposure), instrumentation, or observing practices (WMO, 2009a). Homogeneity means that all the elements of the data series originate from a single population (WMO, 2010). Observation stations (i.e., rain gages) are moved from one location to another and these locational changes may lead to discontinuities in extremes, trends, and observations influenced by local weather and other regional climatic influences. All these factors affect the homogeneous nature of the long-term time series of precipitation data and bias the studies of the extremes. Guidelines on analysis of extreme events developed by WMO (2009b) provide real-life examples of inhomogeneities, and stress the need to have complete station history metadata for resolving these issues. If observations at a station or a set of stations are suspicious, carefully identified reference stations can be used for evaluation. Mass curves can be developed using the observations at the station suspected of problems and the reference station.

Graphical methods of detection of inhomogeneity include moving average plots using smoothing methods (McCuen, 2005). It is also important to collect causal information in the process of analysis to evaluate the reasons for inhomogeneity. Often metadata (data about data) available about observations and information about the stations is extremely helpful to decipher any inhomo-geneity in the data. It is important to associate different rainfall-producing mechanisms (slow-moving frontal systems, hurricane events, and summer convective storms) to rainfall depths in spe-cific years for specific storm durations. Spatial summary statistics for all stations should be used to assess the regional or global vari-ability of rainfall in a region. This analysis will help to establish or confirm if the storm events produced by meteorological processes are similar in nature. This will also strictly satisfy the homogene-ity requirement of statistical analysis of extreme events. Miller (1972b) points out that in the case of extreme precipitation, inho-mogeneity tends to be difficult to decipher and inhomogeneity in yearly precipitation totals is easier to detect.

The Mann–Whitney U statistic is commonly used to decide whether observations of a hydrologic variable are from the same population. In order to apply the Mann–Whitney test, the raw data sequences with n elements should be divided into two sam-ple groups with n_1 and n_2 elements. The raw data set is then ranked from lowest to highest, including tied rank values where appropri-ate. The equations relating to the Mann–Whitney U statistic are as follows (Bobee and Ashkar, 1991; Corder and Foreman, 2009):

$$U_i = n_1 n_2 + \frac{n_i(n_i + 1)}{2} \sum R_i \qquad (7.49)$$

$$\bar{x}_u = \frac{n_1 n_2}{2} \qquad (7.50)$$

$$S_u = n_1 n_2 + \sqrt{\frac{n_1 n_2(n_1 + n_2 + 1)}{12}} \qquad (7.51)$$

$$Z^* = \frac{U_i - \bar{x}_u}{S_u} \qquad (7.52)$$

The variable S_u is the standard deviation, R_i is the rank from the sample of interest, and $\bar{x}_u$ is the mean. The variable Z^* is the z-score for a normal approximation of the data. The null hypothe-sis, which has homogeneity, for the observations is rejected if the calculated Z^* statistic is greater than the selected critical value at 0.05 significance level obtained from the standard normal distribution table.

7.18 PARTIAL DURATION SERIES

Annual extreme value series are often referred to as block maxima values and are generally used for frequency analysis. In the case of an annual extreme series, the maximum value can be obtained from a calendar year or a water year. However, extreme values over a pre-specified threshold are also used for analysis. These series are referred to as partial duration series (PDS). One limitation of an annual series is that it may omit a value that is lower than the maximum value in a specific year but is higher than all other values in the series. The PDS overcomes this limitation by selecting all extreme values above a threshold. In some instances, the number of extreme values adopted from a PDS is the same as the number

of years of data available. Analysis of annual maximum series (AMS) provides estimates of the average period between years when a particular value is exceeded. This is generally referred to as average recurrence interval (ARI). The PDS provides the aver-age period between cases of a specific magnitude. The informa-tion obtained from the PDS is the annual exceedence probability (AEP). Laurenson (1987) provided the definitions for ARI and AEP. In the context of precipitation analysis, ARI refers to the average, or expected, value of the periods between exceedences of a given rainfall total accumulated over a given duration; AEP refers to the probability that a given rainfall total accumulated over a given duration will be exceeded in any one year.

The POT values are first assessed for independence of observa-tions. Methods for identifying independent extreme peak stream-flows are discussed in Chapter 4. Several flood frequency analysis studies (Stedinger, 2000) using PDS in the past have identified the advantages of using such series. One of the difficulties associated with the use of PDS for rainfall frequency analysis is the subjective selection of the threshold.

7.19 INCORPORATING CLIMATE VARIABILITY AND CLIMATE CHANGE INTO RAINFALL FREQUENCY ANALYSIS

Internal modes of climate variability defined as teleconnections often influence precipitation and temperature patterns across the globe. Teleconnections such as ENSO and AMO cycles have influ-ence on rainfall patterns in several parts of the continental USA and other regions in the world. Higher precipitation totals (espe-cially for long temporal durations greater than 24 hours) have been attributed to specific phases of these oscillations. Future rainfall frequency analysis methods should consider these tele-connections to understand regional and temporal variability of rainfall extremes.

Long-term, high-quality, and reliable climate records with a daily (or higher) time resolution are required for assessing changes in extremes. Guidelines for development of data sets of these characteristics are discussed by Brunet et al. (2008). A recently published document by the World Meteorological Organization (WMO, 2009) provides discussion about (1) development of data sets for the analysis of extremes; (2) use of descriptive indices and extreme-value theory concepts to evaluate extreme events; (3) assessment of trends and other statistical approaches for eval-uating changes in extremes; and (4) understanding of observed changes and modeling projected changes in extremes.

7.20 FUTURE DATA SOURCES

Hydrologic design based on design storms will benefit from the emerging sources of rainfall measurement such as radar and satellite. The use of radar data for estimation of PMP is often dismissed due to the shorter length of data compared to long-term data that are available from rain gages (Collier and Hardaker, 2007). However, the utility of the radar data is also being questioned for use in the frequency analysis of extremely rare events due to lack of a reasonable data length of 10 years. Collier and Hardaker (1996) have developed methods to estimate maximum rainfall totals to obtain realistic estimates of PMP. Reliable radar data are now available in many parts of the world at a temporal resolution that is adequate for frequency analysis for hydrologic design. Often the fine temporal resolution radar-based precipitation data are useful in analyzing short duration storms. Satellite data temporal resolution, however coarse it is, still gives the only reasonably accurate rainfall data available for analysis in many regions of the world with extremely low rain gage density and lack of radar-based rainfall.

7.21 STATISTICAL TESTS AND TREND ANALYSIS: EXAMPLE OF EXTREME PRECIPITATION ANALYSIS IN SOUTH FLORIDA

Investigation of climate records to identify changes, trends, or patterns in climate requires data values to be without any gaps and at a fixed time step. Estimation of missing values is often required to obtain complete data sets and to perform statistical analyses. This section discusses an example of extreme precipitation data analysis adopted from a study by Teegavarapu et al. (2011). The following tasks are carried out for evaluation of trends: (1) extraction of wet day frequency (above 25.4 mm and 50.8 mm threshold) data in each year (water year) and month; (2) trend analyses of annual total precipitation, wet season (June–October) total precipitation, wet days (>25.4 mm and 50.8 mm threshold), maximum precipitation intensity (1-day or 5-day in any water year); and (3) precipitation characteristics and AMO comparison.

7.21.1 Statistical analysis of wet day frequency data

The significance of temporal trends present in hydroclimatic observations is evaluated using the Mann–Kendall (MK) statistic (Hirsch and Slack, 1984; Lettenmaier et al., 1994; Yue et al., 2002). The advantage of this statistic is that it tests for consistency in the direction of change and is unbiased by the presence of outliers. The null hypothesis, H_0, is that the variables are independent and randomly distributed. The test statistic is:

$$S = \sum_{i=1}^{n-1} \sum_{j=j+1}^{n} \text{sgn}(X_j - X_i) \quad (7.53)$$

where sgn(X) can be $+1$, 0, or -1 depending on whether X is greater than, equal to, or less than zero, respectively. In the absence of ties, the variance of S is calculated as:

$$\text{Var}[S] = \frac{n(n-1)(2n+5)}{18} \quad (7.54)$$

The test statistic is computed as:

$$Z = \frac{S - \text{sgn}(S)}{\sqrt{\text{Var}[S]}} \quad (7.55)$$

It is assumed that statistically significant trends are present if $|Z| > Z_{1-\alpha}$, at a selected significance level, α. In this study, trends were tested for significance at $\alpha = 0.10$ (90% confidence levels). The value of statistic $Z_{1-\alpha}$ can be found in standard normal distribution statistical tables. Tests based on the following two methods can be used to evaluate statistical significance of trends:

1. Mann–Kendall test with data pre-whitening (PW);
2. Mann–Kendall test with trend-free pre-whitening (TFPW).

7.21.2 Mann–Kendall test with data pre-whitening

One of the problems in detecting trends in hydroclimate data is the presence of serial correlation (i.e., persistence). Zhang et al. (2001) suggested a PW procedure to eliminate the effects of the serial correlation on the statistical testing for the MK test. In this study, the statistical significance of the trends in time series of hydroclimate data is examined by the MK test with PW using the following procedure (Zhang et al., 2001):

1. Compute lag-1 serial correlation coefficient (r_1).
2. If $r_1 < 0.1$ then the MK test is applied to the original time series.
3. Otherwise, the MK test is applied on the PW series

$$(x_2 - r_1 x_1, x_3 - r_1 x_2, \ldots, x_n - r_1 x_{n-1})$$

7.21.3 Mann–Kendall test with trend-free pre-whitening

Yue et al. (2002) found that the PW approach affects the magnitude of slope present in the original data series and reduces the significance of trends present in the original data. They proposed a TFPW approach to address this problem. The steps to implement MK-TFPW to detect significance of trends are given below:

1. Estimate Sen's slope (SS) of the data series.
2. De-trend the data using following equation:

$$X'_t = X_t - SS(t) \quad (7.56)$$

Compute the lag-1 serial correlation coefficient (r_1) for the de-trended data series.

1. If the de-trended data series does not show serial correlation, then the MK test is applied to the original time series.

2. Otherwise, the MK test is applied on the TFPW data series (Y_t) computed as follows:

$$Y_t' = X_t' - (r_1 X_{t-1}) \qquad (7.57)$$

$$Y_t = Y_t' - SS(t) \qquad (7.58)$$

7.21.4 Sen's slope estimator

The magnitude of trends detected by the least squares regression method is affected by the extreme values. To address this problem, a non-parametric method was proposed by Hirsch et al. (1982, 1991) to detect and estimate the magnitude of temporal trends. This method, referred to as Sen's slope (SS) estimator, computes the slope between all data pairs and estimates the slope as the median value among all possible slope values. Upper and lower confidence intervals for SS were also computed for $\alpha = 0.10$ (90% confidence level).

7.22 DIFFERENT TESTS: MOVING WINDOW APPROACHES

Temporal windows can be defined to assess trends in extreme precipitation data. The windows selected may be of constant length or varying length. The former is easier based on fixed chronological data sets.

7.23 IMPLICATIONS OF INFILLED DATA

This section provides an example of long-term precipitation data used for assessment of trends. The data set (Teegavarapu et al., 2012) used in this example provides a unique opportunity to assess the implications of infilled data on the statistical analysis. Improvised deterministic and stochastic interpolation methods to fill the gaps in daily precipitation records at several stations (Table 7.11) are investigated in a previous study. Evaluated methods include four improvised methods as well as three benchmark methods, with implementation variations for all methods. Three benchmark methods are selected to provide a basis against which the evaluated methods' performance can be quantified and compared. Evaluated methods include variations of the inverse distance weighting method (IDWM), correlation coefficient weighting method (CCWM), linear weight optimization method (LWOM), and artificial neural networks (ANNs). Benchmark methods are variations of the single best estimator (SBE) method and spatial and climatological mean estimators (SME and CME). In addition, two classification methods were evaluated in combination with the estimation methods in an attempt to separate zero and

Table 7.11 *List of NOAA stations in Florida*

ID	Station name	Latitude	Longitude
1	Arcadia	27.2181	−81.8739
2	Archbold Biological Stn	27.1819	−81.3511
3	Avon Park	27.5939	−81.5250
4	Bartow	27.8989	−81.8431
5	Belle Glade	26.6931	−80.6711
6	Bradenton	27.4469	−82.5011
7	Brooksville	28.6161	−82.3661
8	Bushnell NWS	28.6619	−82.0831
9	Cedar Key WSW	29.1331	−83.0500
10	Clermont	28.4550	−81.7231
11	Crescent City	29.4169	−81.5131
12	Desoto City NWS	27.3700	−81.5139
13	Everglades	25.8489	−81.3900
14	Fellsmere 7 SSW	27.6831	−80.6500
15	Fort Lauderdale	26.1019	−80.2011
16	Fort Myers Page Field	26.5850	−81.8611
17	Fort Pierce	27.4619	−80.3539
18	Gainesville U. of Florida	29.6500	−82.3500
19	Hialeah	25.8181	−80.2861
20	High Springs	29.8289	−82.5969
21	Hillsborough River St Pk	28.1431	−82.2269
22	Homestead Exp Stn	25.5000	−80.5000
23	Hypoluxo	26.5500	−80.0500
24	Immokalee	26.4611	−81.4369
25	Inverness	28.8031	−82.3131
26	Isleworth	28.4831	−81.5331
27	Key West Wb City	28.4819	−81.5261
28	Kissimmee	28.2761	−81.4239
29	Kissimmee2	28.1701	−81.2459
30	La Belle	26.7431	−81.4319
31	Lake Alfred	28.1039	−81.7139
32	Melbourne WSO	28.0669	−80.6161
33	Miami Beach	25.7800	−80.1300
34	Miami WSO City	25.7139	−80.2769
35	Moore Haven	26.8400	−81.0869
36	Mountain Lake	27.9350	−81.5931
37	Myakka River St Pk NWS	27.2419	−82.3161
38	Naples	26.1689	−81.7161
39	Ocala	29.0800	−82.0781
40	Okeechobee	27.2200	−80.8000
41	Plant City	28.0239	−82.1419
42	Pompano Beach	26.2331	−80.1411
43	St Leo	28.3381	−82.2700
44	St Petersburg	27.7831	−82.6331
45	Stuart 1 S	27.2000	−80.1639
46	Tamiami Trail 40 Mi Bend	25.7611	−80.8239
47	Tampa Intl Airport	27.9611	−82.5400
48	Tarpon Springs	28.1500	−82.7500
49	Tavernier	25.0069	−80.5211
50	Venice NWS	27.1011	−82.4361
51	Wauchula	27.5481	−81.7989
52	West Palm Beach Int Ap	26.6850	−80.0989
53	Winter Haven	28.0150	−81.7331

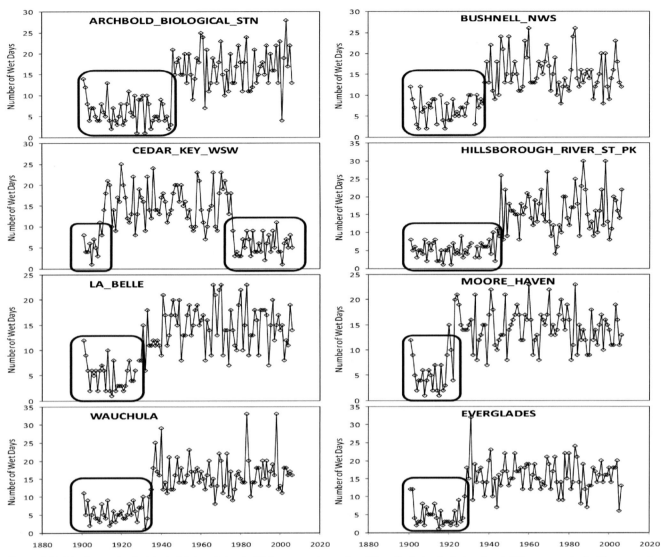

Figure 7.21 Bias in wet day (>1 inch (25.4 mm) threshold) frequency due to data filling. Periods of missing data infilling are marked with rectangular boxes (Teegavarapu *et al.*, 2010).

non-zero precipitation estimates. The evaluated classifiers are support vector machine (SVM) and variations of a single best classifier (SBC). The LWOM was finally adopted for infilling of missing precipitation data. Figure 7.21 shows the bias in wet day (>25.4 mm threshold) frequency due to data filling. Periods of missing data infilling are marked with rectangular boxes.

Figures 7.22 and 7.23 show the cumulative density plots from two stations: Moore Haven and Hillsborough.

The inclusion of zero values of precipitation along with positive values provided CDF plots as shown in Figure 7.24. Figure 7.25 shows the cumulative density plots of precipitation data from a station with the least amount of missing data. The series with 1.75% of missing data and the filled series have the same empirical cumulative density plots, suggesting that the extent of missing data is so small that it does not affect the site statistics.

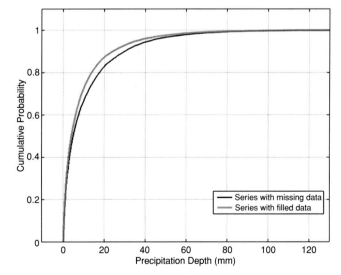

Figure 7.22 Cumulative density plots of daily non-zero precipitation data from Moore Haven station.

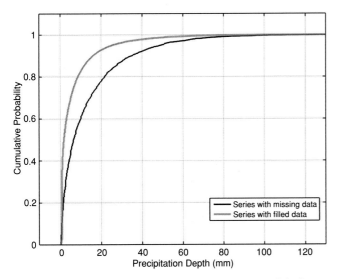

Figure 7.23 Cumulative density plots of daily non-zero precipitation data from Hillsborough station.

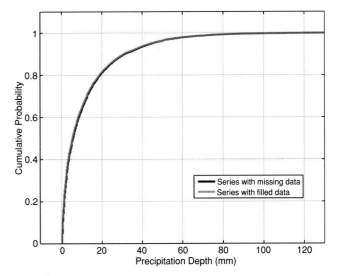

Figure 7.25 Cumulative density plots of daily non-zero precipitation data from Bartow station.

Only trends towards increase in frequency of wet days were found statistically significant, at both 1- and 2-inch thresholds. However, because the majority of stations do not show statistically significant trends, and many stations have trends towards decrease in frequency of wet days, it seems that there is no specific change in frequency of wet days.

7.23.2 Trends in intensity of extreme events

To investigate the changes in magnitude of extreme rainfall, averages of five extreme daily rainfall amounts for each water year at each station are calculated. Table 7.14 presents a statistical analysis of the magnitude of extreme rainfall. MK-TFPW analyses show that the intensity of maximum precipitation days has decreased at 11 stations and increased at 5 stations. However, only two stations (Moore Haven and Ocala) show a statistically significant decrease in intensity of extreme events.

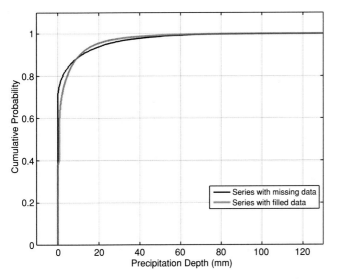

Figure 7.24 Cumulative density plots of daily precipitation data from Hillsborough station.

7.23.1 Trends in frequency of wet days

Tables 7.12 and 7.13 present statistical analysis of frequency of wet days at 25.4 mm (1 inch) and 50.8 mm (2 inches) threshold. Statistics based on MK-TFPW analysis show that the frequency of wet days (>1 inch) has increased at eight stations and decreased at eight stations; however, only two stations (Fort Myers and Myakka River St Park) show a statistically significant increase in frequency of wet days (>1 inch). Similarly, the frequency of wet days (>2 inches) has increased at six stations and decreased at ten stations; however, trends towards increase in frequency of wet days (>2 inches) at only three stations (Clermont, Fort Myers, and Myakka River St Park) were found to be statistically significant.

7.24 DESCRIPTIVE INDICES FOR PRECIPITATION EXTREMES

Descriptive indices for precipitation extremes were developed by the Expert Team on Climate Change Detection and Indices (ETCCDI) and are described by WMO (2009a). These indices describe specific characteristics of extremes, including frequency, amplitude, and persistence (WMO, 2009a). A total of 11 indices for precipitation developed by ETCCDI are provided in Table 7.15. The indices can be used for evaluation of long-term precipitation extremes. The descriptive indices can be used for assessment of trends by defining two separate windows.

Table 7.12 *Statistical analysis results of wet days frequency (>25.4 mm) for the period 1945–2006 using filled data*

Rain gage	Sen slope	LL	UL	Statistical significance	MKPW_S	MKPW_Z	Statistical significance	MK_DPW_S	MK_DPWW_Z	Statistical significance
Avon Park	0.019	− 0.022	0.080	No	126	0.7593	No	126	0.7593	No
Bushnell NWS	− 0.022	− 0.083	0.000	Yes	− 160	− 0.9894	No	− 194	− 1.2010	No
Clermont	0.000	− 0.051	0.031	No	− 73	− 0.4373	No	− 73	− 0.4373	No
Fort Myers Page Field	0.071	0.000	0.143	Yes	187	1.1575	No	309	1.8708	Yes
Moore Haven	− 0.023	− 0.067	0.000	Yes	− 179	− 1.0812	No	− 179	− 1.0812	No
Mountain Lake	0.000	− 0.053	0.063	No	− 3	− 0.0124	No	− 3	− 0.0124	No
Myakka River St Pk NWS	0.083	0.000	0.161	Yes	343	2.0773	Yes	343	2.0773	Yes
Ocala	− 0.042	− 0.095	0.000	Yes	− 217	− 1.3120	No	− 217	− 1.3120	No
Plant City	0.000	− 0.043	0.073	No	77	0.4616	No	77	0.4616	No
St Petersburg	− 0.054	− 0.135	0.000	Yes	− 212	− 1.2816	No	− 212	− 1.2816	No
Tampa Intl Airport	− 0.048	− 0.125	0.000	Yes	− 260	− 1.6117	No	− 226	− 1.3667	No
Tarpon Springs	0.000	− 0.042	0.075	No	70	0.4191	No	70	0.4191	No
Wauchula	0.000	− 0.023	0.059	No	99	0.5953	No	99	0.5953	No
Naples	0.000	− 0.045	0.071	No	48	0.2855	No	48	0.2855	No
Tamiami Trail 40 Mi Bend	0.000	− 0.063	0.043	No	22	0.1307	No	22	0.1307	No
West Palm Beach Int Ap	0.000	− 0.060	0.056	No	− 8	− 0.0425	No	− 8	− 0.0425	No

Table 7.13 *Statistical analysis results of wet days frequency (>50.8 mm) for the period 1945–2006 using filled data*

Rain gage	Sen slope	LL	UL	Statistical significance	MKPW_S	MKPW_Z	Statistical significance	MK_DPW_S	MK_DPWW_Z	Statistical significance
Avon Park	0.000	− 0.043	0.000	Yes	− 199	− 1.2027	No	− 199	− 1.2027	No
Bushnell NWS	0.000	− 0.032	0.000	Yes	− 162	− 1.0019	No	− 162	− 1.0019	No
Clermont	0.026	0.000	0.059	Yes	293	1.7736	Yes	293	1.7736	Yes
Fort Myers Page Field	0.028	0.000	0.057	Yes	205	1.2695	No	274	1.6989	Yes
Moore Haven	0.000	− 0.050	0.000	Yes	− 256	− 1.5489	No	− 256	− 1.5489	No
Mountain Lake	0.000	− 0.023	0.000	Yes	− 29	− 0.1742	No	− 29	− 0.1742	No
Myakka River St Pk NWS	0.048	0.000	0.083	Yes	397	2.4053	Yes	397	2.4053	Yes
Ocala	0.000	− 0.036	0.000	Yes	− 155	− 0.9354	No	− 155	− 0.9354	No
Plant City	0.000	− 0.036	0.000	Yes	− 136	− 0.8200	No	− 136	− 0.8200	No
St Petersburg	− 0.020	− 0.057	0.000	Yes	− 213	− 1.2877	No	− 213	− 1.2877	No
Tampa Intl Airport	− 0.020	− 0.053	0.000	Yes	− 239	− 1.4456	No	− 239	− 1.4456	No
Tarpon Springs	0.000	− 0.036	0.000	Yes	− 124	− 0.7471	No	− 124	− 0.7471	No
Wauchula	0.000	0.000	0.028	Yes	97	0.5831	No	97	0.5831	No
Naples	0.000	0.000	0.040	Yes	174	1.0508	No	174	1.0508	No
Tamiami Trail 40 Mi Bend	0.000	0.000	0.027	Yes	40	0.2369	No	40	0.2369	No
West Palm Beach Int Ap	0.000	− 0.041	0.000	Yes	− 120	− 0.7228	No	− 120	− 0.7228	No

Table 7.14 *Trends in average of five highest annual rainfall events for the period 1945–2006 using filled data*

Rain gage	Sen slope	LL	UL	Statistical significance	MKPW_S	MKPW_Z	Statistical significance	MK_DPW_S	MK_DPWW_Z	Statistical significance
Avon Park	−0.008	−0.018	0.000	No	−267	−1.616	No	−267	−1.616	No
Bushnell NWS	−0.004	−0.010	0.003	No	−112	−0.691	No	−149	−0.899	No
Clermont	0.003	−0.004	0.010	No	135	0.814	No	135	0.814	No
Fort Myers Page Field	0.004	−0.003	0.012	No	176	1.063	No	176	1.063	No
Moore Haven	−0.007	−0.013	0.000	Yes	−287	−1.737	Yes	−287	−1.737	Yes
Mountain Lake	−0.007	−0.014	0.000	Yes	−78	−0.479	No	−182	−1.126	No
Myakka River St Pk NWS	0.004	−0.004	0.011	No	131	0.790	No	131	0.790	No
Ocala	−0.007	−0.013	−0.001	Yes	−324	−1.962	Yes	−324	−1.962	Yes
Plant City	−0.003	−0.010	0.003	No	−142	−0.856	No	−142	−0.856	No
St Petersburg	−0.004	−0.013	0.004	No	−150	−0.905	No	−150	−0.905	No
Tampa Intl Airport	−0.005	−0.012	0.002	No	−188	−1.136	No	−188	−1.136	No
Tarpon Springs	−0.005	−0.013	0.002	No	−175	−1.057	No	−175	−1.057	No
Wauchula	−0.007	−0.015	0.001	No	−232	−1.403	No	−232	−1.403	No
Naples	0.000	−0.007	0.007	No	6	0.030	No	6	0.030	No
Tamiami Trail 40 Mi Bend	0.000	−0.007	0.007	No	−60	−0.367	No	−60	−0.367	No
West Palm Beach Int Ap	0.005	−0.004	0.015	No	152	0.917	No	152	0.917	No

Table 7.15 *Descriptive indices for precipitation extremes*

Index	Description
RX1day	Maximum 1-day precipitation
RX5day	Maximum 5-day precipitation
SDII	Simple daily intensity index
R10mm	Count of precipitation days with RR greater than 10 mm
R20mm	Count of precipitation days with RR greater than 20 mm
Rnnmm	Count of days with RR greater than a threshold value
CDD	Consecutive dry days (RR < 1 mm)
CWD	Consecutive wet days (RR ≥ 1 mm)
R95pTOT	Total precipitation due to wet days (>95th percentile)
R99pTOT	Total precipitation due to extremely wet days (>99th percentile)
PRCPTOT	Total precipitation in wet days (>1 mm)

RR: Recorded daily rainfall

7.24.1 Climate normals and indices

According to WMO technical regulations, standard normals are averages of climatological data computed for the following consecutive periods of 30 years: January 1, 1901 to December 31, 1930; January 1, 1931 to December 31, 1960, and so forth. Climate normals are required for calculation of a few of the indices provided in Table 7.15. These climatological baseline periods are non-overlapping. The current WMO normal period is 1961–90 (Lu, 2006).

The following definitions for the indices are adopted from WMO (2009a).

- RX1day

  $RX1day_j$ is the maximum 1-day precipitation (i.e., highest precipitation amount in 1-day period). If RR_{ij} is the daily precipitation amount on day i in period j, the maximum 1-day value for period j is given by the following equation:

$$RX1day_j = \max\{RR_{ij}\} \qquad (7.59)$$

- RX5day

  $RX5day_j$ is the maximum 5-day precipitation (i.e., highest precipitation amount in 5-day period). If RR_{kj} is the precipitation amount for the 5-day interval k in period j, where k is defined by the last day, the maximum of 5-day values for period j is given by the following equation:

$$RX5day_j = \max\left\{\sum_{n=1}^{k} RR_{nj}\right\} \qquad (7.60)$$

- SDII

  SDII is the simple daily intensity index (i.e., mean precipitation amount on a wet day. Let RR_{wj} be the daily precipitation amount on wet day w (RR $\geq$ 1mm) in period j. If w represents the number of wet days in j, then the simple precipitation intensity index is given by the following equation:

$$\text{SDII}_j = \frac{1}{w} \left\{ \sum_{n=1}^{w} RR_{wj} \right\} \quad (7.61)$$

- R10mm

  R10mm is the number of heavy precipitation days (i.e., count of days where RR (daily precipitation amount) $\geq$ 10 mm). Let RR_{ij} be the daily precipitation amount on day i in period j. The index R10mm is given by the following equation:

$$\text{R10mm}_j = \sum_{i=1}^{nT} RC_i$$
$$\left\{ RC_i = 1, \text{ if } RR_{ij} > 10 \text{ mm}; \text{ else } RC_i = 0 \right\}$$
$$(7.62)$$

- R20mm

  R20mm is the number of heavy precipitation days (i.e., count of days where RR (daily precipitation amount) $\geq$ 20 mm). Let RR_{ij} be the daily precipitation amount on day i in period j. The index R20mm is given by the following equation:

$$\text{R20mm}_j = \sum_{i=1}^{nT} RC_i$$
$$\left\{ RC_i = 1, \text{ if } RR_{ij} > 20 \text{ mm}; \text{ else } RC_i = 0 \right\}$$
$$(7.63)$$

- Rnnmm

  Rnnmm is the number of heavy precipitation days (i.e., count of days where RR (daily precipitation amount) $\geq$ user defined precipitation threshold (nn) expressed in mm). Let RR_{ij} be the daily precipitation amount on day i in period j. The index Rnnmm is given by the following equation:

$$\text{Rnnmm}_j = \sum_{i=1}^{nT} RC_i$$
$$\left\{ RC_i = 1, \text{ if } RR_{ij} > nn; \text{ else } RC_i = 0 \right\}$$
$$(7.64)$$

- CDD

  CDD index is the number of consecutive dry days (i.e., maximum length of dry spell; RR < 1 mm). Let RR_{ij} be the daily precipitation amount on day i in period j. The count of the largest number of consecutive days where RR_{ij} < 1 mm is given by the following equation:

$$\text{CDD}_j = \sum_{i=1}^{nT} RCC_i$$
$$\left\{ RCC_i = 1, \text{ if } RR_{ij} < 1 \text{ mm}; \text{ else } RCC_i = 0 \right\}$$
$$(7.65)$$

- CWD

  CWD index is the consecutive wet days (i.e., maximum length of wet spell; RR > 1 mm). Let RR_{ij} be the daily precipitation amount on day i in period j. The count of the largest number of consecutive days where RR_{ij} > 1 mm is given by the following equation:

$$\text{CWD}_j = \sum_{i=1}^{nT} RCC_i$$
$$\left\{ RCC_i = 1, \text{ if } RR_{ij} > 1 \text{ mm}; \text{ else } RCC_i = 0 \right\}$$
$$(7.66)$$

- R95pTOT and R99pTOT

  The indices R95pTOT and R99pTOT are calculated 95th and 99th percentiles of rainfall depths based on the normal (base period 1961–90).

$$\text{R95pTOT} = \sum_{j=1}^{nT} RR_j \left\{ RR_j > RRn95 \right\} \quad (7.67)$$

$$\text{R99pTOT} = \sum_{j=1}^{nT} RR_j \left\{ RR_j > RRn99 \right\} \quad (7.68)$$

- PRCPTOT

  PRCPTOT is the total precipitation in wet days (i.e., RR (daily precipitation amount) $\geq$ 1 mm). Let RR_{ij} be the daily precipitation amount on day i in period j. The index PRCPTOT is given by the following equation:

$$\text{PRCPTOT} = \sum_{i=1}^{nT} RR_{ij} \left\{ \text{if } RR_{ij} > 1\text{mm} \right\} \quad (7.69)$$

RC is a binary variable used to count the number of days; RCC is a binary variable used to count the largest number of consecutive days.

An example of calculation of one of the indices for a station with missing precipitation data is provided here. The calculations are based on a recent study by Teegavarapu *et al.* (2012). Based on the example discussed previously, indices are developed for extreme precipitation index (RX1day) at Moore Haven with missing data and data with gaps filled. Figures 7.26 and 7.27 show the differences in the cumulative density plots of the index (RX1day) for these two data sets for two different time periods. It is evident from these figures that the infilling process can bias the results based on indices.

Another index (Rnnmm) for extreme precipitation at Moore Haven with infilled data and data with gaps is shown in

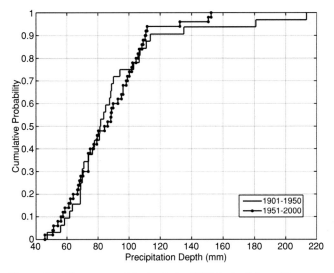

Figure 7.26 Extreme precipitation index (RX1day) at Moore Haven with missing data.

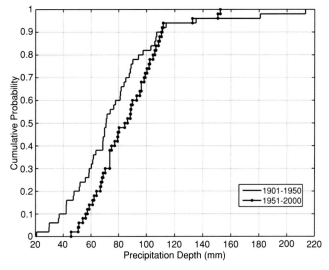

Figure 7.27 Extreme precipitation index (RX1day) at Moore Haven with infilled missing data.

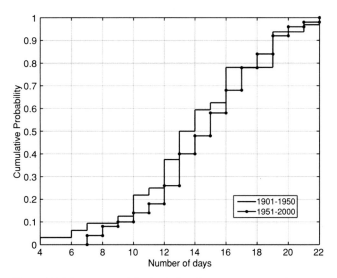

Figure 7.28 Extreme precipitation index (Rnnmm) at Moore Haven with infilled missing data.

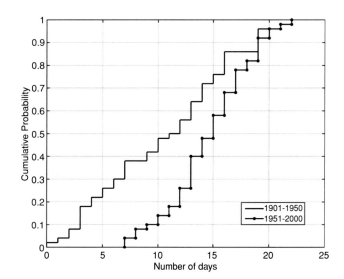

Figure 7.29 Extreme precipitation index (Rnnmm) at Moore Haven with data with gaps.

Figures 7.28 and 7.29 respectively. The threshold value used for calculation of the index Rnnmm is 25.4 mm. Again, the results suggest the influence of infilled data on calculation of indices compared to those from data with gaps.

Several publications are available from WMO that relate different aspects of statistical analysis of extremes. These include climatological practices (WMO, 1983, 2009a); normals (WMO, 1989, 1996, 2007); statistical analysis of observations (WMO, 1990); and guide to hydrologic practices (WMO, 2008). These documents provide an exhaustive review of statistical methods applicable for assessments of extreme precipitation trends. WMO (2011) recommends the use of regional-averaging of indices as

opposed to evaluation of single-site indices relevant to precipitation extremes.

7.25 RARE EXTREMES

Extreme value theory can be used to address rare extremes. An example of a rare extreme may be landfall of a hurricane (or cyclone) with a large amount of precipitation realized along the path of the hurricane. Two main issues to be addressed when dealing with rare extreme events are changes and non-stationarity. The extreme quantiles can be calculated for different periods of time (both in the past and in the future) assuming that non-stationarity within the time periods is sufficiently small, or using more advanced methods in which the parameters in the statistical models vary over time to describe the temporal evolution of the extremes (WMO, 2009a). New methods are now available to incorporate a linear trend (or other covariate) into any of the three parameters of the GEV distribution (WMO, 2009a). Detection of changes in rare extremes is difficult, especially at a specific location in a region. Data pooling (augmentation of a data set based on observations from a number of stations) and spatial pooling are commonly employed in hydrology for evaluation of trends.

7.26 TRENDS BASED ON GCM MODEL SIMULATIONS

Hydrologic variable trends based on downscaled GCM model simulations can be evaluated using the statistical concepts discussed in this chapter. One important and critical assumption of stationarity used in frequency analysis is no longer valid when analysis is carried out with such data sets. In general, precipitation frequency analysis methods always require the assumption of stationary climate. Several statistical tests are used to see if statistically significant trends are present in the AMS or PDS of observations used for frequency analysis. For example, the precipitation frequency analysis carried out by NOAA for Atlas 14 uses the assumption of stationary climate.

Precipitation frequency analysis studies now focus on the evaluation of climate change impact on trends of extreme rainfall. If detailed statistical analysis shows no or little observable change in trends over time, the assumption of stationary climate is not detrimental to precipitation frequency analysis. This is a valid approach as the impact of potential changes in climate on precipitation frequency estimates is uncertain. The uncertainty is mainly due to the documented large differences among different climate model projections with respect to the expected changes in extreme precipitation frequencies and magnitudes. Further research is also needed to determine how to adjust precipitation frequency estimates for future climate change.

7.27 SOFTWARE FOR EVALUATION OF EXTREME PRECIPITATION DATA

Statistical analysis of extreme precipitation data can be conducted using the software on the website for this book. The software is developed by the author of this book and is referred to as statistical evaluation of extreme precipitation (SEEP) and provides the following tools or utilities.

• A suite of exploratory data analysis (EDA) methods and advanced statistical methods.

• EDA methods include histograms, cumulative density function plots, kernel density estimations, scatter-plots, and others.

• Inputs: Temporal precipitation data at a location or multiple locations.

• Outputs: EDA, statistical analysis of extreme annual precipitation (block maxima), parameter estimation using maximum likelihood estimation (MLE) and L-moments approaches for seven different distributions.

• Extreme precipitation indices based on long-term historical precipitation data using WMO's ETCDI document.

The software can also be used for analysis of extremes of other hydrologic variables. A users' manual along with sample data are provided on the website.

7.28 CONCLUSIONS AND SUMMARY

Statistical analysis of precipitation data and trends in precipitation data are the main focal points of this chapter. In setting the stage for precipitation data analysis, exploratory data analysis (EDA) techniques are discussed first. The availability of continuous precipitation data without any gaps and missing extreme observations is critical for the statistical analysis of data to obtain design rainfall. Statistical analysis of extreme precipitation data is clearly explained and the procedures that lead to development of intensity–duration–frequency (IDF) curves and development of iso-pluvial contours based on spatial interpolation are discussed. Assessment of extreme precipitation data and uncertainties associated with frequency analysis of these extremes are discussed. Precipitation frequency analysis in the context of changing climate is addressed.

EXERCISES

7.1 Use the SEEP software provided on the website to evaluate different statistical distributions to characterize extreme precipitation data for different durations provided in Table 7.16.

7.2 Evaluate the impact of missing precipitation data using any of the methods discussed in Chapter 3, to assess the changes in distributions of filled and unfilled data.

7.3 Develop IDF curves using extreme precipitation data for different durations provided in Table 7.17 using the frequency method (Section 7.4.7) discussed in this chapter.

7.4 Evaluate if the pre-specified inter-event time definition (IETD) is changing over time in a region based on precipitation time series available.

Table 7.16 *Extreme precipitation data for different durations*

				Hours					
Year	1	2	6	12	24	48	72	96	120
1	24	37	63	100	106	119	129	129	129
2	46	50	88	88	90	111	117	124	132
3	57	61	67	99	105	153	156	157	161
4	66	73	84	84	122	124	124	124	124
5	30	39	70	75	75	89	96	97	100
6	33	55	84	87	88	88	88	97	104
7	66	91	123	124	124	153	155	155	155
8	59	72	83	86	87	116	116	116	116
9	36	59	91	119	123	123	123	141	160
10	45	67	99	109	115	119	127	127	127
11	36	51	72	81	81	98	110	123	135
12	57	98	102	105	109	150	151	153	165
13	43	58	73	73	85	106	127	186	188
14	51	72	72	72	77	92	99	126	145
15	85	109	152	169	199	219	219	222	225
16	22	33	49	85	85	86	86	90	93
17	40	41	46	50	53	57	60	70	99
18	47	54	88	135	142	142	142	142	142
19	42	43	71	106	113	114	114	114	114
20	35	44	104	181	238	300	301	301	301
21	11	15	17	19	26	33	33	33	33
22	47	51	52	52	69	87	100	130	141
23	47	93	109	109	109	109	109	109	109
24	44	77	84	94	108	113	134	134	137
25	34	45	76	96	135	192	234	252	279
26	30	47	74	80	81	120	122	127	133
27	13	19	38	61	84	113	113	113	115
28	32	37	49	91	141	143	143	143	143
29	64	86	112	122	124	124	124	124	140
30	58	66	69	71	79	109	119	119	124
31	33	61	94	102	112	117	130	135	150
32	53	56	61	61	66	84	84	84	99
33	30	38	48	51	58	69	76	79	89
34	48	61	71	71	91	99	99	114	155
35	36	58	112	142	208	246	249	254	254
36	69	94	109	109	130	142	142	157	157
37	38	61	69	69	71	74	84	86	86
38	69	86	104	104	104	109	109	109	109
39	25	38	56	84	99	99	102	102	102
40	46	51	53	71	91	109	109	109	127
41	51	66	104	104	122	122	147	152	185
42	51	66	76	76	81	86	109	142	142
43	36	48	64	104	112	150	160	188	221
44	46	46	56	58	69	71	119	132	140
45	58	71	91	107	160	173	173	173	173
46	56	71	76	81	81	107	114	119	163
47	48	76	84	84	91	119	132	132	135
48	41	53	69	71	71	71	114	114	114
49	61	66	91	135	173	191	191	191	193

(*cont.*)

				Hours					
Year	1	2	6	12	24	48	72	96	120
50	43	58	61	74	79	79	94	97	114
51	56	91	127	127	165	178	178	178	185
52	33	53	79	102	178	180	183	193	193
53	46	79	89	91	91	91	102	112	114
54	51	56	58	69	99	132	135	135	135
55	38	53	84	97	130	130	191	193	208
56	33	61	104	135	137	150	157	160	163

Table 7.17 *Extreme precipitation data for different durations*

				Hours					
Year	1	2	6	12	24	48	72	96	120
1	52	89	121	167	189	199	207	241	251
2	27	43	55	80	86	122	122	122	122
3	55	81	115	149	172	183	183	183	188
4	68	71	72	72	74	105	115	135	137
5	62	63	94	126	144	211	230	230	235
6	37	59	140	206	283	422	444	464	465
7	46	74	119	155	191	192	192	192	192
8	38	70	122	127	131	151	152	157	158
9	33	43	87	96	104	106	114	116	118
10	43	57	60	60	73	115	115	115	115
11	56	77	80	84	103	116	139	158	159
12	43	77	108	108	108	110	110	156	158
13	80	98	102	102	120	137	148	168	182
14	32	38	65	108	123	123	123	123	134
15	57	57	59	67	86	103	114	138	182
16	44	53	102	150	238	275	304	304	304
17	28	47	52	58	60	66	79	118	123
18	36	52	54	55	87	91	115	123	127
19	36	52	60	60	65	69	78	81	92
20	46	60	94	103	120	161	213	224	224
21	41	60	74	82	101	158	167	183	201
22	47	56	61	61	61	110	123	133	153
23	85	105	105	105	105	133	155	155	168
24	56	70	97	123	127	134	158	158	158
25	48	50	51	84	95	104	107	112	117
26	39	44	58	66	122	129	129	129	139
27	95	95	95	95	126	126	144	144	144

7.5 Evaluate spatial and temporal precipitation trends in your region using the methods described in this chapter.

7.6 Obtain the gridded data sets described in Chapter 3 and evaluate trends associated with extreme precipitation values.

7.7 Define new extreme precipitation indices and evaluate them for any specific region.

7.8 Generalized extreme value distribution is considered appropriate for extreme precipitation data for different durations. Evaluate the scale and location parameter variability for precipitation data provided in Table 7.16 for different durations.

7.9 Develop sampling adjustment factors (or Weiss factors) for rainfall data available in your region and compare them with the factors available from the literature. Note that to develop the Weiss factors fine resolution temporal rainfall data are required.

7.10 Use the statistical evaluation of extreme precipitation (SEEP) available on the website to fit distributions to extreme precipitation data.

7.11 Evaluate the variability of the extreme precipitation indices provided by WMO as discussed in the chapter for different climatic regions and assess the variability also across different temporal windows.

USEFUL WEBSITE

http://maps.grida.no/go/graphic/precipitation_changes_trends_over_land_from_1900_to_2000.

8 Hydrologic modeling and design

8.1 PRECIPITATION AND CLIMATE CHANGE: IMPLICATIONS ON HYDROLOGIC MODELING AND DESIGN

8.1.1 GCM-based model results and implications

The single most essential input to hydrologic design is precipitation. Generally, single-event models and, rarely, continuous modeling approaches are used for hydrologic design in the USA and elsewhere in the world. The main limitations of GCM-based simulations are the availability of the temperature and precipitation data sets only at coarser resolutions that are essentially not particularly useful for hydrologic design.

8.1.2 Non-stationarity issues

Trends in hydrologic time series are generally evaluated based on historical data available at a site or a region. Data selection depends on the temporal slice of historical data time series and is shown in Figure 8.1. Several data selection issues need to be addressed based on the following questions:

1. How great a data length is required to process the hydrologic data for design purposes?
2. Are long-term historical data used for calculations of intensities sufficient for hydrologic design purposes?
3. Should only the most recent extreme events be included in the design process?
4. Which temporal slice (data window) of the time series data should be selected?
5. Should only data based on future climate changes be used for design?

Detection of trends relies on homogeneous and undisturbed observational records. Many meteorological observation stations have changed position, sensor type, calibration, or surrounding environment, often undocumented. Augmenting missing data will need spatial interpolation methods for infilling missing data as

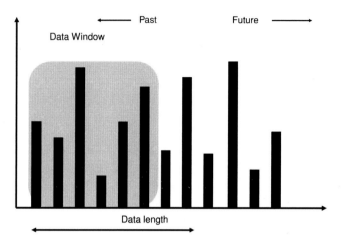

Figure 8.1 Schematic of data window and data length.

discussed in Chapter 3. Estimation of missing data may introduce bias depending on the spatial interpolation algorithm implemented in the estimation process.

8.2 EMERGING TRENDS IN HYDROLOGIC DESIGN FOR EXTREME PRECIPITATION

Future changes in climate that may alter precipitation intensity or duration would likely have consequences for urban stormwater discharge, particularly where stormwater detention and conveyance facilities were designed under assumptions that may no longer be correct. The social and economic impact of increasing the capacity of undersized stormwater facilities, or the disabling of key assets because of more severe flooding, could be substantial. Figures 8.2 and 8.3 show schematic diagrams of possible future increases in precipitation depth for a given duration or increase in the number of events and also increase or decrease in the magnitudes of extreme events. The future hydrologic design should deal with different possible scenarios of changing precipitation patterns.

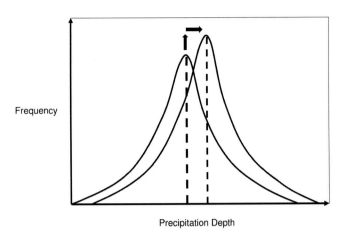

Figure 8.2 Schematic showing variations of extremes in frequency and change in mean.

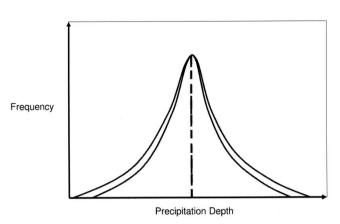

Figure 8.3 Schematic showing variations of extremes with no change in the mean.

8.3 METHODOLOGIES FOR HYDROLOGIC DESIGN

Methodologies for future hydrologic design should consider the cascade of uncertainties propagating from the first step till the end when design intensities are obtained. Figure 8.4 shows the steps in the process of developing a more compromise hydrologic design considering the uncertainties associated with future climate change projections. The design can also be referred to as sustainable design as it considers the current and future climate information to address the issues of uncertainties and also preferences of hydrologists in trusting the climate change model results. The future stormwater infrastructure designed in this compromise design approach will deal with changing storm IDF relationships over time, evolving temporal rainfall distributions, and economic options to minimize cost and improve the reliability of the structure under consideration to adapt to and to resist failure.

An example of such compromise design is presented in Section 8.4.

8.3.1 Data length and design storms

Design storms have been widely employed for design of stormwater conveyance systems in engineering practice. They are generally derived using IDF relationships that are developed using historical rainfall data. Data length is a very important factor that affects the intensity values. Historical rainfall data collected from a rain gaging station (located at Lexington, Kentucky) are used to illustrate this point. It is evident from Figure 8.5 that in general the average intensity over the last 15 years has increased. Many climatologists assume that under warmer conditions the hydrologic cycle will become more intense, stimulating rainfall of greater intensity and longer duration, causing longer floods and droughts (IPCC, 1992). The reasons behind recent increases in rainfall intensities observed at Lexington, Kentucky, are not

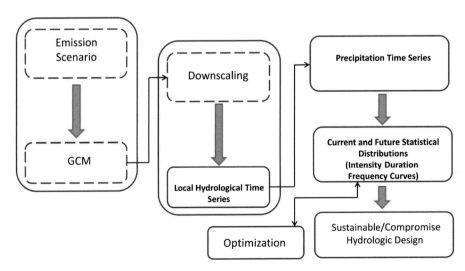

Figure 8.4 Schematic showing various steps of hydrologic design incorporating climate change.

Table 8.1 *Maximum rainfall depths for different return periods and data lengths for Lexington, Kentucky*

Year [Data length]	Return period	Rainfall depth (24 hrs) mm	Rainfall depth (24 hrs) mm [NRCS*]
1971–2004	5	96	97
	10	114	109
1971–1985	5	90	97
	10	105	109
1985–2000	5	111	97
	10	135	109
1985–2004	5	104	97
	10	126	109

* NRCS: Natural Resource Conservation Service, Technical Release-55 Document

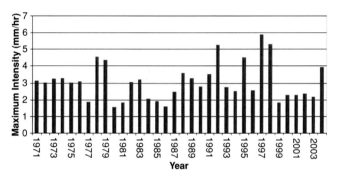

Figure 8.5 Maximum rainfall intensities from the observed data for different years at Lexington, Kentucky.

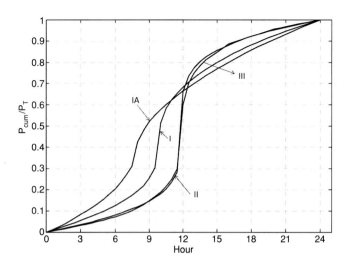

Figure 8.6 Rainfall distributions used in the USA developed by NRCS.

completely known or may be attributed to climate change at this point of time.

Daily rainfall depths are calculated for different return periods using traditional frequency analysis of annual rainfall depths and adopting an extreme value Type I (Gumbel) distribution, and they are reported in Table 8.1. Minimum data (n) required for a given return interval (T) are given by

$$n = mT - 1 \qquad (8.1)$$

where m is the rank. Several data lengths are used for calculation of rainfall depths for different return intervals. It is interesting to note that the values obtained using complete data are close to NRCS (Natural Resources Conservation Service) recommended values. However, when the data length is changed, the depths are different from the recommended values. Two important parameters are affected by the rainfall depths and they are the data length and data window of the rainfall time series.

A number of important questions need to be addressed in the context of climate change and these are: (1) How great a data length is required to obtain IDF curves? (2) Are long-term data

used for calculations of intensities sufficient for hydrologic design purposes ? (3) Should only the most recent extreme events be included in the design process? No concrete answers are available to address all the issues indicated above, especially under climate change scenarios. However, there is a need for a conceptual basis for selection of data length for design purposes considering future climate change.

One of the solutions suggested by Adams and Howard (1986) and Adams and Papa (2000) is viable to address these issues. The solution is to adopt continuous simulation approaches rather than relying on average intensities obtained from IDF curves for design. Adams and Howard (1986) prove that peak design runoff frequency is not equal to average rainfall intensity frequency considering cases of identical rainfall and different antecedent moisture conditions, and identical antecedent moisture conditions and different rainfalls by examining the pathology of design storms. Currently no design considerations are available that can handle future climate changes. Sustainable hydrologic design that attaches importance to climate change is very much needed.

8.4 HYDROLOGIC DESIGN

Hydrologic design methods (e.g., the rational method) aimed at developing stormwater management design plans and synthetic storm distributions (e.g., NRCS 24-hour distributions) mainly rely on rainfall intensities for a given duration and return interval. The four NRCS distributions that are applicable for different parts of the USA are shown in Figure 8.6. The synthetic distributions are based on historical data and need to be constantly revised based on the most recent data or temporal window approaches which result in the highest (conservative) estimates of intensities for hydrologic design.

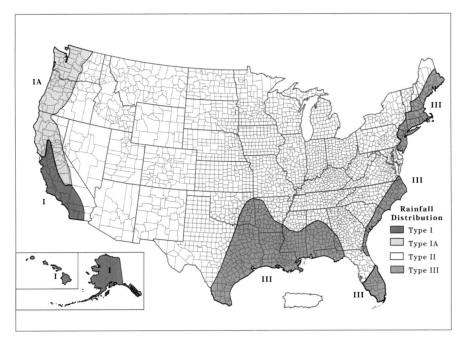

Figure 8.7 Geographical boundaries for NRCS distributions (source: TR-55).

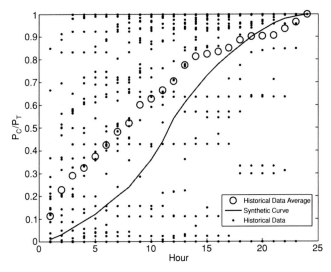

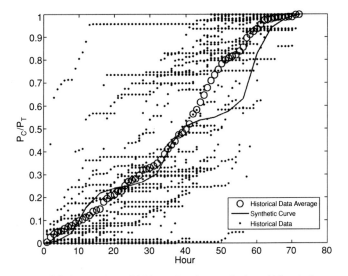

Figure 8.8 Comparison of 72-hour duration synthetic and historical rainfall distributions (rain gage: West Palm Beach Airport, Florida, USA).

Figure 8.9 Comparison of 24-hour duration synthetic and historical rainfall distribution (rain gage: West Palm Beach Airport, Florida, USA).

Validity of synthetic rainfall distributions (i.e., Types I, IA, II, and III) for different parts of the USA is well defined. The regions where these distributions are applicable are shown in Figure 8.7.

8.4.1 Observed and synthetic rainfall distributions

Comparisons of real-life rainfall distributions and synthetic curves based on long-term historical precipitation data at a rain gage station in Florida are shown in Figures 8.8 and 8.9. The distributions based on historical data deviate from the standard curves for both 24- and 72-hour durations. These deviations are possible due to several reasons including climate variability and climate change.

8.5 ADAPTIVE HYDROLOGIC INFRASTRUCTURE DESIGN

Adaptive hydrologic design especially for stormwater infrastructure requires the elements of compromise discussed in Section 8.3.

IDF curves serve as the main design aids for stormwater management systems using the rational approach for estimating peak discharge and hydraulic elements of the pipe infrastructure. As the IDF curves are expected to change in the future due to climate change and variability, these changes need to be reflected in the design or usage of the new IDF curves available from future climate simulations. Considering the uncertainties involved in the climate change model results, incorporation of future changes to precipitation regimes in design is a contentious issue. A fuzzy mathematical programming framework is presented in the next section, which considers the variability in IDF curves and preferences of decision makers towards possible future changes. The framework involves solving three optimization models to obtain a compromise solution that minimizes the cost of the infrastructure (i.e., stormsewer system piping network and others). The decision makers or agencies responsible for design can provide their preferences attached to specific changes in the magnitude and direction of climate change.

8.5.1 Hydrologic design incorporating uncertainties of climate change

The uncertainties associated with IDF curves and preferences of the decision makers are incorporated via fuzzy membership functions and are ultimately used in an optimization model. Details of fuzzy mathematical programming model formulations under uncertain decision-making environments can be found in Teegavarapu and Simonovic (1999) and Teegavarapu (2010).

8.5.2 Fuzzy mathematical programming model for hydrologic design

In an unsymmetrical decision-making environment in the context of fuzzy mathematical programming formulations, the objective function or the constraint space or both are not crisp. In these situations, three different formulations are required for obtaining the solution of fuzzy mathematical programming formulations. These formulations are referred to as original, intermediate, and final and are described in Steps I, II, and III below. Details of these formulations can be found in Teegavarapu and Simonovic (1999). The steps refer to formulations of optimization models in standard form. Step I refers to the original formulation with objective function and constraints. Step II refers to a modified original formulation with tolerances added to constraints that are considered fuzzy. Step III is essentially the final formulation with the objective of maximizing the membership function value using the Bellman and Zadeh (1970) criterion as explained by Zimmermann (1991). Membership functions are required to be defined a priori for the objective function and fuzzy constraints indicating the preferences of the decision maker.

Step I: Original formulation

Maximize

$$f(x) = C^T x \tag{8.2}$$

subject to:

$$Ax \leq b \tag{8.3}$$

$$Dx \leq b' \tag{8.4}$$

$$x \geq 0 \tag{8.5}$$

$$c, x \in \Re^n, \quad b \in \Re^m, \quad A \in \Re^{m \times n}$$

Equations and inequalities defined in 8.2–8.5 refer to classical linear mathematical programming mode formulation in standard form. The coefficients A, b, c, and D are crisp numbers. The notation for this formulation is adopted from Zimmermann (1991). The objective function obtained in this formulation is referred to as f_o, obtained when the original mathematical formulation without any tolerances added to the constraints is solved.

Step II: Intermediate formulation

Maximize

$$f(x) = C^T x \tag{8.6}$$

subject to:

$$Ax \leq b + t_o \tag{8.7}$$

$$Dx \leq b' \tag{8.8}$$

$$x \geq 0 \tag{8.9}$$

The objective function obtained in this formulation is referred to as f_1 obtained when tolerance, t_o, is added to the limit in the constraint 8.7.

Step III: Final formulation

Maximize

$$L \tag{8.10}$$

subject to:

$$L(f_1 - f_o) + C^T x \leq f_1 \tag{8.11}$$

$$Lt_o + Ax \leq b + t_o \tag{8.12}$$

$$Dx \leq b' \tag{8.13}$$

$$L, x \geq 0 \tag{8.14}$$

$$L \leq 1 \tag{8.15}$$

The objective function values, f_o and f_1 obtained in the previous two formulations (Steps I and II) are used to develop a fuzzy

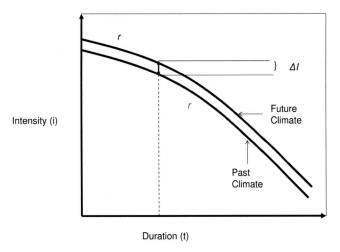

Figure 8.10 IDF curves based on past and future climate.

membership function. Membership function value, L, for the objective function is given by:

$$L = \begin{cases} 1 & \text{if } f_o < C^T x \\ \dfrac{C^T x - f_1}{f_o - f_1} & \text{if } f_1 < C^T x < f_o \\ 0 & \text{if } C^T x \leq f_1 \end{cases} \quad (8.16)$$

8.5.3 Application to stormsewer design problem

The formulations discussed in the previous section are applied to a stormsewer optimization model with the objective of minimizing the overall cost of infrastructure. Fuzzy membership functions are developed for preferences attached to future increases in precipitation intensities and cost. The functional form of a typical IDF curve is used for development of a mathematical equation and also reflects the change (Δi) in the intensities for a given duration. Changes assumed for the IDF curve are shown in Figure 8.10 and the membership functions for intensity and cost are shown in Figures 8.11 and 8.12 respectively.

The formulation is aimed at minimizing the cost considering the length of the stormsewer network. The formulation is given by Equations 8.17–8.21.

Minimize

$$\psi = f(d_k, L_k, \eta_k) \quad (8.17)$$

subject to:

$$t_k = f(L_k, S_k) \quad (8.18)$$

$$d_k = f(Q_k, n_k) \quad (8.19)$$

$$Q_k = f(C_k, i_k, A_k) \quad (8.20)$$

$$i_k = f(r, t_k) \quad (8.21)$$

The formulation defined by Equations 8.17–8.21 minimizes the cost (ψ) associated with different design elements of the stormsewer system. The cost is a function of the length of a sewer pipe

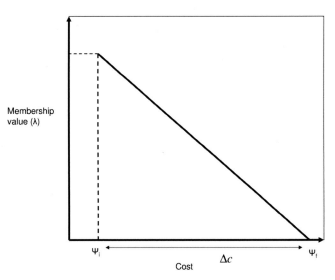

Figure 8.11 Membership function indicating preference for increased intensity for design.

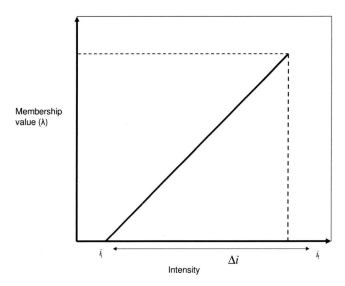

Figure 8.12 Membership function indicating preference for decreased cost for design.

in the system (L_k), diameter of the pipe (d_k) for each sub-region (k) and Manning's roughness coefficient (η_k) of the pipe. The variable t_k is the time of concentration and is a function of length, slope, and other characteristics of the region. The diameter of the pipe depends on the peak discharge, slope of the sewer, and Manning's roughness coefficient. Equation 8.20 is the rational peak discharge equation that is a function of precipitation intensity (i_k), rational coefficient (C_k), and the area of the region. The precipitation intensity (i_k) is based on a specific return period (r) and duration, which is generally equal to time of concentration (t_k). The return period is generally fixed by local drainage authorities or water management agencies that oversee the design of stormsewer management systems.

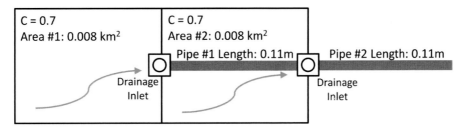

Figure 8.13 Layout of hypothetical stormsewer system.

The current (Equation 8.22) and future (Equation 8.23) IDF curves are defined in functional forms by the following equations, with future increase in the intensity added as a tolerance (Δi) to the original IDF curve.

$$i_k = \left(\frac{\delta}{t_k^\alpha + \beta} \right) \tag{8.22}$$

$$\Delta i_k = \left(\frac{\delta}{t_k^{\alpha o} + \beta^o} - \frac{\delta}{t_k^\alpha + \beta} \right) \tag{8.23}$$

Equations for the membership function for intensity changes are defined as follows:

$$\lambda = 1 - \left\{ 1 - \left(\frac{i_f - i_i}{\Delta i} \right) \right\} \tag{8.24}$$

$$\lambda \Delta i = i - i_i \tag{8.25}$$

The final formulation (Step III), which provides the compromise design elements, is given by the following objective function (8.26) and all the other constraints defined earlier (Equations 8.18–8.25).

Maximize

$$\lambda \tag{8.26}$$

subject to:

$$\psi_f - \Delta \psi \lambda \le \psi_i \tag{8.27}$$

The final formulation results in a compromise design of hydrologic and hydraulic components based on the decision maker's preferences attached to future climate change (magnitude and direction).

8.6 HYDROLOGIC DESIGN EXAMPLE

A hydrologic design problem (numerical example) is illustrated here. Stormsewer design problems are discussed by standard texts on hydrology or stormwater management. A typical stormsewer design problem with a system showing two different areas is illustrated in Figure 8.13.

The design problem considered requires estimation of pipe diameters for storm sewers using a rational approach. Hydraulic elements that are part of a stormsewer system are not considered

as part of the design. The peak discharge at a point (drainage inlet) is given by

$$Q = kCiA \tag{8.28}$$

where Q is the peak discharge, C is the rational coefficient, i is the intensity, A is the area of the watershed or region of interest, and k is unit conversion coefficient. The diameter of the pipe can be designed based on the assumption that the sewer is flowing full and Manning's equation is applicable.

$$D = \left(\frac{2.16Q\eta}{\sqrt{S}} \right)^{\frac{3}{8}} \tag{8.29}$$

Time of concentration is assumed to be known for the two regions in this example and the travel time is calculated based on peak discharge and velocity evaluations. Restriction of velocity is placed as a constraint in the formulation.

$$V = \frac{Q}{D^2 \pi / 4} \tag{8.30}$$

$$t_{total} = t_k + t_t \tag{8.31}$$

$$t_t = \frac{L}{V} \tag{8.32}$$

Equation 8.30 is used to calculate the velocity of flow associated with peak flow (Q) and the diameter obtained from Equation 8.29. The total duration or maximum possible time of concentration is based on time of concentration (t_k) and travel time (t_t) in any pipe given by Equation 8.32. The travel time is calculated using information about the velocity (V) and length (L) of the pipe.

Precipitation intensities are generally obtained using IDF curves available for a region and a specific return period is selected based on guidance from local drainage ordinances. The cost curves used for economic optimization using three different formulations are shown in Figures 8.14 and 8.15.

The first two formulations (i.e., original and intermediate) require minimization of cost functions associated with pipe diameter and smoothness of the pipe indirectly linked to Manning's roughness coefficient. The total cost (C^o) is minimized using the following optimization formulation with several constraints. Using the formulation based on Equations 8.33–8.35, the original

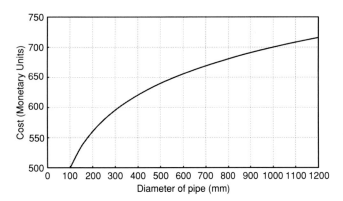

Figure 8.14 Hypothetical cost curve related to diameter of pipe.

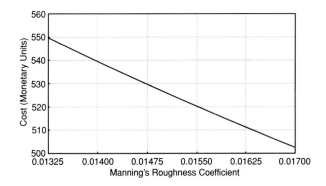

Figure 8.15 Hypothetical cost curve related to smoothness of pipe.

and intermediate formulations are solved, and objective function values (f_o, f_1) are estimated.

Minimize

$$C^o = \sum_{np=1}^{tnp} C(D_{np}, \eta_{np}) \tag{8.33}$$

subject to:

$$\eta_{min} \leq \eta_{np} \leq \eta_{max} \quad \forall np \tag{8.34}$$

$$V_{min} \leq V_{np} \leq V_{max} \tag{8.35}$$

$$Q_j = kC_j i_j A_j j = 1 \tag{8.36}$$

$$D_{np} = \left(\frac{2.16 Q_{np} \eta_{np}}{\sqrt{S_{np}}} \right)^{\frac{3}{8}} \quad \forall np \tag{8.37}$$

$$Q_j = k \left(\frac{\sum_{j=1}^{np} C_j A_j}{\sum_{j=1}^{np} A_j} \right) i_j \sum_{j=1}^{np} A_j \quad \forall j \geq 2 \tag{8.38}$$

The variable C^o is the total cost, which is a function of costs associated with number of pipes (np) considered, η_{min} and η_{max} are the minimum and maximum values of Manning's roughness coefficient, and V_{min} and V_{max} are the minimum and maximum allowable velocities of flow in the pipes. Intensity (i_j)

Table 8.2 *Precipitation intensities based on current and future climate*

Time (minutes)	Intensity (mm/hr)	
	Current	Future
8	239	263
10	229	251
20	183	201
30	150	165
40	130	142
50	114	126
60	104	115
120	68	75
180	51	56

calculations are based on the time of concentration and travel time in pipes.

8.6.1 Solution

Intensity values provided in Table 8.2 are transformed into functional forms with a structure given in Equation 8.22. The values of constants δ, α, and β for current and future intensity–duration relationships are 24.42, 0.43, 0 and 26.52, 0.43, 0 respectively. These values are based on intensities with units of inches/hour. The values are obtained by solving a simple optimization formulation using the Excel solver. The minimum and maximum velocities values used in the formulation are 0.915 m/s and 1.525 m/s respectively.

The time of concentration, length of the pipe, and area values for region 1 are 10 minutes, 0.11 km, and 0.008 km^2 respectively. Region 2 also has the same values as shown in Figure 8.13. The runoff coefficient is assumed to be 0.7 for both the regions. The formulations are solved using a generalized reduced-gradient-based non-linear programming solver available under the Excel environment. Results related to original, intermediate, and final formulations are provided in Table 8.3. The overall satisfaction value achieved for λ is 0.4941 and the compromise diameter values are 48 and 61 mm respectively. An additional linear membership function is also included in the formulation that enforces decreasing preference to channel roughness. No restriction is placed on the diameter sizes.

The variations in costs and diameters are small in magnitude as the increases in future precipitation intensities considered in this example problem are small. The formulations provided in this example use hypothetical yet realistic values of intensities and other variables. The formulations can be expanded to include additional constraints related to hydraulic elements of the storm-sewer design.

Table 8.3 *Results based on original, intermediate, and final formulations*

Variables	Original	Intermediate (climate change)	Final (compromise)
Intensity for area 1 (mm/hr)	229	251	240
Intensity for area 2 (mm/hr)	217	239	228
Diameter of pipe 1 (mm)	47	49	48
Diameter of pipe 2 (mm)	60	62	61
Cost for pipe 1 (monetary units)	915	918	917
Cost for pipe 2 (monetary units)	936	939	938
Total cost (monetary units)	1,852	1,858	1,855

8.7 EXAMPLE OF WATER BALANCE MODEL

Downscaled precipitation estimates from statistically downscaled and bias-corrected climate change models are used to evaluate a conceptually simple four-parameter water budget model. Xu and Singh (1998) provide a state-of-the-art review of water balance models (WBMs) in which they discuss the advantages and disadvantages of monthly WBMs. Many hydrologic models have been presented in the literature and are used to simulate surface and subsurface flow patterns in drainage basins. It is possible that non-physically based parameters might reproduce observed data reasonably well in many instances. However, it is not possible to use a non-physical model outside its period of characteristics. Jakeman and Hornberger (1993) studied the complexity of rainfall–runoff models as measured by the number of parameters. Their study includes the findings of Hornberger *et al.* (1985), who claim that four parameters are adequate to explain basin hydrology based on rainfall. Beven (1989) reported that three to five parameters were usually adequate to simulate basin hydrology based on rainfall. The inference from these papers is that models with more than six parameters may be over-parameterized, and may be less reliable in reproducing the hydrologic variables.

8.7.1 Water balance model: Thomas model

The Thomas model (Thomas, 1981; Alley, 1984, 1985) is a WBM that can be applied at daily, monthly, and annual time scales for estimation of several hydrologic variables. The main inputs to the model are precipitation and potential evaporation. Two additional inputs, initial soil moisture storage (ϕ_{t-1}) and groundwater storage (γ_{t-1}) values, either measured from the field or derived from calibration, are required for the first time interval of model application. Representation of these variables for a finite soil volume is provided in Figure 8.16.

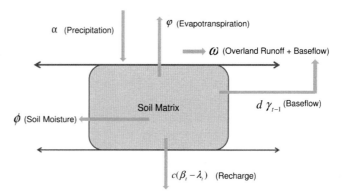

Figure 8.16 Representation of different components of WBM.

The water availability, evapotranspiration, subsurface, and overland flow (runoff) in the Thomas model are given by

$$\beta_t = \alpha_t + \phi_{t-1} \quad \forall t \tag{8.39}$$

$$\lambda_t = \frac{\alpha_t + b}{2a} - \sqrt{\left(\frac{\alpha_t + b}{2a}\right)^2 - \frac{b\alpha_t}{a}} \quad \forall t \tag{8.40}$$

$$\phi_t = \lambda_t e^{\frac{-\varphi_t}{b}} \quad \forall t \tag{8.41}$$

$$\lambda_t = \frac{c\left(\beta_t - \lambda_t\right) + \gamma_{t-1}}{1 + d} \quad \forall t \tag{8.42}$$

$$\omega_t = (1 - c)\left(\beta_t - \lambda_t\right) + d\gamma_{t-1} \quad \forall t \tag{8.43}$$

where β_t is the water available, α_t is the precipitation, ϕ_t is the soil moisture at the beginning of the time interval, λ_t is the evaporation opportunity which is a function of water availability, ϕ_{t+1} is the soil moisture at the end of the time interval which is a function of potential evapotranspiration (ϕ_t). The variable γ_t is groundwater storage at the end of the time interval and is a function of water availability, evaporation opportunity, and groundwater storage at the beginning of the time interval. The variable ω_t is the streamflow (runoff) in the time interval t. All the variables are expressed in units of equivalent water depth and the flow is finally obtained in discharge units using the watershed area.

Table 8.4 *Monthly correction factors for Thornthwaite and Wilm method*

Latitude	Jan	Feb	Mar	Apr	May	Jun	Jul	Aug	Sep	Oct	Nov	Dec
0	1.04	0.94	1.04	1.01	1.04	1.01	1.04	1.04	1.01	1.04	1.01	1.04
10 N	1.00	0.91	1.03	1.03	1.08	1.06	1.08	1.07	1.02	1.02	0.98	0.99
20 N	0.95	0.90	1.03	1.05	1.13	1.11	1.14	1.11	1.02	1.00	0.93	0.94
30 N	0.90	0.87	1.03	1.08	1.18	1.17	1.20	1.14	1.03	0.98	0.89	0.88
40 N	0.84	0.83	1.03	1.11	1.24	1.25	1.27	1.18	1.04	0.96	0.83	0.81
50 N	0.74	0.78	1.02	1.15	1.33	1.36	1.37	1.25	1.06	0.92	0.76	0.70
10 S	1.08	0.97	1.05	0.99	1.01	0.96	1.00	1.01	1.00	1.06	1.05	1.10
20 S	1.14	1.00	1.05	0.97	0.96	0.91	0.95	0.99	1.00	1.08	1.09	1.15
30 S	1.20	1.03	1.06	0.95	0.92	0.85	0.90	0.96	1.00	1.12	1.14	1.21
40 S	1.27	1.06	1.07	0.93	0.86	0.78	0.84	0.92	1.00	1.15	1.20	1.29
50 S	1.37	1.12	1.08	0.89	0.77	0.67	0.74	0.88	0.99	1.19	1.29	1.41

8.7.2 Potential evapotranspiration

The potential evapotranspiration (PE, φ_t) required for the four-parameter WBM is estimated using a widely used method developed by Thornthwaite and Wilm (1944). Applications of this method can be found in Serrano (2010). The PE value for a given month is based on the mean monthly air temperature of that month and an annual air temperature efficiency index, which is defined as the sum of 12 monthly values of heat index given by:

$$\varphi_t = k_t k_1 \left(\frac{10 T_{am}}{J} \right)^{k_2 J^3 + k_3 J^2 + k_4 J + k_5} \quad \forall t \qquad (8.44)$$

where φ_t is the potential evapotranspiration of the month t, k_m is the monthly correction constant (i.e., adjustment factor) and is a function of latitude (Gray, 1973), $k_1 = 16.0$ mm/month, $k_2 = 6.75 \times 10^{-7}\,°C^{-3}$, $k_3 = -7.71 \times 10^{-5}\,°C^{-2}$, $k_4 = 1.792 \times 10^{-2}\,°C^{-1}$, $k_5 = 0.49239$, and T_{am} is the mean monthly air temperature of month t, excluding negative values (°C). The annual heat index, J (°C) is given by

$$J = k_6 \sum_{t=1}^{12} T_{am}^{1.514} \qquad (8.45)$$

where k_6 is equal to $0.0874\,°C^{-0.514}$. Using Equations 8.44 and 8.45 the potential evapotranspiration can be estimated for each month of positive temperature. Monthly correction factors for the Thornthwaite and Wilm method for different latitudes are given in Table 8.4.

The PE values can be obtained at a location using the monthly mean temperature values. The method has been used widely throughout the world; however, the applicability is more appropriate for regions in the eastern USA (Shaw, 1994). If PE values are available from a nearby weather station, those values can be used in lieu of the above discussed temperature-based conceptually simple method for calibration of the Thomas model. The PE estimation method by Hargreaves and Samani (1982) is the most recommended method among all the available methods. However,

since temperature projections are available from climate change models, a conceptually simple temperature-based PE estimation method is recommended in this chapter.

8.7.3 Parameter estimation

The model parameters, a, b, c, and d are physically based and are estimated in general using a trial-and-error method or using an optimization technique. Any optimization technique can be used for obtaining the optimal parameters for the monthly balance model. The optimization formulation is given by Equations 8.46–8.50.

Minimize

$$\sqrt{\frac{1}{no} \sum_{t=1}^{no} (\hat{\omega}_t - \omega_t)^2} \qquad (8.46)$$

subject to:

$$a_l \leq a \leq a_u \qquad (8.47)$$

$$b_l \leq b \leq b_u \qquad (8.48)$$

$$c_l \leq c \leq c_u \qquad (8.49)$$

$$d_l \leq d \leq d_u \qquad (8.50)$$

The variable *no* is the number of time intervals, $\hat{\omega}_t$ is the observed streamflow value, and a_l, b_l, c_l, and d_l are the lower limits on a, b, c, and d parameters respectively. The variables a_u, b_u, c_u, and d_u are the upper limits on a, b, c, and d, respectively. The objective function, mean squared error (MSE), provided by Equation 8.46 can be modified to the following Equation 8.51 to weight high and low flows equally.

$$\frac{1}{no} \sum_{i=1}^{no} (\ln(\hat{\omega}_t) - \ln(\omega_t))^2 \qquad (8.51)$$

The initial guess values for a, b, c, and d provided in any non-linear programming solver should be changed to obtain a set of solutions.

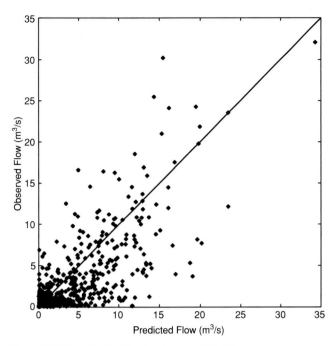

Figure 8.17 Results of calibration (years: 1950–90).

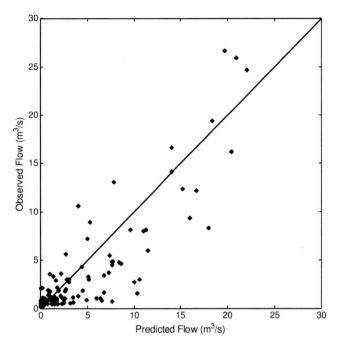

Figure 8.18 Results of validation (years: 1991–2010).

The parameter estimation process (Johnston and Pilgrim, 1976) at one single site is referred to as at-site calibration. Regional calibration of the parameters at multiple sites is also possible using the approach described by Fernandez *et al.* (2000).

8.7.4 Example application of Thomas model to a Florida watershed

The Thomas model is applied for estimation of monthly flows at a site in a watershed in Florida. Future climate change projections from 15 climate change models for precipitation and temperature discussed in Chapter 5 are used for simulating future extreme monthly discharges at a site in the watershed. Model application is discussed in the next section.

8.7.5 Details of the watershed

The case study watershed area used for the Thomas model is located in Florida and identified by USGS hydrologic unit code (HUC) number 03100101, with a drainage area of 341 km². A USGS streamflow gaging station located on the Joshua Creek at Nocatee, Florida (latitude: 27.1664, longitude: −81.8797), is used for historical streamflow observations. The temperature and precipitation data are obtained from a United States Historical Climatology Network (USHCN) station (station ID: 80228; name: Arcadia; location: latitude, 27.218, longitude, −81.874). The Thomas model is provided on the website for this book. The calibration and validation results for the Thomas model are shown in Figure 8.17 and Figure 8.18, respectively. Results based

on GCM-based projections using 15 different models discussed in Chapter 5 are shown in Figures 8.19 and 8.20 respectively. Comparative cumulative density plots for discharges based on 15 different models based on historical climatology and twenty-first century projected temperature and precipitation values are shown in Figure 8.21. Similar plots are used for comparison with data from years 1961–90 (climate normal). These plots are shown in Figure 8.22.

8.8 WATER BUDGET MODEL SOFTWARE ⁻

The Thomas model described in the previous section is available on the website as application software. The model is referred to as the water budget model. The initial screenshot of the model is shown in Figure 8.23.

The main inputs required for the model are long-term precipitation and temperature data sets, along with information about the area of the watershed and observed discharges at the gaging station, which is generally located at the outlet of the watershed. The data available should be without any gaps (i.e., missing data) and the model can be applied to any temporal resolution from a day to a year. Additional inputs that are required for calibration of the model are initial soil moisture and groundwater moisture values. The initial soil moisture storage (ϕ_0) and groundwater storage (γ_0) values are referred to as S_w and S_g in the software, as shown in the input module of the software in Figure 8.24.

The model uses a temperature-based method proposed by Thornthwaite and Wilm (1944) to estimate potential

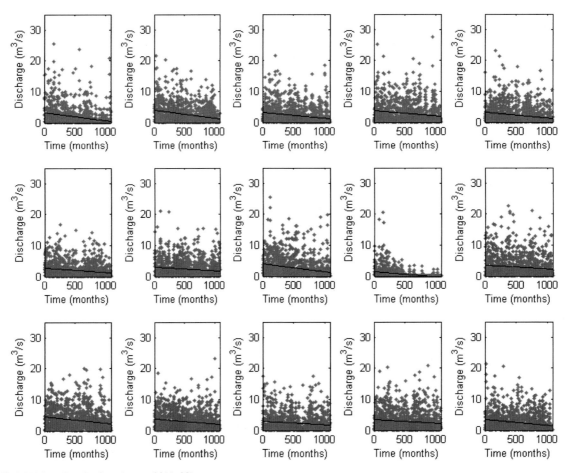

Figure 8.19 GCM-based projections (years: 2011–99).

evapotranspiration values. However, if potential evapotranspiration values are available from any other sources, these values can be provided as inputs to the model. The WBM can be run for several time intervals (i.e., years) and the calibration and validation data sets can be defined by the user. The model provides results with time series, scatter plots, and performance measures for both calibration and validation data periods.

8.9 INFRASTRUCTURAL MODIFICATIONS AND ADAPTATION TO CLIMATE CHANGE

Several modifications and adaptation strategies are suggested and developed by water management agencies around the world to deal with expected increased storm intensity and frequency. Mollenkamp and Kastens (2010) provide a comprehensive review of institutional challenges to adaptation using a case study in Germany. Adaptations to climate change include (1) climate proofing; (2) changes to operations; (3) developing rules to mimic natural hydrologic conditions; (4) demand management; (5) short-term and long-term planning of efficient use of available

water resources; and finally (6) addressing climate change risk and vulnerability assessments in the water management models. The following are some of those modifications suggested by USEPA (2008) that relate to extreme precipitation or changing future precipitation regimes.

1. Emergency plans for drinking water and wastewater infrastructure need to recognize the possibility of increased risk of high-flow and high-velocity events due to intense storms as well as potential low-flow periods.

2. Damage from intense storms may increase the demand for public infrastructure funding and may require re-prioritizing of infrastructure projects.

3. Floodplains may expand along major rivers requiring relocation of some water infrastructure facilities and coordination with local planning efforts; in urban areas, stormwater collection and management systems may need to be redesigned to increase capacity.

4. Combined storm and sanitary sewer systems may need to be redesigned because an increase in storm event frequency and intensity can result in more combined sewer overflows causing increased pollutant and pathogen loading.

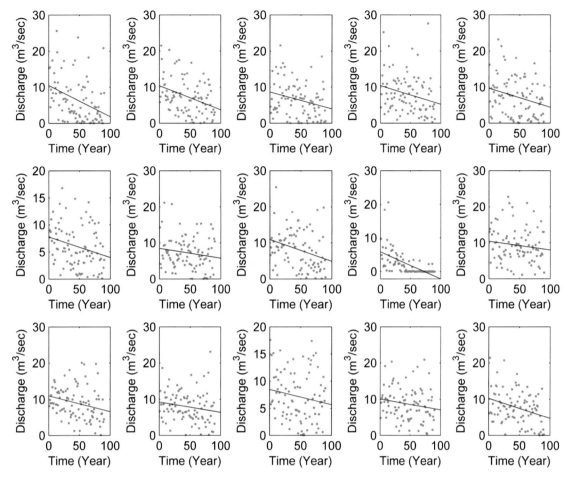

Figure 8.20 GCM-based projections of annual monthly extremes (years: 2011–99).

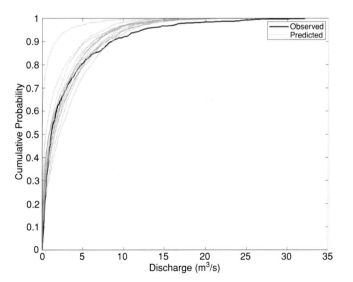

Figure 8.21 Cumulative density plots based on GCM-based projections (years 2011–99) and historical (years 1951–2010) monthly discharges.

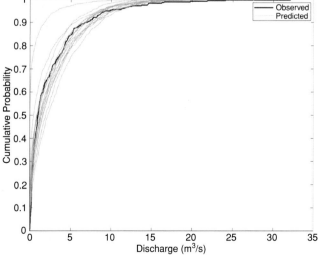

Figure 8.22 Cumulative density plots based on GCM-based projections (years 1961–90) and historical (years 1961–90) monthly discharges.

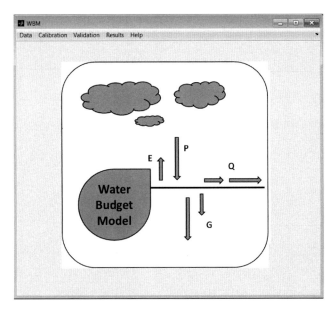

Figure 8.23 Initial screenshot of the WBM.

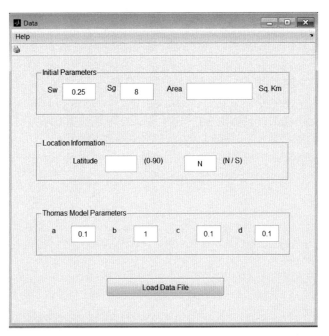

Figure 8.24 Data input section of WBM.

5. Greater use of biological monitoring and assessment techniques will help water resource managers assess system impacts of higher velocities from more intense storms and other climate change impacts.

6. The demand for watershed management techniques that mitigate the impacts of intense storms and build resilience into water management through increased water retention (e.g., green roofs, smart growth techniques) is likely to increase. The management of wetlands for stormwater control purposes and to buffer the impacts of intense storms will be increasingly important.

8.10 CONCLUSIONS AND SUMMARY

Hydrologic design and practice are two focus areas of this chapter considering changing climate and variability. Several issues related to hydrologic design that rely on precipitation extremes are discussed. Adaptive IDF curves, evolving temporal distributions of precipitation events, and limitations of traditional statistical methods in handling non-stationarity issues are evaluated. The chapter also discusses a new method for optimal compromise hydrologic design using fuzzy mathematical programming concepts incorporating current and future IDF curves for stormsewer design problems. A simulation model for future water budgets using a four-parameter WBM, referred to as the Thomas model, is described and implemented for a case study region in the USA. Downscaled precipitation and temperature projections are used in the Thomas model to evaluate future changes to water availability. The chapter also briefly highlights the needs for adaptive hydrologic infrastructure under changing climate.

EXERCISES

8.1 Use the extreme precipitation data from a rain gage in South Florida, USA, provided in Table 8.5 and develop IDF curves. Evaluate the changes in IDF curves when different temporal windows are used for the development of these curves. Use the frequency method discussed in Chapter 7 and an extreme value type distribution.

8.2 Evaluate the impact of the most recent precipitation extremes on IDF curves compared to those curves developed from entire historical data available in your region.

Table 8.5 *Extreme precipitation depths (mm) for different durations*

Year	Duration (hours)								
	1	2	6	12	24	48	72	96	120
1942	112	212	251	385	387	411	433	434	434
1943	52	68	72	72	74	111	116	123	123
1944	31	42	43	70	84	137	158	175	176
1945	63	112	175	226	234	253	263	263	270
1946	67	71	132	150	167	170	182	184	189
1947	58	72	131	198	224	225	258	258	258
1948	54	65	117	139	180	190	190	192	192
1949	56	82	86	117	150	158	175	182	182
1950	34	45	68	88	106	114	163	175	220
1951	51	88	165	181	190	192	192	192	217
1952	63	64	98	99	147	154	193	195	198
1953	35	48	61	62	77	104	114	131	144
1954	37	62	74	80	118	170	202	215	249

Table 8.5 (*cont.*)

	Duration (hours)								
Year	1	2	6	12	24	48	72	96	120
1955	47	55	94	95	116	136	136	136	136
1956	32	51	73	73	88	129	147	164	180
1957	64	94	133	151	162	162	162	162	162
1958	45	87	159	174	179	194	200	205	205
1959	50	65	114	122	162	166	223	252	252
1960	59	99	166	180	217	223	225	228	263
1961	41	44	49	52	56	74	87	95	98
1962	40	54	61	90	91	114	140	145	154
1963	81	116	120	151	169	194	194	196	197
1964	58	69	106	113	180	180	180	239	240
1965	44	56	117	153	243	249	249	249	258
1966	53	53	72	96	122	183	190	190	190
1967	51	72	92	117	122	135	135	135	148
1968	79	86	92	106	117	135	184	193	225
1969	90	109	124	124	124	145	150	150	155
1970	47	72	92	107	124	125	125	133	161
1971	57	95	130	135	138	195	218	240	244
1972	50	71	121	133	157	161	162	162	163
1973	32	52	62	63	71	81	96	120	121
1974	47	54	97	116	129	158	161	162	162
1975	54	64	98	110	110	113	113	139	186
1976	80	113	125	127	139	180	181	181	186
1977	45	85	99	122	134	188	204	221	242
1978	41	75	93	125	135	142	181	185	225
1979	65	99	146	177	186	193	208	226	256
1980	67	76	76	76	88	88	90	106	141
1981	36	58	86	103	106	130	152	157	178
1982	48	83	146	192	224	233	235	239	249
1983	39	60	102	121	132	165	165	168	172
1984	71	87	137	143	195	257	278	288	288
1985	40	45	68	80	80	89	99	128	133
1986	56	71	90	99	105	127	138	141	142
1987	58	61	88	94	102	120	136	154	156
1988	65	110	163	170	171	171	172	172	215
1989	60	84	99	99	99	131	133	133	138
1990	71	99	108	112	114	123	129	134	144
1991	75	98	138	177	215	219	219	226	226
1992	76	114	122	122	124	140	174	194	223
1993	64	100	110	141	142	169	185	188	205
1994	53	82	109	110	155	177	185	193	196
1995	47	60	122	147	161	176	188	215	220
1996	43	46	49	52	61	94	102	109	111
1997	46	46	69	93	98	112	123	123	124
1998	43	74	139	231	232	232	232	232	249
1999	39	58	119	158	181	275	275	288	290
2000	64	98	129	130	131	134	134	134	134
2001	43	74	119	126	148	243	261	263	263
2002	56	76	122	128	134	142	143	169	233
2003	64	67	134	139	140	173	176	179	201
2004	39	59	134	203	278	343	352	394	396

8.3 Qualify and quantify the impact of teleconnections, if any, on your regional precipitation patterns and how they vary from multi-year or multi-decadal oscillations.

8.4 Develop synthetic temporal distributions of precipitation data from historical extremes for specific durations and evaluate any changes in different temporal windows.

8.5 Use the fuzzy mathematical programming framework provided in this chapter to design a sustainable compromise stormsewer of similar layout to that discussed in the example, using details provided in Tables 8.6 and 8.7. Adopt the cost curves provided in this chapter or develop your own cost curves and a specific IDF functional form using the durations and precipitation intensities provided in Table 8.5.

Table 8.6 *Precipitation intensities for different durations*

	Intensity (mm/hr)	
Time (minutes)	Current	Future
8	239	310
10	229	297
20	183	238
30	150	195
40	130	168
50	114	149
60	104	135
120	68	88
180	51	67

Table 8.7 *Parameter values for design*

Parameter	Area 1	Area 2
Time of concentration (minutes)	12	13
Runoff coefficient	0.7	0.8
Area (km^2)	0.008	0.009
	Pipe 1	Pipe 2
Allowable velocity (m/s)	1.5	1.5
Length of pipe (m)	107	100
Slope (%)	1	1

8.6 Design a stormsewer network in your region using the concepts discussed in this chapter for compromise design. Comment on the existing network and the future network if no compromise design is used.

8.7 Using the data provided in Tables 8.8 and 8.9, calibrate and validate the Thomas model using appropriate values for initial soil moisture and groundwater storage and using the watershed area of 30 km^2. Use the data in Table 8.8 for

Table 8.8 *Monthly precipitation, discharge, and potential evapotranspiration values for calibration*

Month	Precipitation (mm)	Discharge (m³/s)	Potential evapotranspiration (mm)
Jan	100.58	3.31	20
Feb	64.52	3.85	45
Mar	86.36	2.40	60
Apr	157.48	3.60	82
May	155.96	4.93	125
Jun	274.57	8.04	146
Jul	202.69	5.64	166
Aug	7.37	1.48	201
Sep	15.49	1.44	154
Oct	61.21	1.07	101
Nov	49.78	0.92	55
Dec	82.04	1.31	43

Table 8.9 *Monthly precipitation and discharge for validation*

Month	Precipitation (mm)	Discharge (m³/s)
Jan	86.11	2.82
Feb	122.17	3.17
Mar	98.81	2.01
Apr	128.78	3.06
May	76.20	4.19
Jun	96.52	7.02
Jul	85.09	4.79
Aug	85.60	1.36
Sep	133.86	1.22
Oct	18.29	0.91
Nov	50.80	0.78
Dec	92.46	1.08

calibration and Table 8.9 for validation assuming that the same potential evapotranspiration values are appropriate. Use any non-linear optimization solver or the Excel-based generalized reduced gradient non-linear solver. Use different objective functions for optimization and different starting values of the soil moisture and groundwater storage.

8.8 Use the Thomas model available on the website within the application software Water Budget Model to evaluate water availability in your region using downscaled future precipitation and temperature projections. Calibrate and validate the model on historical data sets and discuss the limitations of the model.

9 Future perspectives

9.1 FUTURE HYDROLOGIC DESIGN AND WATER RESOURCES MANAGEMENT

The impact of climate change on hydrologic design and management of hydrosystems could be one of the important challenges faced by future practicing hydrologists and water resources managers. Many water resources managers currently rely on the historical hydrologic data and adaptive real-time operations without consideration of the impact of climate change on major inputs influencing the behavior of hydrologic systems and the operating rules. The fourth book in the IFI four-volume series, by Simonovic, addresses the issues of flood risk management under climate change conditions. Issues such as risk, reliability, and robustness of water resources systems under different climate change scenarios have been addressed in the past. However, water resources management with the decision maker's preferences attached to climate change has never been dealt with. The impact of climate change and climate variability on several hydrologic regimes throughout the world and on water resources management is discussed in several works (Dam, 1999; Frederick, 2002; IPCC, 2007d, Brekke *et al.*, 2009) published to date. The main findings, however inconclusive at times, derived from these works indicate that predicted changes will influence the hydrologic cycle in one form or another. General consensus from these studies is that feedback mechanisms within climate are not sufficiently understood to make accurate predictions of the impacts of climate change on hydrologic regimes.

Predictions discussed in the Intergovernmental Panel on Climate Change report (IPCC, 2007d) indicate that Earth may experience a rise of 1–3.5 °C in global surface temperature and changes in spatial and temporal patterns of precipitation. Potential impacts of climate change on runoff mechanisms are also documented in several works (IPCC, 1992, 1995, 2001, 2007d) that suggest future climate change will involve greater extremes of weather, including more high-intensity rainfall events or decreased streamflow conditions. Insufficient hydrologic record lengths, natural variability blended with anthropogenically induced changes, inconclusive results from climate change studies, and the effect of future climatic changes on hydrologic design and water resources management are major issues to be addressed by all future practicing hydrologists and water resources managers.

Climate impact studies generally focus on using three different approaches (referred to as the 3Is) (Arnell, 1996) and they are (1) the impact approach, which deals with cause and effect with emphasis on scenario-based analysis; (2) the interaction approach, which considers the feedbacks; and (3) the integrated approach, which deals with the hierarchies of interactions within each sector and the interactions and feedback between sectors (e.g., global warming impacts on the economy, landscape, and society).

9.2 UNCERTAIN CLIMATE CHANGE MODEL SIMULATIONS

Despite considerable efforts that have been undertaken in research and modeling climate change, the results are still highly uncertain for a number of reasons, ranging from the shortcomings and capabilities of GCMs to the different coupling methods used at different levels of scales in integration of GCMs and hydrologic models. However, researchers suggest that water resources managers may continue to rely on scientists' "best estimates" of future climate change for any long-term planning, and will have to await the anticipated improvements in models and methodologies to obtain better estimates (Leavesley, 1999). A level of uncertainty and distrust associated with the conclusions related to climate change from many of these works prevails among practicing hydrologists and water resources managers. Substantial uncertainty remains in trends of hydrologic variables because of large regional differences, gaps in spatial coverage, and temporal limitations of data (Huntington, 2006). Stainforth *et al.* (2007) discussed uncertainty in the models, model forcing, and initial conditions. Studies by Murphy *et al.* (2004), Tebaldi and Knutti (2007), and Dettinger (2005) have attempted to derive future climate probability distributions from climate projections to characterize uncertainty associated with climate change information. Increasing amounts of literature and studies related to climate change are getting little attention and there exists a cautious reluctance in accepting the

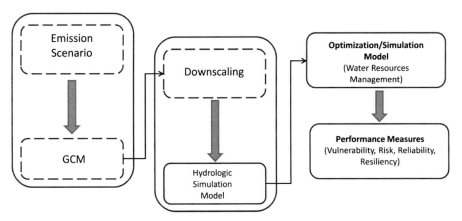

Figure 9.1 Typical steps in water resources system evaluation under changing climate.

possible climate change scenarios. Many earlier studies focusing on impacts of climate change (e.g., Bogardi and Kundzewicz, 2002; Thomas and Bates, 2002; Shreshtha, 2002) discuss criteria for understanding risk, reliability, and robustness of water resource systems considering future climate change scenarios and fall short of discussing the consequences of preferences or beliefs of resource managers attached to climate change that influence water resources management.

Leiserowitz (2006) indicates that risk perceptions associated with climate change are strongly influenced by experiential factors, including effect, imagery, and values, and demonstrate that responses to climate change are influenced by both psychological and socio-cultural factors. While the water resources management agencies and decision makers acknowledge the limitations of the climate change models, their perceptions towards the accuracy of results from these models are generally translated to preferences for predicted future changes to the main hydrologic inputs. To address hydrologic uncertainties associated with stochastic hydrologic inputs, a general-purpose scenario-modeling framework to solve water system optimization problems was presented by Pallottino et al. (2005).

9.3 FUTURE OF HYDROLOGIC DATA FOR DESIGN

Many future climate change studies focus only on mean trends in precipitation or temperature variables at different temporal scales. However, the pathology of storms leading to different flooding events should be assessed. These are especially important in watersheds prone to flash floods. Inter-event time definition (IETD) (Behera et al., 2010) for precipitation events for wet and dry spells and how it affects the future hydrologic design should also be considered. Climate change information (CCI) is available now through several sources and a number of questions remain

about how this information is used for design. Some of these questions are:

1. How do practicing hydrologists and water resources managers conceptualize climate change and react to it?
2. What roles do their expectations of climate change play in their acceptance and use of CCI?
3. How do recipients of CCI deal with forecast uncertainty and implications of this uncertainty in design?

9.3.1 Climate change projections and data sets

Substantial uncertainty remains in trends of hydrologic variables because of large regional differences, gaps in spatial coverage, and temporal limitations of data (Huntington, 2006). Uncertainty in the models, model forcing, and impact of initial conditions are discussed by Stainforth et al. (2007). Studies have attempted to derive future climate probability distributions from climate projections to characterize uncertainty associated with climate change information (Murphy et al., 2004; Tebaldi and Knutti, 2007; Dettinger, 2005).

A schematic of typical steps used in climate change impact studies, especially in water resources management, is shown in Figure 9.1. Uncertainties in different components of the process illustrated in the steps are indicated by boxes with dotted lines (e.g., emissions scenario). Reynard (2007) details several sources of uncertainty and they include (1) future emissions of greenhouse gases, (2) atmospheric concentrations of these gases, (3) global and large-scale climate response, (4) local/small-scale climate response, (5) natural climate variability, and (6) hydrologic impacts. The serial progression of these uncertainties is referred to as a "cascade of uncertainty" by Reynard. Climate change predictions based on GCMs are dependent on the land-surface hydrology in the GCM and computational constraints; large-scale GCM grids limit the possibilities of effectively linking/coupling hydrologic models directly to GCMs. Also, climate change predictions depend on the rainfall–runoff model or hydrologic simulation model used and the climate scenario generated. Studies now attempt to provide estimates of uncertainty associated with the impacts of climate change.

A recent study has focused on understanding and presenting uncertainty about climate change to water resources managers (Groves *et al.*, 2008). They document three different approaches: (1) standard scenarios, (2) probability-weighted scenarios, and (3) policy-relevant scenarios. Their study indicated that a level of uncertainty and distrust associated with the conclusions of many of the climate change studies prevails. Also, they indicated the existence of little attention to uncertainties associated with climate change model simulations and cautious reluctance in accepting the possible climate change scenarios.

9.4 TOOLS FOR CLIMATE-SENSITIVE MANAGEMENT OF WATER RESOURCES SYSTEMS

A number of recommendations can be made to handle the issues of climate change in management models. General consensus among modelers and water resources management personnel is needed to agree on and to utilize the results from several climate change studies/investigations that use a wide variety of methodologies and a range of assumptions regarding the magnitude and direction of the change. Conceptually acceptable practical approaches (Teegavarapu and Simonovic, 2002; Teegavarapu, 2010) should be devised to model the decision maker/resource manager's preferences in accepting the magnitude and direction of climate change within the framework of management models.

Compromise operating policies and water resource management models that are climate-sensitive should be devised based on long-term effects of climate change on major hydrologic processes that influence the quantity and frequency of major inputs to the hydrosystems. Sustainable operation of hydrosystems based on short-term and long-term policies must be derived. The long-term policies can be easily derived based on operations of the systems using new hydrologic inputs derived from climate change scenarios. Through the case study example discussed in this chapter and the recommendations provided, the author intends to motivate the water resources community and practicing hydrologists to look into soft-computing modeling approaches, such as fuzzy set theory, for handling the human perceptions associated with future climate change in operational models. Sensitivity analyses based on different climate change scenarios can only inform the risks associated with operation of water resource systems. However, operating rules that are derived based on the preferences of water resources managers are more indicative of how the future changes are perceived by the managers.

9.5 EXAMPLE: GENERATION OF COMPROMISE OPERATING POLICIES FOR FLOOD PROTECTION

The main objective of presenting this example application in this chapter is to discuss issues related to the influence of climate change on water resources management especially related to flood protection. The example also highlights the need for operational policies that bring regulated flows closer to natural flows to sustain river ecosystems. A soft-computing approach, fuzzy set theory (Zadeh, 1965), for handling the preferences attached by the decision makers to magnitude and direction of climate change in water resources management models is discussed. A case study of a multi-purpose reservoir operation is used to address the above issues within an optimization framework and is adopted from the study by Teegavarapu (2010).

9.5.1 Fuzzy set based reservoir operation model

A fuzzy linear programming (FLP) formulation is used for solving a reservoir operation problem and also to address the preferences attached to the direction and magnitude of climate change by the reservoir managers or decision makers. No specific GCM-based scenario directly linked to the case study area is used but an overall reduction in streamflows is considered following the conclusions of a climate change study (Mulholland and Sale, 2002) conducted for the southeastern part of the USA. To demonstrate the utility of the FLP model in the current context, a short-term operation model is developed for Green Reservoir, Kentucky, USA. The primary objective of the reservoir is flood control in the Green River basin as well as in the downstream areas of the Ohio River. Secondary objectives include recreation and low-flow augmentation. An optimization model formulation is developed and solved to fulfill these objectives.

The original FLP formulation (Teegavarapu and Simonovic, 1999) that uses a piecewise linearized non-linear loss (penalty) function defined for storage and release is modified for this purpose. The penalty or loss function, and the penalty values (monetary points) used in this study and within the optimization formulation are shown in Figure 9.2. Higher penalty values are assigned for storage once the value of storage in any time interval is above the maximum storage (i.e., 892.02×10^6 m^3) or below the ideal target value (i.e., 200.99×10^6 m^3). Considering the recreational benefits and flood protection objectives, the reservoir level was maintained high in the summer (reducing releases) and lowered in winter (increasing releases). The penalty values provided in Figure 9.2 are for winter months. Ideal target storage values are emphasized through penalty values to accommodate late winter flows into the reservoir due to spring runoff. The inflow scheme for the time periods for which optimal operation

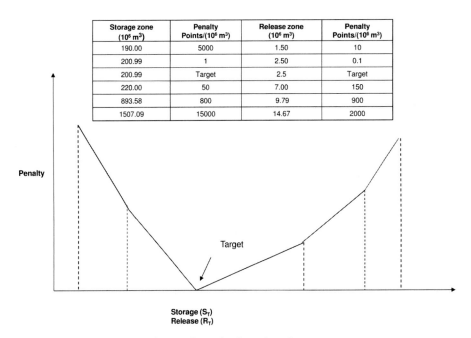

Storage zone (10^6 m^3)	Penalty Points/(10^6 m^3)	Release zone (10^6 m^3)	Penalty Points/(10^6 m^3)
190.00	5000	1.50	10
200.99	1	2.50	0.1
200.99	Target	2.5	Target
220.00	50	7.00	150
893.58	800	9.79	900
1507.09	15000	14.67	2000

Figure 9.2 Linearized penalty function for storage or release and associated penalty values.

rules are required is known. The objective is to minimize the sum of under-achievements or over-achievements (reflected in penalty values) in meeting the storage and release target requirements over a specific time horizon with time interval of one day. Constraints are defined for release and storage zones using upper and lower bounds of these variables along with the best possible target values. Fuzzy constraints are developed for inflows and the problem formulation (original) is solved first, then revised and re-solved twice (intermediate and final) to handle the preferences. The penalty values for storage and release zones and details of the case study can be obtained from earlier work by Can and Houck (1984).

9.5.2 Fuzzy membership functions: preferences towards climate change

Membership functions are generally defined on fuzzy sets to describe the degrees of truth on a scale [0 to 1] for a physical quantity or a linguistic variable. The scale is used to characterize the uncertainty or vagueness associated with the definition of the variable as perceived by humans. Fuzzy membership functions (Zadeh, 1965; Zimmermann, 1987) are used to model the preferences of the decision maker's or reservoir manager's preferences attached to possible variations in hydrologic inputs due to climate change. The nature of the function suggests that the decision maker is more certain about the future reduction in inflow values and therefore increasing preference is attached. The linear membership functions adopted in the current study are used to illustrate the applicability of fuzzy sets in an uncertain decision-making environment. Although they are conceptually

simple and comprehensible, non-linear membership functions can be derived using actual surveys (Fontane et al., 1997). Derivation of non-linear membership functions and their appropriateness for different management problems are discussed by Cox (1999). Practical methods to derive membership functions are discussed in an earlier work by Teegavarapu and Simonovic (1999). Teegavarapu and Elshorbagy (2005) discuss the development and use of membership functions for hydrologic model evaluation. An excellent review of the development of fuzzy membership functions and aggregation operators was provided by Despic and Simonovic (2000). In general, sigmoidal or s-shaped functions are more appropriate to define smooth transitions in the degree of importance (Zimmermann, 1987).

The use of FLP in developing compromise operating policies is illustrated by using an experiment in which a 15% decrease in the daily streamflow values is considered for the operation period. The exact value of percentage decrease in streamflow values is not derived from any specific GCM-based simulation. An arbitrary value of 15% is fixed based on general agreement related to reduction of runoff in the southeastern part of the USA (Mulholland and Sale, 2002) considering a special case of doubled carbon dioxide alone. The assumption regarding the reduction of inflows is appropriate as the case study area is located in the southeastern USA. Three formulations, original, intermediate, and final, are required for solution of the FLP model. The original formulation is solved for normal conditions (constraints and objective function) and the intermediate formulation is solved for increased/decreased values of variables (e.g., inflows). The final formulation is the fuzzy optimization model that considers the membership functions for variables and the objective function. The final formulation

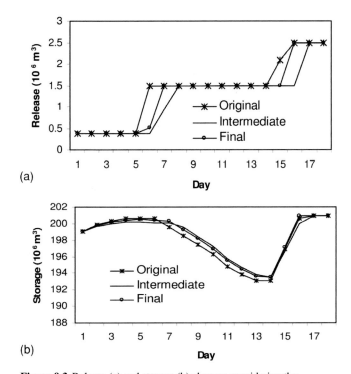

(a)

(b)

Figure 9.3 Release (a) and storage (b) changes considering the decreased inflow scenario for three different formulations.

provides the compromise solution based on fuzzy constraints relevant to decrease in inflow values with an objective of maximizing the overall membership value. More details of these three formulations are available elsewhere (Teegavarapu and Simonovic, 1999).

The final formulation provides the compromise solution based on fuzzy constraints relevant to preferences attached to inflow decrease. The solution of the FLP formulation yields three objective function values, 108.56, 122.19, and 113.86 penalty points/10^6 m^3, and corresponding operating rules. A decrease in the inflow values increases the penalty function (objective) value as less water is available for release to meet storage and release targets. This is reflected in the higher value of the objective function from the intermediate formulation when the inflow values are decreased. In the final formulation, the objective function value obtained is in between that of the initial and intermediate formulations, suggesting that a compromise is achieved based on the conflicting nature of membership functions. The release and storage variations based on three different formulations are shown in Figures 9.3a and 9.3b respectively. It is evident from these figures that release values are reduced to achieve the storage target as the penalty values for under-achieving and over-achieving the storage target are higher than those associated with the release target.

The final membership function value (the level of satisfaction) obtained from the final formulation is 0.63, with a maximum possible value being equal to 1. The final membership value

reflects a compromise based on conflicting behavior of the objective function value that is being minimized and the membership function for inflow variables, with higher preference attached to lower than original values. The membership function for inflow variables increases the penalty value. It can be concluded from the variations observed in storage and release from original values that the reservoir operating rules are sensitive to the decision maker's preferences attached to the magnitude and direction of climate change. Sensitivity analysis of operating rules for a variety of conditions is not equivalent to what is achieved by the fuzzy optimization model discussed in this chapter. A major limitation of the traditional sensitivity analysis approach is its inability to handle the preferences. The conflicting nature of the fuzzy constraints in dealing with inflow values or zones and the objective (monetary units) is captured in this fuzzy optimization framework.

9.6 IMPACTS OF CLIMATE CHANGE ON RESERVOIR OPERATIONS: EXAMPLE FROM BRAZIL

According to IPCC (2007), changes in precipitation patterns and temperature will eventually lead to changes in runoff and water availability. Runoff is predicted to decrease by 10 to 30% over some dry regions at mid latitudes and in the dry tropics. Due to decreases in rainfall and higher rates of evapotranspiration along with spatially expanding drought-affected areas, the potential for adverse impacts on multiple sectors, such as agriculture, water supply, energy production, and health, increases. Energy production and demand are heavily influenced by climate change, and regions with heating and cooling needs are affected the most (USGS, 2009). The effects of climate change on hydropower plant operation are explored here as the important economic value of hydropower justifies the need to obtain an optimal operational policy suitable to future scenarios (Larijani, 2009). An example of reservoir operation in the southern part of Brazil is used to illustrate the impacts of reduced inflows on power generation. According to IPCC (2007), the region where the case study power plant is located is prone to suffer a 5 to 10% reduction in runoff by the end of the twenty-first century (2090–9) in relation to the climatologic normal observed between years 1980 and 1999. In areas affected by drought conditions, a reduction in hydropower generation potential is expected. Details of the hydropower reservoir are provided in the study by Ferreira and Teegavarapu (2012).

While maintaining the specific water quality requirements associated with nutrients, a critical scenario is evaluated by reducing the average reservoir inflows by 10%. In comparison to actual average inflows, a decrease of 10% in reservoir inflows represents a reduction of 9.68% in the overall energy production for

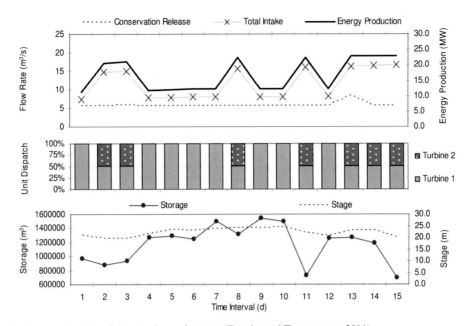

Figure 9.4 Operational policy as a function of climate change forecasts (Ferreira and Teegavarapu, 2011).

a 15-day period (251.83 MW). The results are summarized in Figure 9.4.

The findings of Larijani (2009) in evaluating the effects of climate change on high-elevation hydropower reservoirs indicate that under a drought condition scenario, where runoff is reduced by 20%, revenues decrease by less than 14%. Negative effects on hydropower generation should not be related only to reduced reservoir inflow rates. Actual operating policies in place are derived from complex factors incorporating physical, socioeconomic, and environmental components. Due to an increasing population and higher demands on a resource with decreasing availability, ever greater conflicts between such components of management of water resources systems are expected to arise. The importance of reviewing the operational policies of essential water resource systems as a function of climate change is reinforced by Burn and Simonovic (1996).

9.7 CLIMATE CHANGE AND FUTURE HYDROLOGIC ENGINEERING PRACTICE

It is important to note that optimality alone is not a sufficient characteristic of acceptable system design. Water resources managers can be expected to be "surprised" in the future because of the increasing complexity of water resources management. Surprises are counter-expected or unexpected events, and water resources planners should minimize the likelihood of surprises (Fiering and Kindler, 1987). Adaptive management is an approach that makes decisions sequentially over time and allows adjustments to be made as more information is known. This approach may be useful

in dealing with the additional uncertainty introduced by potential climate change (USGS, 2009).

Flood frequency analysis under a changing climate scenario is one of the most difficult and controversial aspects of future hydrologic engineering practice. Several alternatives are offered by researchers to avoid or get around the enforcement of the stationarity assumption in the hydrologic time series analysis. One method is to use an appropriate factor (i.e., enlargement factor) for multiplying the design flood estimate to reflect future climate changes and non-stationary conditions (DEFRA, 2002). Use of rainfall–runoff models to simulate current and future hydrologic conditions with the idea of using perturbation factors for future climate simulations using current climatic time series is recommended in many studies (Reynard et al., 1999; Prudhomme et al., 2001, 2005; Reynard, 2007). Direct use of outputs from regional climate models to estimate flood frequency was documented by Kay et al. (2005). Inter-event time definition (i.e., minimum dry period between storm events, or IETD) (Behera et al., 2010) is a very crucial parameter that is often neglected in the analysis of rainfall time series for hydrologic design. Based on an IETD, a long-term rainfall record can be analyzed for different storm event characteristics, such as rainfall event volume, duration, intensity, and inter-event time. The design storms adopted considering IETD can influence the stormwater storage decisions within urban watersheds. Moreover, current interests in low-impact development (LID) (Watson and Adams, 2011) require careful analysis of rainfall and runoff volume in the analysis and design of best management practices. Due to change in the future climate, IETD is expected to change, and impacts on hydrologic infrastructure developed for flood protection will need to be addressed.

9.8 FLOODS: STATIONARITY AND NON-STATIONARITY ISSUES

Hydrologic design and practice continue to rely on the assumption that the past climate represents the future climate. There seems to be a gap existing in communication and differences of opinion between climate change scientists and hydrologists in addressing the issue of stationarity. Almost any study attempted today for hydrologic infrastructural design is faced with the question of resolving the stationarity issue and every possible statistical test is being conducted to confirm the existence or non-existence of stationarity in hydrologic time series. Galloway (2011) asks two questions: (1) Is stationarity dead? (2) If it is dead, what do we do now with existing hydrologic design practices? A summary and review of several research papers (Hodgkins and Dudley, 2006; Kiang *et al.*, 2011; Kundzewicz, 2011; Lins and Cohn, 2011; Ouarda and El-Adlouni, 2011; Villarini *et al.*, 2011; Vogel *et al.*, 2011) provide a number of questions raised by these researchers and they are:

- Have the flood-generating mechanisms (e.g., drivers) changed over the past century?
- Can the causes for changes in these mechanisms be easily identified?
- Can the influence of anthropogenic influences on hydrologic processes be confirmed?
- What is the temporal window size to evaluate the existence and non-existence of trends in data?
- How do we incorporate climate change uncertainties into the definition of *n*-year floods?
- How do we evaluate flood risks in future?
- What are the new or emerging statistical methods that are available for hydrologic frequency analysis in the presence of non-stationarity?

Villarini *et al.* (2011) and Vogel *et al.* (2011), based on their analysis of peak discharges in the USA, indicate that it is difficult to pinpoint the causes of changing flood characteristics and generating mechanisms over time. Land-use changes are documented to be the major cause of changing flood characteristics along with contributing factors from climate change. Periodic changes from long-term persistence and natural variations in the hydrologic time series make the process of determining the anthropogenic influences difficult (Lins and Cohn, 2011). Use of the Hurst–Kolmogorov framework to explicitly consider the long-term persistence in hydrologic time series analysis was promoted by Koutsoyiannis (2011). Stedinger and Griffis (2011) indicate that when flood-generating causes are not clearly understood, interpretation of trends becomes extremely difficult. Ouarda and El-Adlouni (2011) use a Bayesian approach to address the issue of non-stationarity. Galloway (2011) suggests that some of the issues related to stationarity are critical to future hydrologic

design of infrastructure for flood protection, and more research is required to understand the implications on non-stationarity and new approaches for planning future water resources projects.

9.9 EXTREME PRECIPITATION: ISSUES FOR THE FUTURE

Studies relating to future extreme precipitation events based on climate change projections conditioned on different emissions scenarios provide information about overall trends in average precipitation values. However, several critical issues need to be addressed for climate-change sensitive and sustainable water resources infrastructure design to handle future climate change, especially related to extremes. Future research should focus on the following:

1. Improvement of the downscaling methods to obtain accurate finer temporal resolution (e.g., daily or hourly) precipitation projections for the future.
2. Development of probabilistic IDF curves to address the uncertainty in projection of precipitation extremes by different climate change models.
3. Improvement of site-specific and regional frequency analysis methods to address spatially variable and temporally evolving distribution parameters of statistical distributions that characterize precipitation extremes.
4. Development of statistical downscaling models for precipitation needs to be improved so that precipitation extremes are accurately reproduced.
5. Modification of hydrologic design procedures based on short- and long-term climate variability especially due to teleconnections.

9.10 INSTITUTIONAL CHANGES AND ADAPTATION CHALLENGES

Adaptation of existing water programs and infrastructure to climate change will be a long and iterative process. The understanding of the impacts of climate change on water that is now emerging from scientific studies, however, provides a sufficient basis for defining an initial set of preliminary steps to adapt water programs to climate change (USEPA, 2008). Several adaptation frameworks are provided by different world agencies, and they differ in their approaches based on the target audience. Lu (2010) provides an in-depth assessment of adaptation strategies for application of climate change information in decision making.

Incorporation of adaptation programs to counter the impacts of climate variability and change should be carried out by collecting historical climate data and sources of climate data, utilizing regional climate change models and modeling scenarios for the major climate parameters, and developing knowledge related to climate variability/climate change impacts tailored to the region and relevant development sectors. Examples of adaptation strategies are discussed for different case study regions across the world in a comprehensive document by USAID (2007). The main challenges of developing adaptation and vulnerability assessment programs are lack of monitoring data, non-availability of region-specific climate change information, and non-existence of capacity-building mechanisms.

9.11 CONCLUSIONS AND SUMMARY

Substantial changes in the operations of water resources systems that are climate-change sensitive are required for future water resources planning and management. Sustainable operations that consider long- and short-term objectives along with consideration of uncertainties in the climate change modeling results are discussed in this chapter. A few ideas are presented on how to deal with uncertainties in climate change scenarios and they are addressed in a compromise operations policy framework. An example case study problem using a fuzzy set theory approach in dealing with the decision maker's preferences attached to climate change (magnitude and direction) is presented. The approach deals with the development of compromise operating policies. Future water resources management heavily relies on the abilitiy to evaluate extreme precipitation events, how these events are characterized by techniques dealing with non-stationary conditions, and appropriate assessments of vulnerability and risk of hydrologic and water resources infrastructure and adaption to climate change.

EXERCISES

9.1 Define issues that need to be considered for adaptive and sustainable hydrologic design and operation of water resources systems under changing climate conditions.

9.2 Develop optimization models for operation of water resources systems (especially reservoir systems) and address uncertainties associated with the future inflows based on any climate change model-specifc scenarios in your region.

9.3 Use the website provided below to obtain daily and monthly future precipitation projections for a region in the USA and discuss the skill of climate change models. http://gdo-dcp. ucllnl.org/downscaled_cmip3_projections/dcpInterface.html #Welcome.

9.4 Develop operation models for operation of water resources systems (scenario-based operation models) to address uncertainties associated with the future inflows based on general trends of magnitude and direction of climate change in the region of interest.

9.5 Develop a fuzzy mathematical programming formulation to address uncertainties in climate change projections of streamflow and precipitation series within an operation model.

Glossary

Accumulated rainfall – Accumulated rainfall over a specific temporal domain.

Alternative hypothesis – A hypothesis that is opposite to a null hypothesis.

Anisotropy – Directional variation of a property of a process.

Anomalous propagation – The abnormal bending of a radar beam as it propagates or passes through the atmosphere.

Artificial neural network – A universal function approximator model built based on the concept of the processing capabilities of neurons in the brain.

Attenuation – The reduction in power of a signal due to refraction or scattering of energy.

Autocorrelation – A characteristic of time series of any variable that exhibits time dependence. The value of correlation can be calculated for different temporal lags.

Base reflectivity – Reflectivity value at a specific elevation which has been obtained directly from the base reflectivity data.

Beam blockage – Obstruction of a radar beam, usually close to the radar unit, caused by any physical object in the vicinity.

Below-beam effect – Inaccurate radar measurement caused by incomplete sampling of the atmosphere. Examples of such an effect include evaporation or growth of precipitation below the beam that is not detected by radar.

Bias – Difference between radar-based rainfall estimation and actual rain gage observation.

Binning – The process of defining the lagged distances for calculation of semi-variogram models.

Bivariate normal distribution – A two-variable distribution with a linear dependence relationship where each variable follows a normal distribution.

Bootstrap samples – Samples generated from a fixed sample set by replacement.

Climate change – Climate change refers to a statistically significant variation in either the mean state of the climate or in its variability, persisting for an extended period (typically decades or longer) (definition according to WMO). A long-term alteration in the climate.

Climate variability – Variations in the mean state and other statistics (variations, the occurrence of extremes, frequency) of the climate on all temporal and spatial scales beyond that of individual episodic events (definition according to WMO). The inherent variability in the climate.

Clutter suppression – A process by which the ground clutter contamination caused by non-meteorological ground return is removed.

Control points – The points in space where observations are available for spatial interpolation.

Correlation – A measure of the association between two variables. A linear form of dependence between variables under analysis.

Correlation coefficient – A measure of correlation between two variables reported as a numerical value, with a range between -1 and 1, where 1 means complete positive dependence, -1 negative dependence, and 0 independence.

Cross-validation – The procedure of using a subset of available (observed) data for validation of a model.

Data mining – The process of extracting useful and non-trivial information about data.

dBZ – A unit commonly used to specify reflectivity. It is referred to as a decibel of radar reflectivity value and is given by $dBZ = 10 \log_{10}(\text{reflectivity})$.

Dependence – Numerical relationship between variables and the extent to which one can be predicted entirely from knowledge of the others.

Desk study approach – A joint probability analysis approach used by DEFRA (2005).

Discharge (or flow) – Rate of water movement past a reference point, measured as volume per unit time.

Disdrometer – An instrument used to measure the drop size distribution and velocity of falling hydrometeors.

Doppler effect – The observed change in the frequency of sound or electromagnetic waves due to the relative motion of the source and observer.

Ducting – A special condition of super-refraction in which the radar beam becomes trapped within a specific layer of the atmosphere.

Echo – Energy backscattered from a target and received by radar.

Elevation angle – The angle between the horizon and a point above the horizon. Usually referred to the angle of the radar antenna.

Events – Episodic occurrences often identified objectively as records in which a threshold of interest is exceeded amongst the environmental or response variables.

Exact interpolation – Interpolation in which the observation values at the control points are strictly honored.

Exceedence probability – Probability of exceedence for a particular value of a variable.

Exploratory data analysis – Preliminary quantitative and visual analysis of data to develop a basic understanding of the data.

Extreme – Unusually high value of a variable, which is used to calculate the exceedence probability.

Extreme events or X-events – Values that are substantially different from the ordinary or usual. (Floods and droughts are extreme hydrologic events.)

Fuzzy membership function – A linear or non-linear function used to define preferences.

Global interpolation – An interpolation scheme that does not limit the number of control points used in the interpolation process.

Ground clutter – Radar echoes that result from physical obstructions.

Ground truth – A term used in cartography, meteorology, and allied disciplines referring to an observation made at a location.

Heterogeneity – Variability in samples or a process.

Heteroscedasticity – Non-constant variance in data.

Homogeneity – Data sample exhibiting no variability.

Homoscedasticity – Indicates that the variance of the data is constant.

Hydrometeor – In general, any form of precipitation. A product from condensation of atmospheric vapor formed in the free atmosphere or at the Earth's surface.

Hyporheic zone – A subsurface layer where mixing of shallow groundwater and surface water takes place.

Independence – Status of no dependence between any two variables.

Inexact interpolation – Interpolation in which the observation values at the control points are not strictly honored.

Interpolation – The process of obtaining a missing value in space or time based on observations available at other control points in space or time. Estimating values of a variable in time and space based on available observations.

Inverse distance weighting method – A deterministic exact interpolation method in which the observations from control points are distance-weighted for generation of spatially interpolated maps or missing values at points of interest.

Joint exceedence – The probability that two related variables will simultaneously exceed specified values, e.g., rainfall and discharge are higher than specific values.

Joint probability – A probability referring to the distribution and extremes of two or more related variables.

Joint probability analysis – An approach used for analysis of joint probabilities with the evaluation of dependence and prediction of extreme conditions.

Kriging – A stochastic inexact interpolation method in which the spatial autocorrelation is modeled using the semi-variance at difference distance lags.

Kurtosis – A measure of the peakedness of a distribution. The kurtosis coefficient is equal to 3 for Gaussian distributions.

Lag – A term used in time series methods. A lag 1 observation is an observation from the previous time step. A lag 2 observation is an observation from two time steps before.

Lagged dependence – Dependence between two variables, based on a temporal lag.

Local interpolation – An interpolation scheme that limits the number of control points used in the interpolation process.

Maximum likelihood – A common method for estimation of distribution parameters by maximizing likelihood function.

Mean – Average value of a series.

Median – The middle ranking value of a series.

Multiple regression function – the relationship between several independent or predictor variables and the dependent variable.

Multivariate analysis – A method for simultaneous analysis of a number of dependent variables (of time and/or space), including links between these variables.

Neighborhood – A term used in spatial interpolation to define surrounding observation points or grids around one specific point or grid.

Non-parametric test – A test that does not involve estimation of parameters.

Non-stationarity – Indicates that the distribution of a random variable is changing with time (e.g., increasing mean).

Normal distribution – a symmetrical probability distribution (Gaussian distribution), specified by mean and standard deviation parameters.

Nugget – Random variance associated with zero lag distance (applicable in kriging).

Null hypothesis – A hypothesis to be tested.

Partial beam filling – The effect that occurs when a target fills only a small portion of the radar beam.

Peaks over threshold – A method of identifying and collecting independent maxima above a given threshold to prepare data for extremes analysis.

Persistence – Property of long memory of the system, whereby high or low values (excursions to high or low states) are clustered over longer time periods. Persistence is also referred to as autocorrelation or serial correlation.

Point estimation – Estimation of a value at a point in space.

Probability density function – Function describing the distribution of data; it expresses the relative likelihood that a random variable attains different values.

Quantitative precipitation estimate – An estimate of the total precipitation that has fallen during a specified interval of time.

Quaternary period – A period that extends from about 1.8 million years ago to the present.

Radar – An acronym for radio detection and ranging. A radar unit is used to measure reflectivity values.

Radome – A structure that houses weather radar.

Random error – An error mainly due to measurement processes that is not consistent and not predictable.

Record – Temporal record of observations for one or more variables in space.

Records per year – Average number of records per year, needed to assign exceedence probabilities to high values.

Refraction – The process in which the direction of energy propagation is changed as the result of a change in density. Large changes in the refractive index of the atmosphere can cause sub-refraction and super-refraction.

Regression analysis – A technique for modeling and analyzing the relationship between a dependent variable and one or many independent variables.

Regression model – A mathematical model that relates dependent and independent variables to develop a function relationship for prediction.

Return period – Estimate of the average period between successive exceedences (events) of a specific threshold.

Runs test – A non-parametric test (refer to Wald–Wolfowitz test).

Sample size – Number of data elements in the sample.

Seasonality – Seasonal (periodic) behavior of variable.

Segmentation – Method of finding abrupt changes in the time series by fitting a step-wise function (parameters: amplitude and time instant of a change).

Semi-variance – Variance based on observations at two locations.

Skewness – A measure of the asymmetry.

Smoothing – Replacement of the raw series by a more regular function of time that has less variability.

Stage – Height of a water surface above an established reference point (datum or elevation).

Standardization – A data transformation method that is achieved by subtracting the mean of the data and dividing by the standard deviation.

Stationarity – Property of constant or non-variable statistical moments over time.

Step change – A jump or sudden change in a time series.

Structure – Man-made pump stations, reservoirs, channel improvements, canals, levees, and diversion channels.

Super-refraction – Bending of a radar beam in the vertical that is greater than under standard refractive conditions. This causes the beam to be lower than indicated, and often results in extensive ground return as well as an overestimation of cloud top heights.

Surface generation – Generation of a field of values in space using uniform gridded observation points or non-uniform point observations.

Systematic error – An error mainly due to instrumental and measurement errors that are consistent and predictable.

Teleconnections – Internal modes of climate variability defined by spatially varying climate anomalies.

Thiessen polygons – These polygons define the areas of influence around each point in such a way that the polygon boundaries are equidistant from neighboring points, and each location within a polygon is closer to its contained point than to any other point.

Threshold – A particular value (sometimes specified by exceedence probability) of a variable, above which it is regarded as an extreme value needing special analysis.

Transfer function (downscaling) – A linear or non-linear functional form linking predictors and predictand.

Transfer function (neural networks) – A function used in the input, hidden, and output layer neurons in artificial neural networks to transform inputs and outputs. Sigmoidal functions are generally used as transfer functions.

Trend – A gradual and noticeable temporal change in the mean value of a variable.

Type I error – Error that occurs when a null hypothesis is rejected when it is true.

Type II error – Error that occurs when a null hypothesis is accepted when it is false.

Validation – A procedure used to evaluate model performance based on parameters established through the calibration process.

Virga – Refers to precipitation that evaporates before reaching the Earth's surface.

Volume coverage pattern – The particular scanning strategy employed by the WSR-88D in the USA.

Voronoi polygons – Another name for Thiessen polygons constructed from point observations.

Wald–Wolfowitz test – A non-parametric test to check the randomness of a series.

Weather generator – A stochastic mathematical model that can be used to generate weather variable values at different temporal scales.

Weather Surveillance Radar – 88 Doppler (WSR-88D) – A meteorological surveillance radar deployed by the National Weather Service of the USA.

X-events – Extremes or extreme events in general that are natural and unnatural.

Note: Some of the definitions are adopted from WMO and NOAA web-based publications.

References

Abdul Aziz, O. I. and Burn, D. H. (2006). Trends and variability in the hydrological regime of the Mackenzie River Basin. *Journal of Hydrology*, **319**, 282–294.

Abramowitz, M. and Stegun, I. A. (1965). *Handbook of Mathematical Functions: with Formulas, Graphs, and Mathematical Tables*, Applied Mathematics Series 55. New York: Dover Publications.

Adams, B. J. and Howard, C. D. D. (1986). Pathology of design storms. *Canadian Water Resources Journal*, **11**(3), 49–55.

Adams, B. J. and Papa, F. (2000). *Urban Stormwater Management Planning with Analytical Probabilistic Models*. New York: John Wiley and Sons.

Adler, R. F., Huffman, G. J., Chang, A., *et al.* (2003). The Version 2 Global Precipitation Climatology Project (GPCP) monthly precipitation analysis (1979–present). *Journal of Hydrometeorology*, **4**, 1147–1167.

Agrawal, R., Imielinski, T., and Swami, A. (1993). Mining association rules between sets of items in large databases. *Proceedings of the ACM SIGMOD International Conference on Management of Data*, Washington, DC, 207–216.

Ahrens, B. (2006). Distance in spatial interpolation of daily rain gage data. *Hydrology and Earth System Sciences*, **10**, 197–208.

Ahuja, P. R. (1960). Planning for precipitation network for water resources development in India. *WMO Flood Control Series*. Report No. **15**, 106–112.

Akaike, H. (1974). A new look at the statistical model identification. *IEEE Transactions on Automatic Control*, **19**(6), 716–723.

Alexandersson, H. (1986). A homogeneity test applied to precipitation data. *Journal of Climatology*, **6**, 661–675.

Alia, Y. (2000). Regional rainfall depth-duration frequency equations for Canada. *Water Resources Research*, **36**, 1767–1778.

Alley, W. M. (1984). On the treatment of evapotranspiration, soil moisture accounting, and aquifer recharge in monthly water balance models. *Water Resources Research*, **20**(8), 1137–1149.

Alley, W. M. (1985). Water balance models in one-month-ahead stream flow forecasting. *Water Resources Research*, **21**(4), 597–606.

Allison, P. D. (1998). *Multiple Regression: A Primer*. Thousand Oaks: Pine Forge Press.

Aly, A., Pathak, C., Teegavarapu, R. S. V., Ahlquist, J., and Fuelberg, H. (2009). Evaluation of improvised spatial interpolation methods for infilling missing precipitation records. *World Environmental and Water Resources Congress*, ASCE, 1–10.

Amitai, E. (2000). Systematic variation of observed radar reflectivity-rainfall rate relations in the tropics. *Journal of Applied Meteorology*, **39**, 2198–2208.

Andrea, I. P. and Depetris, P. J. (2007). Discharge trends and flow dynamics of South American rivers draining the Southern Atlantic Seaboard: an overview. *Journal of Hydrology*, **333**, 385–399.

Andronova, N. G. and Schlesinger, M. E. (2000). Causes of global temperature changes during 19th and 20th centuries. *Geophysical Research Letters*, **27**, 2137–2140.

Angstrom, A. (1935). Teleconnections of climate changes in present time. *Geografiska Annaler*, **17**, 242–258.

AOML (2011). http://www.aoml.noaa.gov/phod/amo_faq.php (accessed December, 2011).

Arnell, N. (1996). *Global Warming, River Flows and Water Resources*. New York: John Wiley and Sons.

ASCE (1996). *Hydrology Handbook*, second edition. New York: American Society of Civil Engineers.

ASCE (2001a). Task committee on artificial neural networks in hydrology. I. Preliminary concepts. *Journal of Hydrologic Engineering*, ASCE, **52**, 115–123.

ASCE (2001b). Task committee on artificial neural networks in hydrology. II. Hydrologic applications. *Journal of Hydrologic Engineering*, ASCE, **52**, 124–137.

Ashkar, F. (1996). Extreme floods. In *Hydrology of Disasters*, V. P. Singh (editor), Water Science and Technology Library, Dordrecht: Kluwer Academic Publishers.

Ashraf, M., Loftis, J. C., and Hubbard, K. G. (1997). Application of geostatisticals to evaluate partial weather station network. *Agricultural Forest Meteorology*, **84**, 255–271.

Atlas, D. (editor) (1990). *RADAR in Meteorology*. Boston, MA: American Meteorological Society.

Atlas, D., Ulbrich, C. W., Marks, F. D., Amitai, E., and Williams, C. R. (1999). Systematic variation of drop size and radar–rainfall relations. *Journal of Geophysical Research*, **104**, 6155–6169.

Austin, P. M. (1987). Relation between measured radar reflectivity and surface rainfall. *Monthly Weather Review*, **115**, 1053–1070.

Baldonado, F. D. (1996) Experiments in determination of radar reflectivity – rainfall intensity (Z-R) relations in the Philippines. *Philippine Atmospheric, Geophysical and Astronomical Services Administration*, AGSSB Abstracts.

Barnes, R. J. and Johnson, T. B. (1984). Positive kriging. In *Geostatistics for Natural Resources Characterization, Part 1*, G. Verly (editor), Dordrecht: Reidel.

Barnes, S. L. (1964). A technique for maximizing details in numerical weather-map analysis. *Journal of Applied Meteorology*, **3**(4), 396–409.

Barsugli, J., Anderson, C., Smith, J. B., and Vogel, J. M. (2009). *Options for Improving Climate Modeling to Assist Water Utility Planning for Climate Change*. Available online at http://www.wucaonline.org/assets/pdf/actions_whitepaper_120909.pdf (accessed December, 2011).

Basilevsky, A. (1983). *Applied Matrix Algebra in the Statistical Sciences*. Mineola: Dover Publications.

Battan, L. J. (1973). *Radar Observation of the Atmosphere*. Chicago: The University of Chicago Press.

BCSD (2011). http://gdo-dcp.ucllnl.org/downscaled_cmip3_projections/dcpInterface.html (accessed September, 2011).

Bedient, P. B., Huber, W. C., and Vieux, B. E. (2009). *Hydrology and Floodplain Analysis*. Upper Saddle River: Prentice Hall.

Beek, E. V. (2010). Managing water under current climate variability. In *Climate Change Adaptation in Water Sector*, F. Ludwig, P. Kabat, H. Schaik, and M. Valk (editors), London: Earthscan.

Behera, P., Guo, Y., Teegavarapu, R. S. V., and Branham, T. (2010). Evaluation of antecedent storm event characteristics for different climatic regions based on inter-event time definition (IETD). *Proceedings of the World Environmental and Water Resources Congress 2010*, ASCE, doi:10.1061/41114(371)251.

Bell, F. C. (1969). Generalized rainfall-duration-frequency relationships. *Journal of Hydraulic Division*, ASCE, **95**(HY1), 311–327.

Bellman, R. E. and Zadeh, L. A. (1970). Decision-making in a fuzzy environment. *Management Science*, **17**, 141–164.

Benestad, R. E., Hanssen-Bauer, I., and Chen, D. (2008). *Empirical-Statistical Downscaling*. Singapore: World Scientific.

Betancourt, J. (2009). Coping with non-stationarity in water and ecosystem management. *9th Annual SAHRA Meeting*, September 23–24.

Beven, K. (1989). Changing ideas in hydrology – the case of physically-based models. *Journal of Hydrology*, **105**, 159.

Bivand, R., Pebesma, E., and Rubio, V. (2008). *Applied Spatial Data Analysis with R*. Use R Series, Heidelberg: Springer.

Blanchard, D. O. and Lopez, R. E. (1985). Spatial patterns of convection in South Florida. *Monthly Weather Review*, **113**, 1282–1299.

Blom, G. (1958). *Statistical Estimates and Transformed Beta-Variables*. New York: John Wiley and Sons.

Bobee, B. and Ashkar, F. (1991). *The Gamma Family and Derived Distributions Applied in Hydrology*. Littleton: Water Resources Publications.

Bogardi, J. J. and Kundzewicz, Z. W. (2002). *Risk, Reliability, Uncertainty and Robustness of Water Resources Systems*, International Hydrology Series. Cambridge: UNESCO and Cambridge University Press.

Bonnin, G. M., Martin, D., Lin, B., *et al.* (2006). *Precipitation-Frequency Atlas of the United States, Atlas 14*. NOAA.

Boots, B. N. (1985). *Voronoi (Thiessen) Polygons*, Concepts and Techniques in Modern Geography, No. 45. Norwich: Geo Books.

Bortot, P., Coles, S. G., and Tawn, J. A. (2000). The multivariate Gaussian tail model: an application to oceanographic data. *Applied Statistics*, **49**, 31–49.

Bourke, P. (1987). http://paulbourke.net/geometry/insidepoly/.

Box, G. E. P. and Cox, D. R. (1964). An analysis of transformations. *Journal of the Royal Statistical Society, Series B*, **26**, 211–252.

Brekke, L. D., Kiang, J. E., Olsen, J. R., *et al.* (2009). *Climate Change and Water Resources Management – A Federal Perspective*. US Geological Survey Circular 1331.

Brimicombe, A. (2003). *GIS, Environmental Modeling and Engineering*. London: Taylor and Francis.

Brooke, A., Kendrik, D., and Meeraus, A. (1996). *GAMS: A User's Guide*. GAMS Development Corporation.

Brown, J., Dingman, S. L., and Lewellen, R. J. (1968). *Hydrology of a Drainage Basin on the Alaskan Coastal Plains*. US Army Cold Regions Research and Engineering Lab., Hanover, Research Report 240.

Brunet, M., Saladié, O., Jones, P., *et al.* (2008). *A Case-Study/Guidance on the Development of Long-Term Daily Adjusted Temperature Datasets*. WCDMP-66, WMO, Geneva.

Buishand, T. A. (1982). Some methods for testing the homogeneity of rainfall records. *Journal of Hydrology*, **58**, 11–27.

Burger, G. (1996). Expanded downscaling for generating local weather scenarios. *Climate Research*, **7**, 111–128.

Burn, D. H. and Elnur, H. (2002). Detention of hydrologic trend and variability. *Journal of Hydrology*, **255**, 107–122.

Burn, D. H. and Simonovic, S. P. (1996). Sensitivity of reservoir operation performance to climatic change. *Water Resources Management*, **10**(6), 463–478.

Burrough, P. A. and McDonnell, R. A. (1998). *Principles of Geographical Information Systems*. New York: Oxford University Press.

Byers, H. R. and Rodebush, H. R. (1948). Causes of thunderstorms of the Florida peninsula. *Journal of Meteorology*, **5**, 275–280.

Can, E. K. and Houck, M. H. (1984). Real-time reservoir operation by goal programming. *Journal of Water Resources Planning and Management*, ASCE, **110**, 297–309.

Castellarin, A., Burn, D. H., and Brath, A. (2001). Assessing the effectiveness of hydrological similarity measures for flood frequency analysis. *Journal of Hydrology*, **241**, 270–285.

Chakrabarti, S., Cox, E., Frank, E., *et al.* (2009). *Data Mining: Know It All*. Burlington: Morgan Kaufmann.

Chang, K.-T. (2009). *Introduction to Geographic Information Systems*, fifth edition. New York: McGraw Hill.

Chang, K.-T. (2010). *Introduction to Geographic Information Systems*, sixth edition. New York: McGraw Hill.

Chang, K.-T. and Li, Z. (2000). Modeling snow accumulation with a geographic information system. *International Journal of Geographical Information Science*, **14**(7), 693–707.

Changnon, S. A. and Kunkel, K. E. (1995). Climate-related fluctuations in Midwestern flooding. *Journal of Water Resources Planning and Management*, **121**, 326–334.

Chatters, J. C. and Hoover, K. A. (1986). Changing late holocene flooding frequencies on the Columbia River, Washington. *Quaternary Research*, **26**, 309–320.

Chen, C. L. (1983). Rainfall intensity-duration-frequency formulas. *Journal of Hydraulic Engineering*. **109**, 1603–1621.

Chen, M. S., Han, J., and Yu, P. S. (1996). Data mining: an overview from a database perspective. *IEEE Transactions on Knowledge and Data Engineering*, **8**, 866–883.

Chernick, M. R. (1999). *Bootstrap Method: A Practitioner's Guide*. New York: Wiley.

Chiew, F. H. A., Piechota, T. C., Dracup, J. A., and McMahon, T. A. (1998). El Niño southern oscillation and Australian rainfall, streamflow and drought: links and potential for forecasting. *Journal of Hydrology*, **204**, 138–149.

Chiew, F. H. S., Zhaou, S. L., and McMahon, T. A. (2003). Use of seasonal streamflow forecasts in water resources management. *Journal of Hydrology*, **270**, 135–144.

Chiles, J.-P. and Delfiner, P. (1999). *Geostatistics: Modeling Spatial Uncertainty*. New York: Wiley.

Chin, D. (2006). *Water Resources Engineering*. Upper Saddle River: Prentice Hall.

Chou, Y. H. (1997). *Exploring Spatial Analysis in Geographic Information Systems*. New York: OnWord Press.

Chow, V. T. (1951). A general formula for hydrologic frequency analysis. *Transactions, American Geophysical Union*, **32**(2), 231–237.

Christensen, J. H., Risnen, J., Iversen, T., *et al.* (2001). A synthesis of regional climate change simulations: a Scandinavian perspective. *Geophysics Research Letters*, **28**(6), 1003–1006.

Christensen, N. S., Wood, A. W., Voisin, N., Lettenmaier, D. P., and Palmer, R. N. (2004). The effects of climate change on the hydrology and water resources of the Colorado River basin. *Climatic Change*, **62**, 337–363.

CICS (2011). Canadian Institute for Climate Studies, http://www.cics.uvic.ca/scenarios/index.cgi?More_Info-Downscaling_Background (accessed September, 2011).

CIRIA (1996). *Beach Management Manual*. CIRIA Report 153.

Clark, M. P., Serreze, M. C., and McCabe, G. J. (2001). Historical effects of El Niño and La Niña events on the seasonal evolution of the montane snowpack in the Columbia and Colorado River Basins. *Water Resources Research*, **37**(3), 741–757.

Clement, P. C. (1995). A comparison of radar-derived precipitation and rain gage precipitation in Northeastern Colorado. Master of Science Thesis, Colorado State University, Fort Collins, USA.

Coles, S. G., Heffernan, J., and Tawn, J. (1999). Dependence measures for extreme value analyses. *Extremes*, **2**, 339–365.

Collier, C. G. (1989). *Applications of Weather Radar Systems: A Guide To Uses Of Radar Data In Meteorology and Hydrology*. New York: John Wiley.

Collier, C. G. and Hardaker, P. J. (1996). Estimating probable maximum precipitation using a storm model approach. *Journal of Hydrology*, **183**, 277–306.

Collier, C. G. and Hardaker, P. J. (2007). Using radar in hydrometeorology. In *Weather Radar: Principles and Advanced Applications*, P. Meischner (editor), Berlin: Springer.

Collins, W. D., Bitz, C. M., Blackmon, M. L., *et al.* (2006). The Community Climate System Model Version 3 (CCSM3). *Journal of Climate*, **19**(11), 2122–2143.

Congalton, R. G. and Green, K. (1999). *Assessing the Accuracy of Remotely Sensed Data: Principles and Practices*. Boca Raton: Lewis.

Conger, D. H. (1978). Method for determining baseflow adjustments to synthesized peaks produced from the US Geological Survey rainfall-runoff model. *Hydrological Sciences*, **23**(4), 401–408.

Corder, W. G. and Foreman, D. I. (2009). *Nonparametric Statistics for Nonstatisticians*. Hoboken: Wiley.

Cox, E. (1999). *The Fuzzy Systems Handbook: A Practitioner's Guide to Building, Using, and Maintaining Fuzzy Systems*. San Diego: AP Professional.

Cromwell, J. B., Labys, W. C., and Terraza, M. (1994). *Univariate Tests for Time Series Models*, 07-99. London: Sage Publications.

Cronin, T. M. (2009). *Paleoclimates: Understanding Climate Change Past and Present*. New York: Columbia University Press.

CRU (2011). http://www.cru.uea.ac.uk/ (accessed December, 2011).

Cullen, A. C. and Frey, H. C. (1999). *Probabilistic Techniques in Exposure Assessment*. New York: Plenum Press.

Cullen, H. M. and deMonocal, P. B. (2000). North Atlantic influence on Tigris-Euphrates stream flow. *International Journal of Climatology*, **20**, 853–863.

Cunderlik, J. M. and Burn, D. H. (2002). Analysis of the linkage between rain and flood regime and its application to regional flood frequency estimation. *Journal of Hydrology*, **261**, 115–131.

Cunderlik, J. M., Ouarda, T. B. M. J., and Bobee, B. (2004). Determination of flood seasonality from hydrological records. *Hydrological Sciences (Journal des Sciences Hydrologiques)*, **49**(3), 511–526.

Dai, A., Fung, J., and Del Genio, A. (1997). Surface observed global land precipitation variations during 1900–1988. *Journal of Climate*, **11**, 2943–2962.

Daly, C., Neilson, R. P., and Phillips, D. L. (1994). A statistical topographic model for mapping climatological precipitation over mountainous terrain. *Journal of Applied Meteorology*, **33**, 140–158.

Daly, C., Gibson, W. P., Taylor, G. H., Johnson, G. H., and Pasteris, P. (2002). A knowledge-based approach to the statistical mapping of climate. *Climate Research*, **22**, 99–113.

Dam, J. C. (1999). *Impacts of Climate Change and Climate Variability on Hydrological Regimes*, International Hydrology Series. Cambridge: Cambridge University Press.

Dawod, M. A. A. and El-Rafy, M. A. (2002). Towards long range forecast of Nile Flood. In *Proceedings of the Fourth Conference on Meteorology and Sustainable Development*, Meteorologist Specialist Association, Cairo, Egypt.

Day, S. (1953). Horizontal convergence and the occurrence of summer shower precipitation at Miami, Florida. *Monthly Weather Review*, **81**, 155–161.

DEFRA (2002). *Catchment Flood Management Plans: Development of a Modelling and Decision Support Framework*. Report EX4495.

DEFRA (2005a). *Dependence between Extreme Sea Surge, River Flow and Precipitation: A Study in South and West Britain*. Defra/Environment Agency R&D Technical Report FD2308/TR3.

DEFRA (2005b). *Joint Probability: Dependence Mapping and Best Practice*. Defra/Environment Agency Flood and Coastal Defence R&D Programme.

DEFRA (2005c). *Use of Joint Probability Methods in Flood Management: A Guide to Best Practice*. Defra/Environment Agency R&D Technical Report FD2308/TR2 (HR Wallingford Report SR 653).

Delworth, T. L. *et al.* (2006). GFDL's CM2 global coupled climate models. Part 1: Formulation and simulation characteristics. *Journal of Climate*, **19**, 643–674.

Deraisme, J., Humbert, J., Drogue, G., and Frelson, N. (2001). Geostatistical interpolation of rainfall in mountainous areas. In GeoENV III – Geostatistics for Environmental Applications, P. Monestiez, D. Allard, and R. Froidevaux (editors), Dordrecht: Kluwer Academic Publishers.

Dery, S. J. and Wood, E. F. (2004). Teleconnection between the Arctic Oscillation and Hudson Bay River discharge. *Geophysical Research Letters*, **31**, L18205, doi: 10.1029/2004GL020729.

Dery, S. J. and Wood, E. F. (2005). Decreasing river discharge in Northern Canada. *Geophysical Research Letters*, **32**, L10401, doi: 10.1029/2005GL022845.

Despic, O. and Simonovic, S. P. (2000). Aggregation operators for soft decision making in water resources, *Fuzzy Sets and Systems*, **115**(1), 11–33.

DETR (2000). *Guidelines for Environmental Risk Assessment and Management*. Environment Agency, Institute for Environmental and Health, London: The Stationery Office.

Dettinger, M. D. (2005). From climate change spaghetti to climate change distributions for 21st century: San Francisco Estuary Watershed. *Science*, **3**(1), 1–14.

Dettinger, M. D. and Diaz, H. F. (2000). Global characteristics of streamflow seasonality and variability. *Journal of Hydrometeorology*, **1**, 289–310.

Deutsch, C. V. and Journel, A. G. (1998). *Geostatistical Library Software and User's Guide*. New York: Oxford University Press.

Diansky, N. A. and Volodin, E. M. (2002). Simulation of present-day climate with a coupled atmosphere-ocean general circulation model. *Izvestiya, Atmospheric and Oceanic Physics* (English translation), **38**(6), 732–747.

Dibike, Y. B. and Coulibaly, P. (2005). Hydrologic impact of climate change in the Saguenay watershed: comparison of downscaling methods and hydrologic models. *Journal of Hydrology*, **307**, 144–163.

Dice, L. R. (1945). Measures of the amount of ecologic association between species. *Ecology*, **26**, 297–302.

Dingman, S. L. (2008). *Physical Hydrology*. Long Grove: Waveland Press.

Dingman, S. L., Barry, R. G., Weller, G., *et al.* (1980). Climate, snow cover, microclimate, and hydrology. In *An Arctic Ecosystem: The Coastal Tundra At Barrow, Alaska*, J. Brown, P. C. Miller, L. L. Tieszen, and F. Bunnell (editors), Stroudsburgh: Dowden, Hutchinson and Ross, 30–65.

Doswell, C. A. and Bosart, L. F. (2001). Extratropical synoptic-scale processes and severe convection. *Meteorological Monographs*, **28**, 27–70.

Douglas, E. M., Vogel, R. M., and Kroll, C. N. (2000). Trends in floods and low flows in the United States: impact of spatial correlation. *Journal of Hydrology*, **240**, 90–105.

Doviak, R. J. and Zrnic, D. S. (1993). *Doppler Radar and Weather Observations*. San Diego: Academic Press, 223–231.

Draper, R. and Smith, H. (1998). *Applied Regression Analysis*. New York: Wiley.

Drosg, M. (2007). *Dealing with Uncertainties: A Guide to Error Analysis*. Berlin: Springer-Verlag.

Dungan, J. L., Perry, J. N., Dale, M. R. T., *et al.* (2002). A balanced view of scale in spatial statistical analysis. *Ecography*, **25**, 626–640.

Dunham, M. (2002). *Data Mining: Introductory and Advanced Topics*. Upper Saddle River: Prentice Hall.

Dunne, T. (1983). Relation of field studies and modeling in the prediction of storm runoff. *Journal of Hydrology*, **65**, 25–48.

Eagleson, P. S. (1972). Dynamics of flood frequency. *Water Resources Research*, **8**, 878–898.

Earthgauge (2011). http://www.earthgauge.net/wp-content/CF_Oscillations%20and%20Teleconnections.pdf (accessed December, 2011).

Ebert, E. E., Janowiak, J. E., and Kidd, C. (2007) Comparison of near real-time precipitation estimates from satellite observations and numerical models. *Bulletin of the American Meteorological Society*, **88**(1), 47–64.

Efron, B. (1979). Bootstrap methods: another look at the jackknife. *Annals of Statistics*, **7**, 1–26.

Efron, B. and Gong, G. (1983). A leisurely look at the bootstrap, the jackknife, and cross validation. *American Statistician*, **37**, 36–48.

Efron, B. and Tibshirani, R. (1993). *An Introduction to the Bootstrap*. London: Chapman and Hall.

Einfalt, T., Arnbjerg-Nielsen, K., Faure, D., *et al.* (2004). Towards a roadmap for use of radar rainfall data in urban drainage. *Journal of Hydrology*, **299**, 186–202.

Eischeid, J. K., Pasteris, P. A., Diaz, H. F., Plantico, M. S., and Lott, N. J. (2000). Creating a serially complete, national daily time series of temperature and precipitation for the western United States. *Journal of Applied Meteorology*, **39**, 1580–1591.

Ellis, D., Furner-Hines, J., and Willett, P. (1993). Measuring the degree of similarity between objects in text retrieval systems. *Perspectives in Information Management*, **3**(2), 128–149.

El Sharif, H. and Teegavarapu, R. S. V. (2011). Spatial statistics preserving interpolation methods for estimation of missing precipitation data. *AGU Fall Meeting*, H41G-1132.

El Sharif, H. and Teegavarapu, R. S. V. (2012). Evaluation of spatial interpolation methods for missing precipitation data: preservation of spatial statistics. *ASCE-EWRI Proceedings*, May 2012.

Ely, L. L. (1997). Response of extreme floods in the southwestern United States to climatic variations in the Late Holocene. *Geomorphology*, **19**, 175–201.

Ely, L., Enzel, Y., Baker, V. R., and Cayan, D. R. (1993). A 5000-year record of extreme floods and climate change in the southwestern United States. *Science*, **262**, 410–412.

Emanuel, K. (2005). Increasing destructiveness of tropical-cyclones over the past 30 years. *Nature*, **436**, 686–688.

Emery, W. J. and Thomson, R. E. (1997). *Data Analysis Methods in Physical Oceanography*. Amsterdam: Elsevier.

Enders, C. (2010). *Applied Missing Data Analysis*, Methodology in the Social Sciences Series. New York: The Guilford Press.

Enfield, D. B., Mestas-Nunez, A. M., and Trimble P. J. (2001). The Atlantic multidecadal oscillation and its relation to rainfall and river flows in the continental U.S. *Geophysical Research Letters*, **28**(10), 2077–2080.

ESRI (2001). *Using ArcGIS Geostatistical Analyst 9*. Environmental Sciences Research Institute, Redlands, CA, USA.

Farris, G. S., Smith, G. J., Crane, M. P., *et al.* (editors) (2007). *Science and the Storms: The USGS Response to the Hurricanes of 2005*. US Geological Survey Circular 1306.

Fealy, R. and Sweeney, J. (2007). Statistical downscaling of precipitation for a selection of sites in Ireland employing a generalized linear modeling approach. *International Journal of Climatology*, **22**, 843–866.

Fernandez, W., Vogel, R. M., and Sankarasubramanian, A. (2000). Regional calibration of a watershed model. *Hydrological Sciences Journal*, **45**(5), 689–707.

Ferreira, A. and Teegavarapu, R. S. V. (2012). Optimal and adaptive operation of a hydropower system with unit commitment and water quality constraints. *Water Resources Management*, **26**, 707–732.

Ferreira, A., Teegavarapu, R. S. V., and Pathak, C. (2009). Evaluation of optimal reflectivity-rainfall (Z-R) relationships for improved precipitation estimates. EOS Abstract H31D-0805, *AGU Fall Meeting*, San Francisco.

Field, C. B., Mortsch, L. D., Brklacich, M., *et al.* (2007). North America. In *Climate Change 2007: Impacts, Adaptation and Vulnerability. Contribution of Working Group II to the Fourth Assessment Report of the Intergovernmental Panel on Climate Change*, M. L. Parry, O. F. Canziani, J. P. Palutikof, P. J. van der Linden, and C. E. Hanson (editors), Cambridge: Cambridge University Press.

Fielding, A. H. (2007). *Cluster and Classification Techniques for the Biosciences*. New York: Cambridge University Press.

Fiering, M. and Kindler, J. (1987). Surprise in water resources design. *International Journal of Water Resources Development*, **2**(4), 1–10.

Flato, G. M. and Boer, G. J. (2001). Warming asymmetry in climate change simulations. *Geophysics Research Letters*, **28**, 195–198.

Fleming, M. and Neary, V. (2004). Continuous hydrologic modeling study with the hydrologic modeling system. *Journal of Hydrologic Engineering*, **9**(3), 1084–1099.

Florida (2011). http://www.floridadisaster.org/EMTOOLS/elnino/elnino.htm (accessed October, 2011).

Floudas, C. A. (1995). *Nonlinear and Mixed-Integer Optimization: Fundamentals and Applications*. New York: Oxford University Press.

Fontane, D. G., Gates, T. K., and Moncada, E. (1997). Planning reservoir operations with imprecise objectives. *Journal of Water Resources Planning and Management*, ASCE, **123**(3), 154–162.

Fortin, M. J. and Dale, M. (2005). *Spatial Analysis: A Guide for Ecologists*. Cambridge: Cambridge University Press.

Fotheringham, A. S., Brunsdon, C., and Charlton, M. (2002). *Geographically Weighted Regression: The Analysis of Spatially Varying Relationships*. Chichester: Wiley.

Frank, N. L. and Smith, D. L. (1968). On the correlation of radar echoes over Florida with various meteorological parameters. *Journal of Applied Meteorology*, **7**, 712–714.

Franke, R. (1982). Smooth interpolation of scattered data by local thin plate splines. *Computers and Mathematics with Applications*, **8**(4), 237–281.

Frederick, D. K. (2002). *Water Resources and Climate Change*. Cheltenham: Edward Elgar Publishing.

Frederick, R. H., Myers, V. A., and Auciello, E. (1977). *Five-To-60-Minute Precipitation Frequency for the Eastern and Central United States*. NOAA, HYDRO-35 Report, National Weather Service, Silver Spring, MD.

Freeman, J. A. and Skapura, D. M. (1991). *Neural Networks: Algorithms, Applications and Programming Techniques*. Reading: Addison-Wesley.

French, M. N., Krajewski, W. F., and Cuykendal, R. R. (1992). Rainfall forecasting in space and time using a neural network, *Journal of Hydrology*, **137**, 1–37.

FREND (1989). *I: Hydrological Studies. II: Hydrological data. Flow Regimes from Experimental and Network Data*. Wallingford: Institute of Hydrology.

Fritsch, J. M. and Maddox, R. A. (1981). Convectively-driven mesoscale pressure systems aloft, Part I: Observations. *Journal of Applied Meteorology*, **20**, 9–19.

Fritsch, J. M., Kane, R. J., and Chelius, C. H. (1986). The contribution of mesoscale convective weather systems to the warm season precipitation in the United States. *Journal of Applied Meteorology*, **25**,1333–1345.

Fulton, R. A., Breidenbach, J. P., Seo, D. J., Miller, D. A., and O'Bannon, T. (1998). The WSR-88D rainfall algorithm. *Weather Forecasting*, **13**, 377–395.

Furevik, T., Bentsen, M., Drange, H., *et al.* (2003). Description and evaluation of the Bergen climate model: ARPEGE coupled with MICOM. *Climate Dynamics*, **21**, 27–51.

Gagin, A., Rosenfeld, D., and Lopez, R. E. (1985). The relationship between height and precipitation characteristics of summertime convective cells in south Florida. *Journal of Atmospheric Science*, **42**, 84–94.

Galloway, G. E. (2011). If stationarity is dead, what do we do now? *Journal of the American Water Resources Association*, **47**(3), 563–570.

Ganguli, M. K., Rangarajan, R., and Panchang, G. M. (1951). Accuracy of mean rainfall estimates: data of Domodar catchment. *Irrigation and Power Journal*, **8**.

Garbrecht, J. D. and Piechota, T. C. (editors) (2006). *Climate Variations, Climate Change and Water Resources Engineering*. Reston: ASCE, EWRI.

GCOS (2010). http://unfccc.int/resource/docs/2010/sbsta/eng/misc09.pdf (accessed December, 2011).

GCOS (2011). http://www.wmo.int/pages/prog/gcos/ (accessed December, 2011).

Genton, M. G. and Furrer, R. (1998a). Analysis of rainfall data by simple good sense: is spatial statistics worth the trouble? *Journal of Geographic Information and Decision Analysis*, **2**, 11–17.

Genton, M. G. and Furrer, R. (1998b). Analysis of rainfall data by robust spatial statistics using S+Spatial Stats. *Journal of Geographic Information and Decision Analysis*, **2**, 126–136.

Gentry, R. C. and Moore, P. L. (1954). Relation of local and general wind interaction near the sea coast to time and location of air mass showers. *Journal of Meteorology*, **11**, 507–511.

Gershunov, A. and Barnett, T. P. (1998). ENSO influence on intraseasonal extreme rainfall and temperature frequencies in the contiguous United States: observations and model results. *Journal of Climate*, **11**, 1575–1586.

Getis, A. and Ord, J. K. (1992). The analysis of spatial association by use of distance statistics. *Geographical Analysis*, **24**, 189–206.

Ghosh, S. and Mujumdar, P. P. (2006). Future rainfall scenario over Orissa with GCM projections by statistical downscaling. *Current Science*, **90**(3), 396–404.

Ghosh, S. and Mujumdar, P. P. (2008). Statistical downscaling of GCM simulations to streamflow using relevance vector machine. *Advances in Water Resources*, **31**(1), 132–146.

Gibbons, J. D. (1997). *Nonparametric Methods for Quantitative Analysis*. Columbus: American Sciences Press.

Gilbert, R. O. (1987). *Statistical Methods For Environmental Pollution Monitoring*. New York: Van Nostrand Reinhold.

Glantz, M. H. (2001). *Currents of Change: Impacts of El Niño and La Niña on Climate and Society*. Cambridge: Cambridge University Press.

Goldenberg, S. B., Landsea, C. W., Mestas-Nuñez, A. M., and Gray, W. M. (2001). The recent increase in Atlantic hurricane activity: causes and implications. *Science*, **293**(5529), 474–479.

Goly, A. and Teegavarapu, R. S. V. (2011). Spatial and temporal variation of precipitation extremes in Florida: influence of teleconnections. H21D-1139, *AGU, Fall Meeting*, San Francisco, CA.

Goodchild, M. F. (1986). *Spatial Autocorrelation, Concepts and Techniques in Modern Geography*. Norwich: GeoBooks.

Goodrich, G. B. and Ellis, A. W. (2008). Climatic controls and hydrologic impacts of a recent extreme seasonal precipitation reversal in Arizona. *Journal of Applied Meteorology and Climatology*, **47**, 498–508.

Goovaerts, P. (2000). Geostatistical approaches for incorporating elevation into the spatial interpolation of rainfall. *Journal of Hydrology*, **228**(1), 113–129.

Gordon, C., Cooper, C., Senior, C. A., *et al.* (2000). The simulation of SST, sea ice extents and ocean heat transports in a version of the Hadley Centre coupled model without flux adjustments. *Climate Dynamics*, **16**, 147–168.

Gordon, H. B., Rotstayn, L. D., McGregor, J. L., *et al.* (2002). *The CSIRO Mk3 climate system model*. CSIRO Atmospheric Research Technical Paper No. 60, Division of Atmospheric Research, CSIRO, Victoria, Australia.

Goswami, B. N., Venugopal, V., Sengupta, D., Madhusoodanan, M. S., and Xavier, P. K. (2006). Increasing trend of extreme rain events over India in a warming environment. *Science*, **314**, 1442–1445.

Govindaraju, R. S. and Rao, A. R. (2000). *Neural Networks in Hydrology*, Dordrecht: Kluwer Academic Publishers.

Gower, J. C. and Legendre, P. (1986). Metric and Euclidean properties of dissimilarity coefficients. *Journal of Classification*, **5**, 5–48.

GPCC (2011). http://gpcc.dwd.de/ (accessed December, 2011).

GPM (2011). http://gpm.nasa.gov (accessed September, 2011).

Grantz, K., Rajagopalan, B., Clark, M., and Zagona, E. (2005). A technique for incorporating large-scale climate information in basin-scale ensemble streamflow forecasts. *Water Resources Research*, **41**, W10410, doi:10.1029/2004WR003467.

Gray, D. M. (1973). *Handbook on the Principles of Hydrology*. Water Information Center, Manhasset Isle, Port Washington, New York.

Gray, S. T., Graumlich, L. J., Betancourt, J. L., and Pederson, G. T. (2004). A tree-ring reconstruction of the Atlantic Multidecadal Oscillation since 1567 A.D. *Geophysical Research Letters*, **31**, L12205-1–4.

Grayson, R. and Bloschl, G (2001). *Spatial Patterns in Catchment Hydrology: Observations and Modeling*. Cambridge: Cambridge University Press.

Greenwood, J. A., Landwehr, J. M., Matalas, N. C., and Wallis, J. R. (1979). Probability-weighted moments: definition and relation to parameters of several distributions expressible in inverse form. *Water Resources Research*, **15**, 1049–1054.

Gregory, M. A., Cunningham, B. A., Schmidt, M. F., and Mack, B. W. (1999). Estimating soil storage capacity for storm water modeling applications. *6th Biennial Storm Water Research and Watershed Management Conference*, September 14–17, Tampa, FL.

Griffith, D. A. (1987). *Spatial Autocorrelation: A Primer*. Washington, DC: Association of American Geographers.

Groisman, P. Y. and Easterling, D. R. (1994). Variability and trends of total precipitation and snowfall over the United States and Canada. *Journal of Climate*, **7**(1), 184–205.

Groisman, P. Y., Karl, T. R., Knight, R. W., and Stenchikov, G. L. (1994). Changes in snow cover, temperature, and radiative heat balance over the northern hemisphere. *Journal of Climate*, **7**, 1633–1656.

Groisman, P. Y., Karl, T. R., Easterling, D. R., *et al.* (1999). Changes in the probability of heavy precipitation: important indicators of climatic change. *Climatic Change*, **42**, 243–283.

Groisman, P. Y., Knight, R. W., and Karl, T. R. (2001). Heavy precipitation and high streamflow in the contiguous United States: trends in the twentieth century. *Bulletin of the American Meteorological Society*, **82**(2), 219–246.

Groves, D. G., Knopman, D., Lempert, R. J., Berry, S. H., and Wainfan, L. (2008). *Presenting Uncertainty About Climate Change to Water-Resource Managers: A Summary of Workshops with the Inland Empire Utilities Agency*, Appendix A: Workshop Presentations, Santa Monica, CA, TR-505-NSF.

Grubb, H. and Robson, A. (2000). Exploratory/Visual Analysis. In *Detecting Trend and Other Changes in Hydrological Data*. Z. W. Kundzewicz and A. Robson (editors), WMO/TD-No. 1013.

GTN (2011). http://gtn-h.unh.edu/ (accessed December, 2011).

Gupta, R. (2008). *Hydrology and Hydraulic Systems*. Long Grove: Waveland Press.

Gutierrez, F. and Dracup, J. A. (2001). An analysis of the feasibility of long-range streamflow forecasting for Columbia using El Nino-Southern Oscillation indicators. *Journal of Hydrology*, **246**, 181–196.

Haan, C. T. (2002). *Statistical Methods in Hydrology*. Ames: Iowa Press.

Hagemeyer, B. C. (2006). ENSO, PNA and NAO Scenarios for extreme storminess, rainfall and temperature variability during the Florida dry season. Preprints, *18th Conference on Climate Variability and Change*, American Meteorological Society, Atlanta, GA, CD-ROM P2.4.

Haining, R. P. (1990). *Spatial Data Analysis in the Social and Environmental Sciences*. Cambridge: Cambridge University Press.

Hair, J. F., Black, W. C., Babin, B. J., and Anderson, R. E. (2010). *Multivariate Data Analysis*. Upper Saddle River: Prentice Hall.

Hamming, R. W. (1950). Error detecting and error correcting codes. *Bell System Technical Journal*, **29**(2), 147–160.

Han, J. and Kamber, M. (2006). *Data Mining: Concepts and Techniques*. New York: Morgan Kaufmann.

Handcock, M. S. and Wallis, J. R. (1994). An approach to statistical spatial-temporal modeling of meteorological fields. *Journal of the American Statistical Association*, **89**(426), 368–378.

Hansen, E. M., Schreiner, L. C., and Miller, J. F. (1982). *Application of Probable Maximum Precipitation Estimates: United States East of the 105th Meridian*. Hydrometeorological Report No. 52, US National Weather Service.

Hargreaves, G. H. and Samani, Z. A. (1982). Estimating potential evapotranspiration. *Journal of Irrigation and Drainage Engineering*, **108**(3), 225–230.

Hauser, D., Roux, F., and Amayenc, P. (1988). Comparison of two methods of thermodynamic and microphysical variables from Doppler radar measurements: application of case of tropical squall line. *Journal of Atmospheric Science*, **45**, 1285–1303.

Hawkes, P. J. (2008). Joint probability analysis for estimation of extremes. *Journal of Hydraulic Research*, **46**(2), 246–256.

Hawkes, P. and Svensson, C. (2005). *Joint Probability: Dependence Mapping and Best Practice: Technical Report on Dependence Mapping*. DEFRA/Environment Agency R&D Technical Report FD2308/TR1.

Hawkes, P. J., Gouldby, B. P., Tawn, J. A., and Owen, M. W. (2002). The joint probability of waves and water levels in coastal defence design. *Journal of Hydraulic Research*, **40**(3), 241–251.

Hayhoe, K., Wake, C. P., Huntington, T. G., *et al.* (2007). Past and future changes in climate and hydrological indicators in the US Northeast. *Climate Dynamics*, **28**(4), 381–407.

Haylock, M. and Goodess, C. (2004). Inter-annual variability of European extreme winter rainfall and links with mean large-scale circulation. *Journal of Climatology*, **24**, 759–776.

Heffernan, J. E. (2000). A directory of coefficients of tail dependence. *Extremes*, **3**, 279–290.

Held, I. M. and Soden, B. J. (2006). Robust responses of the hydrological cycle to global warming. *Journal of Climate*, **19**, 5686–5699.

Hellstrom, C. and Malmgren, B. A. (2004). Spatial analysis of extreme precipitation in Sweden 1961–2000. *AMBIO: A Journal of the Human Environment*, **33**(4), 187–192.

Hengl, T. (2009). *A Practical Guide to Geostatistical Mapping*. EN Scientific and Technical Research series report, Office for Official Publications of the European Communities, Luxembourg.

Hense, A. and Friederichs, P. (2006). Wind and precipitation extremes in the Earth's atmosphere. In *Extreme Events in Nature and Science*, S. Albeverio, V. Jentsch, and H. Kantz (editors), New York: Springer.

Henson, R. and Trenberth, K. E. (1998) Children of the Tropics: El Niño and La Niña. http://www.ucar.edu/communications/factsheets/elnino/ (accessed September, 2011).

Hershfield, D. M. (1961). *Rainfall Frequencies Atlas of the United States for Durations from 30 minutes to 24 hours and Return Periods from 1 to 100 years*. Technical Paper No. 40, US Weather Bureau, Washington, DC.

Hershfield, D. M. (1965). Method for estimating probable maximum rainfall. *Journal of the American Waterworks Association*, **57**, 965–972.

Hess, P. and Brezowsky, H. (1977). Katalog der Grosswetterlagen Europas 1881–1976. *Berichtes des Deustcher Wetterdienst*, **113**. Deutscher Wetterdienst, Offenbach am Main.

Higgins, R. W., Yao, Y. P., Yarosh, E. S., Janowiak, J. E., and Mo, K. C. (1997). Influence of the Great Plains low-level jet on summertime precipitation and moisture transport over the central United States. *Journal of Climate*, **10**, 481–507.

Hirsch, R. M. (2011). A perspective on nonstationarity and water management. *Journal of the American Water Resources Association*, **47**(3), 436–446.

Hirsch, R. M. and Slack, J. R. (1984). Nonparametric trend test for seasonal data with serial dependence. *Water Resources Research*, **20**(6), 727–732.

Hirsch, R. M., Slack, J. R., and Smith, R. A. (1982). Techniques of trend analysis for monthly water quality data. *Water Resources Research*, **18**, 107–121.

Hirsch, R. M., Alexander, R. B., and Smith, R. A. (1991). Selection of methods for the detection and estimation of trends in water quality. *Water Resources Research*, **27**(5), 803–813.

Hirschboeck, K. K. (1987a). Hydroclimatically defined mixed distributions in partial duration flood series. In *Hydrologic Frequency Modeling*, V. P. Singh (editor), Dordrecht: D. Reidel Publishing Company, 199–212.

Hirschboeck, K. K. (1987b). Catastrophic flooding and atmospheric circulation anomalies. In *Catastrophic Flooding*, L. Mayer and D. B. Nash (editors), Allen & Unwin, 23–56.

Hirschboeck, K. K. (1988). Flood hydroclimatology. In *Flood Geomorphology*, V. R. Baker, R. C. Kochel, and P. C. Patton (editors), New York: John Wiley & Sons, 27–49.

Hirschboeck, K. K., Ely, L., and Maddox, R.A. (2000). Hydroclimatology of meteorologic floods. In *Inland Flood Hazards: Human, Riparian and Aquatic Communities*, E. Wohl (editor), Cambridge: Cambridge University Press, 39–72.

Hodgkins, G. A. and Dudley, R. W. (2006). Changes in the timing of winter-spring streamflows in eastern North America, 1913–2002. *Geophysical Research Letters*, **33**, L06402.

Holawe, F. and Dutter, R. (1999). Geostatistical study of precipitation series in Austria: time and space. *Journal of Hydrology*, **219**, 70–82.

Holland, P. W. and Welsch, R. E. (1977). Robust regression using iteratively reweighted least-squares. *Communications in Statistics: Theory and Methods*, **A6**, 813–827.

Holliday, J. D., Hu, C. Y., and Willett, P. (2002). Grouping of coefficients for the calculation of intermolecular similarity and dissimilarity using 2D fragment bit-strings. *Combinatorial Chemistry and High Throughput Screening*, **5**(2), 155–166.

Hong, Y., Hsu, K., Gao, X., and Sorooshian, S. (2004). Precipitation estimation from remotely sensed imagery using artificial neural network–cloud classification system. *Journal of Applied Meteorology*, **43**(12) 1834–1853.

Hornberger, G. M., Beven, K. J., Cosby, B. J., and Sappington, D. E. (1985). Shenandoa watershed study: calibration of a topography-based, variable contributing area hydrological model to a small forested catchment. *Water Resources Research*, **21**, 1841–1859.

Hosking, J. R. M. (1990). L-moments: analysis and estimation of distributions using linear combinations of order statistics. *Journal of Royal Statistical Society, B*, **52**(2), 105–124.

Hosking, J. R. M. and Wallis, J. R. (1998). *Regional Frequency Analysis: An Approach Based on l-Moments*. Cambridge: Cambridge University Press.

Hosking J. R. M., Wallis, J. R., and Wood, E. F. (1985). Estimation of the generalized extreme-value distribution by the method of probability-weighted moments. *Technometrics*, **27**(3), 251–261.

House, P. K. and Hirschboeck, K. K. (1997). Hydroclimatological and paleohydrological context of extreme winter flooding in Arizona, 1993. In *Storm-Induced Geological Hazards: Case Histories from the 1992–1993 Winter Storm in Southern California and Arizona*, R. A. Larson and J. E. Slosson (editors), Boulder: Geological Society of America Reviews in Engineering Geology, **XI**, 1–24.

HR Wallingford (2000a). *The Joint Probability of Waves and Water Levels: JOIN-SEA: A Rigorous But Practical New Approach*. HR Report SR 537.

HR Wallingford (2000b). *The Joint Probability of Waves and Water Levels: JOIN-SEA Version 1.0: User Manual*. HR Report TR 71.

Hu, K., Li, B., Lu, Y., and Zhang, F. (2004). Comparison of various spatial interpolation methods for non-stationary regional soil mercury content. *Environmental Science*, **25**(3), 132–137.

Hua, C., Guo, S., Xu, C. Y., and Singh, V. P. (2007). Historical temporal trends of hydroclimatic variables and runoff response to climatic variability and their relevance in water resource management in the Hanjiang Basin. *Journal of Hydrology*, **344**, 171–184.

Hudlow, M. D., Arkell, R., Patterson, V., *et al.* (1979). *Calibration and Intercomparison of the GATE C-band Radars*. NOAA Tech. Report EDIS 31, NTIS, Springfield, VA.

Huff, F. A. (1967). Time distribution of rainfall in heavy storms. *Water Resources Research*, **3**(4), 1007–1019.

Huffman, G. J., Adler, R. F., Stocker, E. F., Bolvin, D. T., and Nelkin, E. J. (2003). Analysis of TRMM 3-hourly multi-satellite precipitation estimates computed in both real and post-real time. Preprints, *Twelfth Conference on Satellite Meteorology and Oceanography*, Long Beach, CA.

Huffman, G. J., Adler, R. F., Bolvin, D. T., *et al.* (2007). The TRMM multi-satellite precipitation analysis: quasi-global, multi-year, combined-sensor precipitation estimates at fine scale. *Journal of Hydrometeorology*, **8**, 38–55.

Hughes, D. A., Hannart, P., and Watkins, D. (2003). Continuous baseflow separation from time series of daily and monthly streamflow data. *Water SA*, **29**(1), 43–48.

Hulme, M. (1992). A 1951–80 global land precipitation climatology for the evaluation of General Circulation Models. *Climate Dynamics*, **7**, 57–72.

Hulme, M. (1994). Validation of large-scale precipitation fields in General Circulation Models. In *Global Precipitations and Climate Change*, M. Desbois and F. Desalmand (editors), NATO ASI Series, Berlin: Springer-Verlag, 387–406.

Hulme, M. and New, M. (1997). The dependence of large-scale precipitation climatologies on temporal and spatial gauge sampling. *Journal of Climate*, **10**, 1099–1113.

Huntington, T. G. (2006). Evidence for intensification of the global water cycle: review and synthesis. *Journal of Hydrology*, **319**, 83–95.

Hutchinson, M. F. (1995). Interpolating mean rainfall using thin-plate smoothing splines. *International Journal of Geographical Information Systems*, **9**, 385–403.

Institute of Hydrology (1980). *Low Flow Studies*. Research Report 1, Institute of Hydrology, Wallingford, UK.

IPCC (1992). *Climate Change 1992; Supplementary Report to the IPCC Scientific Assessment*. J. T. Houghton, B. A. Callender, and S. K. Varney (editors), Cambridge: Cambridge University Press.

IPCC (1995). *Second IPCC Report*. Chapter 10: Hydrology and freshwater ecology, and Chapter 14: Water resources management, Cambridge: Cambridge University Press.

IPCC (2000). Emissions scenarios. In *Special Report of the Intergovernmental Panel on Climate Change*. N. Nakicenovic and R. Swart (editors), Cambridge and New York: Cambridge University Press.

IPCC (2001). *Intergovernmental Panel on Climate Change, Climate Change* (Three reports). Cambridge: Cambridge University Press.

IPCC (2007a). Summary for Policymakers. In *Climate Change 2007: The Physical Science Basis. Contribution of Working Group I to the Fourth Assessment Report of the Intergovernmental Panel on Climate Change*. S.

Solomon, D. Qin, M. Manning, *et al.* (editors), Cambridge and New York: Cambridge University Press.

IPCC (2007b). *Climate Change 2007: The Physical Science Basis. Contribution of Working Group I to the Fourth Assessment Report of the Intergovernmental Panel on Climate Change*. S. Solomon, D. Qin, M. Manning, *et al.* (editors), Cambridge and New York: Cambridge University Press, 1–18; www.ipcc.ch/press/index.htm.

IPCC (2007c). *Climate Change 2007: Climate Change Impacts, Adaptation And Vulnerability, Contribution of Working Group 2 to Fourth Assessment Report of the Intergovernmental Panel on Climate Change*, M. L. Parry *et al.* (editors), Cambridge and New York: Cambridge University Press, 1–16.

IPCC (2007d). *Climate Change 2007: The Physical Science Basis. Contribution of Working Group I to the Fourth Assessment Report of the Intergovernmental Panel on Climate Change*. S. Solomon, D. Qin, M. Manning, *et al.* (editors), Cambridge and New York: Cambridge University Press.

IPCC (2008). *Climate Change and Water*. IPCC Technical Paper VI. http://www.ipcc.ch/pdf/technical-papers/climate-change-water-en.pdf.

IPCC-TGICA (2007). *General Guidelines on the Use of Scenario Data for Climate Impact and Adaptation Assessment, Version 2*. Prepared by T. R. Carter on behalf of the Intergovernmental Panel on Climate Change, Task Group on Data and Scenario Support for Impact and Climate Assessment, http://www.ipcc-data.org/guidelines/TGICA_guidance_sdciaa_v2_final.pdf.

IPSL (2005). *The New IPSL Climate System Model: IPSL-CM4*. Paris: Institut Pierre Simon Laplace des Sciences de l'Environnement Global, 73.

IRI (2011). http://iri.columbia.edu/climate/ENSO/enso.html (accessed December, 2011).

Isaaks, H. E. and Srivastava, R. M. (1989). *An Introduction to Applied Geostatistics*. New York: Oxford University Press.

Iwashima, T. and Yamamoto, R. (1993). A statistical analysis of the extremes events: long-term trend of heavy daily precipitation. *Journal of Meteorological Society of Japan*, **71**, 637–640.

Jaccard, P. (1908). Nouvelles recherches sur la distribution florale. *Société Vaudoise des Sciences Naturelles*, **44**, 223–270.

Jacob, D., Smalley, R., Meighen, J., Xuereb, K., and Taylor, B. (2009). Climate change and probable maximum precipitation. HRS Report No. 12, Bureau of Meteorology, Australia.

Jain, A. K. and Dubes, R. C. (1988). *Algorithms for Clustering Data*. Upper Saddle River: Prentice Hall.

Jakeman, A. J. and Hornberger, G. M. (1993). How much complexity is warranted in a rainfall-runoff model? *Water Resources Research*, **29**, 2637–2649.

Jarvis, C. H., Stuart, N., and Cooper, W. (2003). Infometric and statistical diagnostics to provide artificially-intelligent support for spatial analysis: the example of interpolation. *International Journal of Geographical Information Science*, **17**, 495–516.

Jentsch, V., Kantz, H., and Albeverio, S. (2008). Extreme events: magic, mysteries and challenges. In *Extreme Events in Nature and Society*, S. Albeverio, V. Jentsch, and H. Kantz (editors), Berlin: Springer, 1–18.

Johnson, S. C. (1996). Hierarchical clustering schemes. *Psychometrika*, **32**(2), 241–254.

Johnston, P. R. and Pilgrim, D. H. (1976). Parameter optimization for watershed models. *Water Resources Research*, **12**(3), 477–486.

Jorgensen, D. P. and Willis, P. T. (1982). A Z-R relationship for hurricanes. *Journal of Applied Meteorology*, **21**, 356–366.

Joss, J. and Lee, R. (1995). The application of radar-gage comparisons to operational precipitation profile corrections. *Journal of Applied Meteorology*, **34**, 2612–2630.

Journel, A. G. and Huijbregts, C. J. (1978). *Mining Geostatistics*, New York: Academic Press.

Jungclaus, J. H., Botzet, M., Haak, H., *et al.* (2006). Ocean circulation and tropical variability in the coupled model AOGCM ECHAM5/MPI-OM. *Journal of Climate*, **19**, 3952–3972.

Junninen, H., Niska, H., Tuppurainen, K., Ruuskanen, J., and Kolehmainen, M. (2004). Methods for imputation of missing values in air quality data sets. *Atmospheric Environment*, **38**, 2895–2907.

K-1 model developers (2004). *K-1 Coupled Model (MIROC) Description*. K-1 technical report, 1, H. Hasumi and S. Emori (editors), Center for Climate System Research, University of Tokyo.

Kahya, E. and Dracup, J. A. (1993). U.S. streamflow patterns in relation to the El Niño/Southern Oscillation. *Water Resources Research*, **29**(8), 2491–2503.

Kalnaya, E., Kanamitsua, M., Kistlera, R., *et al.* (1996). The NCEP/NCAR 40-year reanalysis project. *Bulletin American Meteorological Society*, **77**, 437–470.

Kanevski, M. and Maignan, M. (2004). *Analysis and Modeling of Spatial Environmental Data*. Lausanne: EPFL Press (Marcel Dekker, Inc.).

Kaplan, A. M., Cane, A., Kushnir, Y., and Clement, A. C. (1998). Analysis of global sea surface temperatures. *Journal of Geophysical Research*, **103**, 18567–18589.

Karl, T. R. and Knight, R. W. (1998). Secular trends of precipitation amount, frequency, and intensity in the United States. *Bulletin of the American Meteorological Society*, **79**, 231–241.

Karl, T. R., Knight, R. W., and Plummer, N. (1995). Trends in high-frequency climate variability in the twentieth century. *Nature*, **377**, 217–220.

Kay, A. L., Jones, R. G., and Reynard, N. S. (2005). RCM rainfall for UK flood frequency estimation. 1. Method and validation. *Journal of Hydrology*, **318**(1–4), 151–162.

Kennedy, M. R., Pearce, H. J., Canterford, R. P., and Mintz, L. J. (1988). The estimation of generalized probable maximum precipitation in Australia. Workshop on spillway design floods, Canberra, 4 February 1988. *Australian National Committee on Large Dams Bulletin*, **79**.

Kerr, R. A. (2000). A North Atlantic climate pacemaker for the centuries. *Science*, **288**, 1984–1985.

Khan, S. (2007). Nonlinear dependence and extremes in hydrology and climate, Ph.D. Thesis, University of South Florida.

Kiang, J. E., Olsen, J. R., and. Waskom, R. M. (2011). Introduction to the featured collection on "nonstationarity, hydrologic frequency analysis, and water management." *Journal of the American Water Resources Association*, **47**(3), 433–435.

Kim, U., Kaluaracchchi, J. J., and Smakhtin, V. U. (2008). Generation of monthly precipitation under climate change for the Upper Blue Nile River Basin, Ethiopia. *Journal of the American Water Resources Association*, 1231–1247.

King, K., Irwin, S., Sarwar, R., and Simonovic, S. P. (2012). The effects of climate change on extreme precipitation events in the Upper Thames River Basin: A comparison of downscaling approaches. *Canadian Water Resources Journal*, **37**(3), 253–274.

Knight, J. R., Folland, C. K., and Scaife, A. A. (2006). Climate impacts of the Atlantic Multidecadal Oscillation. *Geophysical Research Letters*, **33**, L17706.

Knox, J. C. (1993). Large increases in flood magnitude in response to modest changes in climate. *Nature*, **361**, 430–432.

Kothyari, U. C. and Garde, R. J. (1992). Rainfall intensity-duration-frequency formula for India. *Journal of Hydraulic Engineering*, **118**(2), 323–336.

Koutsoyiannis, D. (2011). Hurst–Kolmogorov dynamics and uncertainty. *Journal of the American Water Resources Association*, **47**(3). 481–495.

Krajewski, W. F. (1987). Co-kriging of radar and rain gage data. *Journal of Geophysics Research*, **92D8**, 9571–9580.

Krause, E. F. (1987). *Taxicab Geometry: An Adventure in Non-Euclidean Geometry*. Mineola: Dover.

Kuhn. G. (2006). On dependence and extremes. Ph.D. thesis, Munich University of Technology.

Kuhn, G., Khan, S., Ganguly, A. R., and Branstetter, M. L. (2007). Geospatial-temporal dependence among weekly precipitation extremes with applications to observations and climate model simulations in South America. *Advances in Water Resources*, **30**(12), 2401–2423.

Kundzewicz, Z. W. (2011). Nonstationarity in water resources: central European perspective. *Journal of the American Water Resources Association*, **47**(3), 550–562.

Kundzewicz, Z. W., Mata, L. J., Arnell, N. W., *et al.* (2007). Freshwater resources and their management. In *Climate Change 2007: Impacts, Adaptation and Vulnerability. Contribution of Working Group II to the Fourth Assessment Report of the Intergovernmental Panel on Climate Change*, M. L. Parry, O. F. Canziani, J. P. Palutikof, P. J. van der Linden, and C. E. Hanson (editors), Cambridge and New York: Cambridge University Press, 173–210.

Kundzewicz, Z. W., Mata, L. J., Arnell, N. W., *et al.* (2008). The implications of projected climate change for freshwater resources and their management. *Hydrological Sciences Journal*, **53**(1), 3–10.

Kunkel, K. E., Andsager, K., and Easterling, D. (1999). Long-term trends in extreme precipitation events over the conterminous United States and Canada. *Journal of Climate*, **12**(8), 2515–2527.

Kyriakidis, P. C., Miller, N. L., and Kim, J. (2004). A spatial time series framework for simulating daily precipitation at regional scales. *Journal of Hydrology*, **297**, 236–255.

Lachniet, M. S., Burns, S. J., Piperno, D. R., *et al.* (2004). A 1500-year El Niño/Southern Oscillation and rainfall history for the isthmus of Panama from speleothem calcite. *Journal of Geophysical Research*, **109**, D20117, doi:10.1029/2004JD004694.

Lagarias, J. C., Reeds, J. A., Wright, M. H., and Wright, P. E. (1998). Convergence properties of the nelder-mead simplex method in low dimensions. *SIAM Journal of Optimization*, **9**(1), 112–147.

Lamb, H. H. (1972). British Isles weather types and a register of daily sequence of circulation patterns, 1861–1971. *Geophysical Memoir*, 116, London: HMSO.

Lance, G. N. and Williams, W. T. (1966). Computer programs for hierarchical polythetic classification. *Computer Journal*, **9**, 60–64.

Landot, T., Sgellari, S., Lima, C., and Lall, U. (2008). *In-filling Missing Historical Daily Rainfall Data Study*. Final Report, South Florida Water Management District, Columbia University, New York.

Langbein, W. B. (1949). Annual floods and the partial duration flood series. *Transactions American Geophysical Union*, **30**(6), 879–881.

Larijani, M. A. (2009). Climate change effects on high-elevation hydropower system in California. Ph.D. thesis, Department of Civil and Environmental Engineering, University of California, Davis.

Larose, T. (2005). *Discovering Knowledge in Data: An Introduction to Data Mining*. Hoboken: Wiley.

Larson, L. W. and Peck, E. L. (1974). Accuracy of precipitation measurements for hydrologic forecasting. *Water Resources Research*, **156**, 1687–1696.

Latif, M. and Barnett, T. P. (1994). Causes of decadal climate variability over the north Pacific and North America. *Science*, **266**, 634–637.

Laurenson, E. M. (1987). Back to basics on flood frequency analysis. *Civil Engineering Transactions*, **29**, 47–53.

Leavesley, G. H. (1999). Assessment of the impacts of climate variability and change on the hydrology of North America. In *Impacts of Climate Change and Climate Variability on Hydrological Regimes*, J. C. van Dam (editor), International Hydrology Series, Cambridge: Cambridge University Press.

Lee, D. I., Jang, M., You, C. H., Kim, K. E., and Suh, A. S. (2002). Kuduck Dwsr-88c radar rainfall estimation and Z-R relationships by poss during 2001 in Korea, *EGS XXVII General Assembly*, Nice, April 21–26, 2002, abstract 2242.

Legates, D. R. and Willmott, C. J. (1990). Mean seasonal and spatial variability in gauge-corrected global precipitation. *International Journal of Climatology*, **10**, 111–127.

Legutke, S. and Voss, R. (1999). *The Hamburg Atmosphere-Ocean Coupled Circulation Model ECHO-G*. Technical report, No. 18, German Climate Computer Centre (DKRZ), Hamburg.

Leiserowitz, A. (2006). Climate change risk perception and policy preferences: the role of affect, imagery, and values. *Climate Change*, **77**, 45–72.

Lettenmaier, D. P., Wood, E. F., and Wallis, J. R. (1994). Hydro-climatological trends in the continental United States 1948–1988. *Journal of Climate*, **7**, 586–607.

Li, D., Harms, S., Goddard, S., Waltman, W., and Deogun, J. (2003). Time-series data mining in a geospatial decision support system. *Proceedings of National Conference on Digital Government Research*, 1–4.

Li, J. and Heap, A. D. (2008). *A Review of Spatial Interpolation Methods for Environmental Scientists*. Canberra: Geoscience Australia.

Li, J. and Heap, A. D. (2011). A review of comparative studies of spatial interpolation methods in environmental sciences. *Performance and Impact Factors*, **6**(3–4), 228–241.

Liao, Y., Zhang, Q., and Chen, D. (2004). Stochastic modeling of daily precipitation in China. *Journal of Geographical Science*, **14**(4), 417–426.

Lindstrom, G. and Berstrom, S. (2004). Runoff trends in Sweden 1807–2002. *Hydrological Sciences*, **49**(1), 69–83.

Lins, H. F. (1997). Regional streamflow regimes and hydroclimatology of the United States. *Water Resources Research*, **33**, 1655–1667.

Lins, H. F. and Cohn, T. A. (2011). Stationarity: wanted dead or alive? *Journal of the American Water Resources Association*, **47**(3), 475–480.

Lins, H. F. and Slack, J. R. (1999). Stream flow trends in the United States. *Geophysical Research Letters*, **26**(2), 227–230.

Linsley, R. K., Kohler, M. A., Paulhus J. L. H., and Wallace, J. S. (1958). *Hydrology for Engineers*. New York: McGraw-Hill.

Little, D. G. and Rubin, D. B. (1987). *Statistical Analysis with Missing Data.* New York: Wiley.

Lloyd, C. D. (2007). *Local Models for Spatial Analysis.* Boca Raton: CRC Press.

Lloyd, C. D. (2010). *Spatial Data Analysis: An Introduction for GIS Users.* New York: Oxford University Press.

Loader, C. (1999). *Local Regression and Likelihood.* New York: Springer Verlag.

Lohmann, U. (2008). Aerosol effects on precipitation locally and globally. In *Climate Variability and Extremes During the Past 100 Years*, Advances in Global Change Research 33, S. Bronnimann, J. Luterbacher, E. Tracy, *et al.* (editors), Dordrecht: Springer.

Loukas, A. and Quick, M. C. (1996). Spatial and temporal distribution of storm precipitation in southwestern British Columbia. *Journal of Hydrology*, **174**, 37–56.

Lu, G. Y. and Wong, D. W. (2008). An adaptive inverse-distance weighting spatial interpolation technique. *Computers and Geosciences*, **34**(9), 1044–1055.

Lu, X. (2006), *Guidance on the Development of Regional Climate Scenarios for Application in Climate Change Vulnerability and Adaptation Assessments within the Framework of National Communications from Parties Not Included in Annex I to the United Nations Framework Convention on Climate Change.* National Communications Support Programme, UNDP-UNEP-GEF New York, USA.

Lu, X. (2010). *Applying Climate Information for Adaptation Decision-Making: A Guidance and Resource Document.* National Communications Support Programme, UNDP.

Ludwig, F. (2009). Using seasonal climate forecasts for water management. In *Climate Change Adaptation in the Water Sector*, F. Ludwig, P. Kabat, H. V. Schaik, and M. Van Der Valk (editors), London: Earthscan Publishers.

Lyne, V. and Hollick, M. (1979). Stochastic time-variable rainfall-runoff modelling. *I.E. Australian National Conference Publication 79/10*, 89–93, Institute of Engineering Australia, Canberra.

Maddox, R. A. (1980). Mesoscale convective complexes. *Bulletin of the American Meteorological Society*, **61** (11), 1374–1387.

Madsen, H., Mikkelsen, P. S., Rosbjerg, D. S., and Harremoes, P. (2002). Regional estimation of rainfall intensity-duration-frequency curves using generalized least squares regression of partial duration series statistics. *Water Resources Research*, **38**(11), 1239.

Maggio, R. C. and Long, D. W. (1991). Developing thematic maps from point sampling using Thiessen polygon analysis. In *Proceedings of GIS/LIS'91, Atlanta, Georgia.*, **1**(10), American Society of Photogrammetry and Remote Sensing, Bethesda, MD.

Maier, H. R. and Dandy, G. C. (1998). The effect of internal parameters and geometry on the performance of back-propagation neural networks: an empirical study. *Environmental Modeling and Software*, **13**(2), 193–209.

Malakpet, C. G., Habib, E., Meselhe, E. A., and Tokay, A. (2007). Sensitivity analysis of variability in reflectivity-rainfall relationships on runoff prediction. In *Proceedings of 21st Conference on Hydrology, 87th AMS Annual Meeting*, San Antonio, TX.

Manivannan, S., Ramaswamy, K., and Shanthi, R. (2001). Predicting runoff from tank catchments using Green–Ampt and SCS runoff curve number models. *International Agricultural Engineering Journal*, **10**(1–2), 57–69.

Manton, M. J., Della-Marta, P. M., Haylock, M. R., *et al.* (2001). Trends in extreme daily rainfall and temperature in Southeast Asia and the South Pacific: 1961–1998. *International Journal of Climatology*, **21**, 269–284.

Marshall, J. S. and Palmer, W. M. (1948). The distribution of raindrops with size. *Journal of Meteorology*, **5**, 165–166.

Martínez-Cob, A. (1996). Multivariate geostatistical analysis of evapotranspiration and precipitation in mountainous terrain. *Journal of Hydrology*, **174**, 19–35.

Matalas, N. C. and Langbein, W. B. (1962). Information content of the mean. *Journal of Geophysics Research*, **67**(9), 3441–3448.

Mather, P. M. (2004). *Computer Processing of Remotely-Sensed Images: An Introduction.* Hoboken: John Wiley and Sons.

Maurer, E. P. (2007). Uncertainty in hydrologic impacts of climate change in the Sierra Nevada, California under two emissions scenarios. *Climatic Change*, **82**(3–4), 309–325.

Maurer, E. P. and Duffy, P. B. (2005). Uncertainty in projections of streamflow changes due to climate change in California. *Geophysical Research Letters*, **32**, L03704.

Mcbean, A. E. and Rovers, F. A. (1998). *Statistical Procedures for Analysis of Environmental Monitoring Data and Risk Assessment.* Upper Saddle River: Prentice Hall.

McCabe, G. J. (1996). Effects of winter atmospheric circulation on temporal and spatial variability in annual streamflow in the western United States. *Hydrological Sciences Journal*, **41**(6), 873–887.

McCabe, G. J., Palecki, M. A., and Betancourt, J. L. (2004). Pacific and Atlantic Ocean influences on multidecadal drought frequency in the United States. *Proceedings of the National Academy of Sciences*, **101**(12), 4136–4141.

McCuen, R. H. (1998). *Hydrologic Analysis and Design*, second edition. Upper Saddle River: Prentice-Hall.

McCuen, R. (2005). *Hydrologic Analysis and Design*, third edition. Upper Saddle River: Pearson-Prentice Hall.

Meischner, P. (2005). *Weather Radar.* Principles and Advanced Applications Series, Physics of Earth and Space Environments, New York: Springer.

Mekonnen, G. and Hossain, F. (2009). *Satellite Rainfall Applications for Surface Hydrology.* New York: Springer.

Millard, S. P. and Neerchal, N. K. (2001). *Environmental Statistics With S-Plus.* Boca Raton: CRC Press.

Miller, J. F. (1964). *Two-to-Ten-Day Precipitation for Return Periods of 2 to 100 Years in the Contiguous United States.* Technical Report- 49. U.S. Weather Bureau.

Miller, J. R. (1972a). A climatic Z-R relationship for convective storms in the northern Great Plains. *Proceedings of the 15th Radar Meteorology Conference*, Boston: AMS, 153–154.

Miller, J. F. (1972b). *Physiographically Adjusted Precipitation Frequency Maps: Distribution of Precipitation in Mountainous Areas.* WMO-No. 326(11), 264–277.

Milly, P. C. D., Betancourt, J., Falkenmark, M., *et al.* (2008). Stationarity is dead: whither water management. *Science*, **319**, 573–574.

Min, S-K., Zhang, X., Zwiers F. W., and Hegerl, G. C. (2011). Human contribution to more-intense precipitation extremes. *Nature*, **470**, 378–381.

Mitas, L. and Mitasova, H. (1988). General variational approach to the interpolation problem. *Computers and Mathematics and Applications*, **16**(12), 983–992.

Molini, A., Lanza, L. G., and La Barbera, P. (2005). The impact of tipping-bucket rain gage measurements errors on design rainfall for urban-scale applications. *Hydrological Processes*, **19**, 1073–1088.

Mollenkamp, S. and Kastens, B. (2010). Institutional adaptation to climate change: current status and future strategies in the Elbe Basin, Germany. In *Climate Change Adaptation in Water Sector*, F. Ludwig, P. Kabat, H. Schaik, and M. Valk (editors), London: Earthscan.

Molnar, P. and Ramirez, J. A. (2001). Recent trends in precipitation and streamflow in the Rio Puerco Basin. *American Meteorological Society*, **14**, 2317–2328.

Montroy, D. L. (1997). Linear relation to central and eastern North American precipitation to tropical Pacific sea surface temperature anomalies. *Journal of Climate*, **10**, 541–558.

Mulholland, P. J. and Sale, M. J. (2002). Impacts of climate change on water resources: findings of the IPCC regional assessment of vulnerability for North America. In *Water Resources and Climate Change*, K. D. Fredrick (editor), Cheltenham: Edward Elgar, 10–14.

Murphy, J. M. (1999). An evaluation of statistical and dynamical techniques for downscaling local climate. *Journal of Climate*, **12**, 2256–2284.

Murphy, J. M., Sexton, D. M., Barnett, D. N., *et al.* (2004). Quantifying uncertainties in climate change using a large ensemble of global climate model predictions. *Nature*, **430**, 768–772.

Myatt, G. J. and Johnson, W. P. (2009). *Making Sense of Data: A Practical Guide to Data Visualization, Advanced Data Mining and Applications.* Hoboken: John Wiley.

Naoum, S. and Tsanis, I. K. (2003). Temporal and spatial variation of annual rainfall on the island of Crete, Greece. *Hydrological Processes*, **17**(10), 1899–1922.

Narayana Rao, T., Narayana Rao, D., Mohan, K., and Raghavan, S. (2001). Classification of tropical precipitating systems and associated Z-R relationships. *Journal of Geophysics Research*, **106**, 17,699–17,771.

Nathan, R. J. and McMahon, T. A. (1990). Evaluation of automated techniques for baseflow and recession analysis. *Water Resources Research*, **26**(7), 1465.

Navone, H. D. and Ceccatto, H. A. (1994). Predicting Indian monsoon rainfall: a neural network approach. *Climate Dynamics*, **10**, 305–312.

NCDC (2011). http://www.ncdc.noaa.gov/oa/climate/climatedata.html (accessed December, 2011).

NCEP (2011). http://www.ncep.noaa.gov/ (accessed December, 2011).

NIST (2011). http://www.itl.nist.gov/div898/handbook/toolaids/index.htm (accessed December, 2011).

NOAA (2010). http://www.srh.noaa.gov/mfl/?n=summer_season (accessed December, 2011).

Nott, J. (2006). *Extreme Events: A Physical Reconstruction and Risk Assessment*. Cambridge: Cambridge University Press.

Novotny, E. V. and Stefan, H. G. (2007). Stream flow in Minnesota: indicator of climate change. *Journal of Hydrology*, **334**, 319–333.

O'Connor, J. E. and Costa, J. E. (2004). *The World's Largest Floods, Past and Present: Their Causes and Magnitudes*. US Geological Survey Circular, 1254.

O'Sullivan, D. and Unwin, D. J. (2010). *Geographical Information Analysis*. Hoboken: John Wiley & Sons.

Ott, W. R. (1995). *Environmental Statistics and Data Analysis*. Boca Raton: CRC Press.

Ouarda, T. B. M. J. and El-Adlouni, S. (2011). Bayesian nonstationary frequency analysis of hydrological variables. *Journal of the American Water Resources Association*, **47**(3), 496–505.

Pagliara, S. and Vitti, C. (1993). Discussion of rainfall intensity-duration-frequency formula for India. *Journal of Hydraulic Engineering*, **119**(8), 962–966.

Pallottino, S., Sechi, G. M., and Zuddas, P. (2005). A DSS for water resources management under uncertainty by scenario analysis. *Environmental Modeling & Software*, **20**, 1031–1042.

Pathak, C. S., Onderlinde, M., and Fuelberg, H. E. (2009). Use of NEXRAD rainfall data to develop climatologically homogeneous rain areas for Central and South Florida. CD-ROM, *Proceedings of World Environmental and Water Resources Congress*, 1–11, doi:10.1061/41036(342)618.

Petterssen, S. (1958). *Introduction to Meteorology*, second edition, New York: McGraw-Hill.

Pettitt, A. N. (1979). A non-parametric approach to the change-point problem. *Applied Statistics*, **28**(2), 126–135.

Pettyjohn, W. A. and Henning, R. (1979). *Preliminary Estimate of Ground-Water Recharge Rates, Related Streamflow and Water Quality in Ohio*. Ohio State University Water Resources Center Project Completion Report No. 552.

Pielke, R. A. (1973). *An Observational Study of Cumulus Convective Patterns in Relation to the Sea Breeze Over South Florida*. NOAA Technical Memo. ERL OD-16, US Department of Commerce, Boulder, CO, 1–81.

Pielke, R. A. (1974). A three-dimensional numerical model of the sea breezes over south Florida. *Monthly Weather Review*, **102**, 115–139.

Poore, R. Z., Quinn, T., Richey, J., and Smith, J. L. (2005). *Cycles of Hurricane Landfalls on the Eastern United States Linked to Changes in Atlantic Sea-surface Temperatures*. USGS Report 130.

Prudhomme, C., Stevenson, C., and Jakob, D. (2001). *Climate Change and Water Management: Managing European Water Resources in an Uncertain Future: Changing Streamflow in the U.K.* EU Report, CEH, Wallingford.

Prudhomme, C., Piper, B., Osborn, T., and Davies, H. (2005). *Climate Change Uncertainty in Water Resources Planning*. Report, UKWIR.

Puma, M. J. and Gold, S. (2011). *Formulating Climate Change Scenarios to Inform Climate-Resilient Development Strategies: A Guidebook for Practitioners*. United Nations Development Programme, New York.

Quenouille, M. H. (1949). Approximate tests of correlation in time series. *Journal of the Royal Statistical Society, Series B*, **11**, 68–84.

Rabinovich, G. S. (2005). *Measurement Errors and Uncertainties: Theory and Practice*. New York: Springer.

Raghavan, S. (2003). *Radar Meteorology*, Atmospheric and Oceanographic Sciences Library, Dordrecht: Springer.

Raghavan, S. and Sivaramakrishnan, T. R. (1982). Radar estimation precipitation around Madras. *Mausam*, **33**(1), 21–28.

Rakhecha, P. R. and Singh, V. P. (2009). *Applied Hydrometeorology*, New Delhi: Springer.

Redmond, K. T. and Koch, R. W. (1991). Surface climate and streamflow variability in the Western United States and their relationship to largescale circulation indices. *Water Resources Research*, **27**, 2381–2399.

Regonda, S. K., Rajagopalan, B., Clark, M., and Zagona, E. (2006). A multi-model ensemble forecast framework: application to spring seasonal flows in the Gunnison River Basin. *Water Resources Research*, **42**(9), W09404.

Reynard, N. S. (2007). *Climate Change, Future Flooding and Coastal Erosion Risks*. London: Thomas Telford.

Reynard, N. S., Prudhomme, C., and Crooks, S. M. (1999). *Climate Change Impacts for Fluvial Flood Defense*, Report, MAFF.

Richardson, C. W. and Wright, D. A. (1984). *WGEN: A Model for Generating Daily Weather Variables*. US Department of Agriculture, Agricultural Research Service, ARS-8.

Robson, A, J., Jones, T. K., Reed, D. W., and Bayliss, A. C. (1998). A study of national trend and variation in UK floods. *International Journal of Climatology*, **18**(2), 165–182.

Rodo, X., Baert, E., and Comin, F. A. (1977). Variations in seasonal rainfall in Southern Europe during the present century: relationships with the North Atlantic Oscillation and the El Niño-Southern Oscillation. *Climate Dynamics*, **13**, 275–284.

Rogers, D. J. and Tanimoto, T. T. (1960). A computer program for classifying plants. *Science*, **132**, 1115–1118.

Rogers, J. C. and Coleman, J. S. M. (2003). Interactions between the Atlantic multi-decadal oscillation, El Niño/La Niña, and the PNA in winter Mississippi Valley stream flow. *Geophysical Research Letters*, **30**(10), 1518, doi:10.1029/2003GL017216.

Ropelewski, C. F. and Halpert, M. S. (1986). North American precipitation and temperature patterns associated with the El Niño/Southern Oscillation (ENSO). *Monthly Weather Review*, **114**, 2352–2362.

Ropelewski, C. F. and Halpert, M. S. (1987). Global and regional scale precipitation patterns associated with the El Niño/Southern Oscillation. *Monthly Weather Review*, **115**, 1606–1626.

Rosenfeld, D., Wolff, D. B., and Amitai, E. (1994). The window probability matching method for rainfall measurements with radar. *Journal of Applied Meteorology*, **33**(6), 689–693.

Rosenzweig, C. and Hillel, D. (2008). *Climate Variability and the Global Harvest: Impacts of El Niño and Other Oscillations on Agroecosystems*. New York: Oxford University Press.

Roudier, P., Sultan, B., Quirion, P., *et al.* (2011). An ex-ante evaluation of the use of seasonal climate forecasts for millet growers in SW Niger. *International Journal of Climatology*, doi:10.1002/joc.2308.

Rudolf, B. and Schneider, U. (2005). Calculation of gridded precipitation data for the global land-surface using in-situ gage observations. *Proceedings of the 2nd Workshop of the International Precipitation Working Group IPWG*, Monterey, October 2004, EUMETSAT, 231–247.

Rudolf, B., Hauschild, H., Rueth, W., and Schneider, U. (1994). Terrestrial precipitation analysis: operational method and required density of point measurements. In *Global Precipitations and Climate Change*, M. Desbois and F. Desalmond (editors), NATO ASI Series I, Vol. 26, Springer-Verlag, 173–186.

Rudolf, B., Fuchs, T., Schneider, U., and Meyer-Christoffer, A. (2003). *Introduction of the Global Precipitation Climatology Centre (GPCC)*. Global Precipitation Climatology Centre, DWD, Germany.

Rudolf, B., Beck, C., Grieser, J., and Schneider, U. (2005). *Global Precipitation Analysis Products*. Global Precipitation Climatology Centre, DWD, Germany, Internet publication, 1–8.

Rummukainen, M. (1997). *Methods for Statistical Downscaling of GCM Simulations*. Swedish Meteorological and Hydrological Institute, Report 71.

Russel, P. F. and Rao, T. R. (1940). On habitat and association of species of anopheline larvae in south-eastern Madras. *Journal of Malaria Institute of India*, **3**, 153–178.

Russell, G. L., Miller, J. R., Rind, D., *et al.* (2000). Comparison of model and observed regional temperature changes during the past 40 years. *Journal of Geophysics Research*, **105**, 14891–14898.

Rycroft, H. B. (1949). Random sampling of rainfall. *Journal of South African Forestry Association*, **18**.

Salas, J. D. (1993). Analysis and modeling of hydrological time series. In *Handbook of Hydrology*, D. R. Maidment (editor), New York: Mc-Graw-Hill.

Salas-Melia, D., Chauvin, F., Deque, M., *et al.* (2005). *Description and Validation of the CNRMCM3 Global Coupled Model*. Note de centre GMGEC, CNRM France.

Salathe, E. P., Steed, R., Mass, C. F., and Zahn, P. H. (2008). A high-resolution climate model for the US Pacific Northwest: mesoscale feedbacks and local responses to climate change. *Journal of Climate*, **21**, 5708–5726.

Sarewitz, D., Pielke, R. A., and Byerly, R. (2000). *Prediction: Science, Decision Making, and the Future of Nature*. Washington, DC: Island Press.

Satterthwaite, F. E. (1946). An approximate distribution of estimates of variance components. *Biometrics Bulletin*, **2**, 110–114.

Schlesinger, M. E. and Ramankutty, N. (1994). An oscillation in the global climate system of period 65–70 years. *Nature*, **367**, 723–726.

Schlesinger, M. E., Ramankutty, N., Andronova, N., and Margolis, M. (2000). Temperature oscillation in the North Atlantic. *Science*, **289**, 547b.

Schmidt, N., Lipp, E. K., Rose, J. B., and Luther, M. E. (1999). ENSO influences on seasonal rainfall and river discharge in Florida. *Journal of Climate*, **14**, 615–628.

Schneider, U., Becker, A., Meyer-Christoffer, A., Ziese, M., and Rudolf, B. (2010). *Global Precipitation Analysis Products of the GPCC*, GPCC status report.

Schuenemeyer, J. H. and Drew, L. J. (2010). *Statistics for Earth and Environmental Scientists*. Hoboken: Wiley.

Schwarz, G. (1978). Estimating the dimension of a model. *Annals of Statistics*, **6**(2), 461–464.

SCS (1972). Hydrology. In *National Engineering Handbook, Supplement A*, Soil Conservation Service, USDA, Washington, DC.

Seber, G. A. F. and Wild, C. J. (2003). *Nonlinear Regression*. Hoboken: Wiley-Interscience.

Semenov, M. A. (2008). Ability of a stochastic weather generator to reproduce extreme weather events. *Climate Research*, **35**, 203–212.

Semenov, M. A. and Barrow, E. M. (1997). Use of a stochastic weather generator in the development of climate change scenarios. *Climatic Change*, **35**, 397–414.

Semenov, M. A., Brooks, R. J., Barrow, E. M., and Richardson, C. W. (1998). Comparison of WGEN and LARS-WG stochastic weather generators for diverse climates. *Climate Research*, **10**, 95–107.

Sene, K. (2008). *Flood Warning, Forecasting and Emergency Response*. New York: Springer.

Sene, K. (2009). *Hydrometeorology: Forecasting and Applications*. New York: Springer.

Seo, D.-J. (1996). Nonlinear estimation of spatial distribution of rainfall: an indicator cokriging approach. *Stochastic Hydrology and Hydraulics*, **10**, 127–150.

Seo, D. J. (1998). Real-time estimation of rainfall fields using radar rainfall and rain gage data. *Journal of Hydrology*, **208**, 37–52.

Seo, D. J. and Smith, J. A. (1993). Rainfall estimation using rain gages and radar: a bayesian approach. *Journal of Stochastic Hydrology and Hydraulics*, **5**(1), 1–14.

Seo, D.-J., Krajewski, W. F., and Bowles, D. S. (1990a). Stochastic interpolation of rainfall data from rain gauges and radar using cokriging. 1. Design of experiments. *Water Resources Research*, **26**(3), 469–477.

Seo, D.-J., Krajewski, W. F., and Bowles, D. S. (1990b). Stochastic interpolation of rainfall data from rain gauges and radar using cokriging. 2. Results. *Water Resources Research*, **26**(5), 915–924.

Serrano, S. E. (2010). *Hydrology for Engineers, Geologists, and Environmental Professionals: An Integrated Treatment of Surface, Subsurface, and Contaminant Hydrology*. HydroScience, Inc.

Sevruk, B. and Geiger, H. (1981). *Selection of Distribution Types for Extremes of Precipitation*. WMO, Operational Hydrology, Report 15.

Sevruk, B. and Klemm, S. (1989). Types of standard precipitation gages. In *WMO/IAHS/ETH International Workshop on Precipitation Measurement*, St. Moritz.

Sharif, M. and Burn, D. (2009). Detection of linkages between extreme flow measures and climate indices. *World Academy of Science, Engineering and Technology*, **60**, 871–876.

Shaw, E. M. (1994). *Hydrology in Practice*. London: Chapman and Hall.

Shawe-Taylor, J. and Cristianini, N. (2000). *An Introduction to Support Vector Machines*. Cambridge: Cambridge University Press.

Shelton, M. L. (2009). *Hydroclimatology: Perspectives and Applications*. Cambridge: Cambridge University Press.

Shepard, D. (1968). A two-dimensional interpolation function for irregularly spaced data, *Proceedings of the Twenty-Third National Conference of the Association for Computing Machinery*, 517–524.

Shepard, D. S. (1984). Computer mapping: the SYMAP interpolation algorithm. In *Spatial Statistics and Models*, G. L. Gaile and C. J. Willmott (editors), Norwell: D. Reidel, 133–145.

Shimazaki, H. and Shinomoto, S. (2007). A method for selecting the bin size of a time histogram. *Neural Computation*, **19**(6), 1503–1527.

Shimazaki, H. and Shinomoto, S. (2010). Kernel bandwidth optimization in spike rate estimation. *Journal of Computational Neuroscience,* **29**(1–2), 171–182.

Shirmohammadi, A., Knisel, W. G., and Sheridan, J. M. (1984). An approximate method for partitioning daily streamflow data. *Journal of Hydrology*, **74**, 335–354.

Shreshtha, B. P. (2002). Uncertainty in risk analysis of water resources systems under climate change. In *Risk, Reliability, Uncertainty and Robustness of Water Resources Systems*, J. J. Bogardi and Z. W. Kundzewicz (editors), International Hydrology Series, Cambridge: UNESCO and Cambridge University Press, 153–161.

Sibson, R. (1981). A brief description of natural neighbor interpolation. In *Interpreting Multivariate Data*, V. Barnett (editor), Chichester: John Wiley, 21–36.

Sloto, R. A. and Crouse, M. Y. (1996). *HYSEP: A Computer Program for Streamflow Hydrograph Separation and Analysis*. US Geological Survey, Water Resources Investigations Report 96–4040.

Smakhtin, V. Y. (2001). Estimating continuous monthly baseflow time series and their possible applications in the context of the ecological reserve. *Water SA*, **27**(2), 213–217.

Small, M. J. (1990). Probability distributions and statistical estimation. In *Uncertainty: A Guide to Dealing with Uncertainty in Quantitative Risk and Policy Analysis*, M. G. Morgan and M. Henrion (editors), New York: Cambridge University Press.

Smith, J. A. (1993). Precipitation. In *Handbook of Hydrology*, D. R. Maidment (editor). New York: McGraw Hill.

Smith, M. D., Goodchild, M. F., and Longley, P. A. (2007). *Geospatial Analysis: A Comprehensive Guide to Principles, Techniques and Software Tools*. Leicester: Winchelsea Press.

Smith, T. M., Reynolds, R. W., Peterson, T. C., and Lawrimore, J. (2008). Improvements to NOAA's historical merged land-ocean surface temperature analysis (1880–2006). *Journal of Climate*, **21**, 2283–2296.

Sneath, P. H. and Sokal, R. R. (1962). Numerical taxonomy. *Nature*, **193**, 855–860.

Sokal, R. R. and Michener, C. D. (1958). A statistical method for evaluating systematic relationships. *University of Kansas Scientific Bulletin*, **38**(22), 1409–1438.

Sokolov, A. A., Rantz, S. E., and Roche, M. (1976). *Floodflow Computation: Methods Compiled from World Experience, Studies and Reports in Hydrology*. Report 22, UNESCO.

Solaiman, T. A., King, L. M., and Simonovic, S. P. (2010). Extreme precipitation vulnerability in the Upper Thames river basin: uncertainty in climate model projections. *International Journal of Climatology*, doi:10.1002/joc.2244).

Soon-Kuk, K. (2006). Rainfall observations in Korea by the world's first rain gage. *Paddy and Environment*, **4**, 67–69.

Sorensen, T. (1948). A method of establishing groups of equal amplitude in plant sociology based on similarity of species content and its application to analyses of the vegetation on Danish commons. *Kongelige Danske Videnski Selskab Biologiske Skrifter*, **5**, 1–34.

Soukup, T. L., Aziz, O. A., Tootle, G. A., Piechota, T. C., and Wulff, S. S. (2009). Long lead-time streamflow forecasting of the North Platte River incorporating oceanic–atmospheric climate variability. *Journal of Hydrology*, **368**, 131–142.

Souza, F. A. and Lall, U. (2003). Seasonal to interannual ensemble streamflow forecasts for Ceara, Brazil: application of a multivariate, semiparametric algorithm. *Water Resources Research*, **39**, 1307–1319.

Spilhaus, A. F. (1948). Raindrop size, shape, and falling speed. *Journal of Meteorology*, **5**, 108–110.

Stainforth, D. A., Allen, M. R., Tredger, E. R., and Smith, L. A. (2007). Confidence, uncertainty and decision support relevance in climate predictions. *Philosophical Transactions of the Royal Society*, **365**, 2145–2161.

Stedinger, J. (2000). Flood frequency analysis and statistical estimation of flood risk. In *Inland Flood Hazards: Human, Riparian, and Aquatic Communities*, E. E. Whol (editor), Cambridge: Cambridge University Press.

Stedinger, J. R. and Griffis, V. W. (2011). Getting from here to where? Flood frequency analysis and climate. *Journal of the American Water Resources Association*, **47**(3), 506–513.

Stedinger, J. R., Vogel, R. M., and Foufoula-Georgiu, E. (1992). Frequency analysis of extreme events. In *Handbook of Hydrology*, D. R. Maidment (editor), New York: Mc-Graw-Hill.

Steiner, M. and Houze, R. A. Jr. (1993). Three-dimensional validation at TRMM ground truth sites: some early results from Darwin, Australia. *Preprints, 26th International Conference on Radar Meteorology*, American Meteorological Society, Norman, OK, 417–420.

Steiner, M., Houze, R. A. Jr., and Yuter, S. E. (1995). Climatological characterization of three-dimensional storm structure from operational radar and rain gage data. *Journal of Applied Meteorology*, **34**, 1978–2007.

Stevenson, S., Fox-Kemper, B., Jochum, M., *et al.* (2011). Will there be a significant change to El Nino in the 21st century? *16th Annual CESM Workshop*, Breckenridge, CO.

Stone, R. C., Hammer, G. L., and Marcussen, T. (1996). Prediction of global rainfall probabilities using phases of the Southern Oscillation Index. *Nature*, **384**, 252–255.

Storch, V. H. and Zwiers, F. W. (1999). *Statistical Analysis in Climate Research*. Cambridge: Cambridge University Press.

Strangeways, I. (2003). *Measuring the Natural Environment*, second edition. Cambridge: Cambridge University Press.

Strangeways, I. (2007). *Precipitation: Theory, Measurement and Distribution*. Cambridge: Cambridge University Press.

Suppiah, R. and Hennessy, K. J. (1998). Trends in total rainfall, heavy rainfall events, and number of dry events in Australia. *International Journal of Climatology*, **18**(10), 1141–1164.

Sutton, R. T. and Hodson, D. L. (2005). Atlantic Ocean forcing of North American and European summer climate. *Science*, **309**(5731), 115–118.

Svensson, C. and Jones, D. (2010). Review of rainfall frequency estimation methods. *Journal of Flood Risk Management*, **3**, 296–313, doi:10.1111/j.1753-318X.2010.01079.x.

Tabios, G. Q., III and Salas, J. D. (1985). A comparative analysis of techniques for spatial interpolation of precipitation. *Water Resources Bulletin*, **21**, 365–380.

Tan, Pang-Ning, Steinbach, M., and Kumar, V. (2006). *Introduction to Data Mining*. New York: Pearson.

Taylor, K. E. (2001). Summarizing multiple aspects of model performance in a single diagram. *Journal of Geophysical Research*, **106**(D7), 7183–7192.

Tebaldi, C. and Knutti, R. (2007). The use of the multi-model ensemble in probabilistic climate projections. *Philosophical Transactions of the Royal Society A*, **365**, 2053–2075.

Teegavarapu, R. S. V. (2007). Use of universal function approximation in variance-dependent interpolation technique: an application in hydrology. *Journal of Hydrology*, **332**, 16–29.

Teegavarapu, R. S. V. (2008). Innovative spatial interpolation methods for estimation of missing precipitation records: concepts and applications. *Proceedings of HydroPredict 2008*, Prague, J. Bruthans, K. Kovar, and Z. Hrkal (editors), 79–82.

Teegavarapu, R. S. V. (2009a). *Comparing NEXRAD Rainfall and Rain Gage Data in South Florida Water Management District*. Report, USGS 104B Grant, Water Resources Research Center, University of Florida.

Teegavarapu, R. S. V. (2009b). Estimation of missing precipitation records integrating surface interpolation techniques and spatio-temporal association rules. *Journal of Hydroinformatics*, **11** (2), 133–146.

Teegavarapu, R. S. V. (2010). Modeling climate change uncertainties in water resources management models. *Environmental Modeling and Software*, **25**(10), 1261–1265.

Teegavarapu, R. S. V. (2012a). Spatial interpolation using non-linear mathematical programming models for estimation of missing precipitation records. *Hydrological Sciences Journal*, **57**, 383–406.

Teegavarapu, R. S. V. (2012b). Spatial interpolation for missing precipitation data: use of proximity metrics, nearest neighbor classifiers and clusters. *Proceedings of World Environmental and Water Resources Conference*.

Teegavarapu, R. S. V. and Chandramouli, V. (2005). Improved weighting methods, deterministic and stochastic data-driven models for estimation of missing precipitation records. *Journal of Hydrology*, **312**, 191–206.

Teegavarapu, R. S. V. and Elshorbagy, A. (2005). Fuzzy set based error measure for hydrologic model evaluation. *Journal of Hydroinformatics*, **7**(3), 199–208.

Teegavarapu, R. S. V. and Goly, A. (2011). *Assessment of Spatial and Temporal Variation of Extreme Precipitation Events in Coastal Basins of SFWMD Region*. SFWMD final report.

Teegavarapu, R. S. V. and Mujumdar, P. P. (1996). Rainfall forecasting using neural networks. *Proceedings of IAHR International Symposium on Stochastic Hydraulics*, **I**, 325–332.

Teegavarapu, R. S. V. and Pathak, C. (2008). Infilling of rain gage records using radar (NEXRAD) data: influence of spatial and temporal variability of rainfall processes. *Proceedings of World Environmental and Water Resources Conference*, R. W. Babcock and R. Walton (editors), EWRI-ASCE, Hawaii, 1–9, doi:10.1061/40976(316)406.

Teegavarapu, R. S. V. and Pathak, C. (2011). Development of optimal Z-R relationships, weather radar and hydrology. *IAHS Red Book*, International Association of Hydrological Sciences, UK, **351**(33).

Teegavarapu, R. S. V. and Simonovic, S. P. (1999). Modeling uncertainty in reservoir loss functions. *Water Resources Research*, **35**(9), 2815–2823.

Teegavarapu, R. S. V. and Simonovic, S. P. (2000). Short-term operation model for coupled hydropower reservoirs. *Journal of Water Resources Planning and Management, ASCE*, **126**(2), 98–106.

Teegavarapu, R. S. V. and Simonovic, S. P. (2002). Optimal operation of water resource systems: trade-offs between modeling and practical solutions. Integrated Water Resources Management, *IAHS Red Book*, **272**, 257–262.

Teegavarapu, R. S. V., Peters, D., and Pathak, C. (2008). Evaluation of functional forms of rain gage – radar (NEXRAD) data relationships. CD-ROM. *Proceedings of World Environmental and Water Resources Congress*, R. W. Babcock and R. Walton (editors), 1–9, doi:10.1061/40976(316)413.

Teegavarapu, R. S. V., Tufail, M., and Ormsbee, L. (2009). Optimal functional forms for estimation of missing precipitation records. *Journal of Hydrology*, **374**, 106–115.

Teegavarapu, R. S. V., Pathak, C., and Chinatalapudi, S. (2010a). Infilling missing precipitation data using NEXRAD data: use of optimal spatial interpolation and data-driven methods. In *Proceedings of World Environmental and Water Resources Congress 2010: Challenges of Change*, ASCE, doi:10.1061/41114(371)473.

Teegavarapu, R. S. V., Senarath, S. U. S., and Pathak, C. (2010b). Sampling schemes for uncertainty assessment of a hydrologic simulation model. In *Proceedings of World Environmental and Water Resources Congress 2010: Challenges of Change*, ASCE, doi:10.1061/41114(371)484.

Teegavarapu, R. S. V., Nayak, A., and Pathak, C. (2011a). Assessment of long-term trends in extreme precipitation: implications of in-filled historical data and temporal window-based analysis. In *Proceedings of the 2011 World Environmental and Water Resources Congress*, doi:10.1061/41173(414)419.

Teegavarapu, R. S. V., Meskele, T., and Pathak, C. (2011b). Geo-spatial grid-based transformations of precipitation estimates. *Computers and Geosciences*, doi:10.1016/j.cageo.2011.07.004.

Thomas, H. A. (1981). *Improved Methods for National Water Assessment*. Report, Contract WR15249270, US Water Resources Council, Washington, DC.

Thomas, J. F. and Bates, B. C. (2002). Responses to the variability and increasing uncertainty of climate in Australia. In *Risk, Reliability, Uncertainty and Robustness of Water Resources Systems*, J. J. Bogardi and Z. W. Kundzewicz (editors), International Hydrology Series, Cambridge: UNESCO and Cambridge University Press, 54–69.

Thome, C. R., Evans, E. R., and Penning-Rowsell, E. C. (2007). *Future Flooding and Coastal Erosion Risks*. London: Thomas Telford.

Thompson, S. A. (1999). *Hydrology for Water Management*. Rotterdam: Balkema Publishers.

Thornthwaite, C. W. and Wilm, H. G. (1944). Report of the committee on transpiration and evaporation, 1943–1944. *Transactions, American Geophysical Union*, **25**, 683–693.

Ting, M., Kushnir, Y., Seager, R., and Li, C. (2011). Robust features of Atlantic multi-decadal variability and its climate impacts. *Geophysical Research Letters*, **38**(L17705), doi:10.1029/2011gl048712.

Tobler, W. R. (1970). A computer movie simulating urban growth in the Detroit region. *Economic Geography*, **46**, 234–240.

Tokay, A., Hartmann, P., Battaglia, A., *et al.* (2008) A field study of reflectivity and Z-R relations using vertically pointing radars and disdrometers. *Journal of Atmospheric and Oceanic Technology*, doi:10.1175/2008JTECHA1163.1.

Tomczak, M. (1998). Spatial interpolation and its uncertainty using automated anisotropic inverse distance weighting (IDW): cross-validation/jackknife approach. *Journal of Geographic Information and Decision Analysis*, **2**, 18–30.

Tootle, G. A., Piechota, T. C., and Singh, A. (2005). Coupled oceanic-atmospheric variability and U.S. streamflow. *Water Resources Research*, doi:10.1029/2005WR004381.

Tootle, G. A., Singh, A. K., Piechota, T. C., and Farnham, I. (2007). Long lead-time forecasting of U.S. streamflow using partial least squares regression. *Journal of Hydrologic Engineering*, **12**(5), 442–451.

Tootle, G. A., Piechota, T. C., and Gutierrez, F. (2008). The relationships between Pacific and Atlantic ocean sea surface temperatures and Colombian streamflow variability. *Journal of Hydrology*, **349**, 268–276.

Torgo, L. (2011). *Data Mining with R: Learning with Case Studies*. Boca Raton: CRC Press.

Trafalis, T. B., Santosa, B., and Richman, M. B. (2005). Learning networks in rainfall estimation. *Computational Management Science*, 2, **3**(7), 229–251.

Trenberth, K. E. (1999). The extreme weather events of 1997 and 1998. *Consequences*, **5**(1), 3–15.

Trenberth, K. E. and Caron, J. M. (2000). The Southern Oscillation revisited: sea level pressures, surface temperatures, and precipitation. *Journal of Climate*, **13**, 4358–4365.

Trenberth, K. E. and Hurrell, J. W. (1994). Decadal atmosphere-ocean variations in the Pacific. *Climate Dynamics*, **9**, 303–319.

Trenberth, K. E. and Stepaniak, D. P. (2001). Indices of El Nino Evolution. *Journal of Climate*, **14**, 1697–1701.

Trenberth, K. E., Caron, J. M., Stepaniak, D. P., and Worley, S. (2002a). Evolution of El Nino–Southern Oscillation and global atmospheric surface temperatures. *Journal of Geophysical Research*, **107**(D8), 4065.

Trenberth, K. E., Stepaniak, D. P., and Caron, J. M. (2002b). Interannual variations in the atmospheric heat budget. *Journal of Geophysical Research*, **107**(D8), 4066.

Trenberth, K. E., Jones, P. D., Ambenje, P., *et al.* (2007). Observations: surface and atmospheric climate change. In *Climate Change 2007: The Physical Science Basis. Contribution of Working Group I to the Fourth Assessment Report of the Intergovernmental Panel on Climate Change.* S. Solomon, D. Qin, M. Manning, *et al.* (editors), Cambridge and New York: Cambridge University Press.

TRMM (2011). http://trmm.gsfc.nasa.gov/ (accessed December, 2011).

Tufail, M. and Ormsbee, L. E. (2006). A fixed functional set genetic algorithm (FFSGA) approach for functional approximation. *IWA Journal of Hydroinformatics*, **3**, 193–206.

Tukey, J. W. (1958). Bias and confidence in not quite large samples (Abstract). *Annals of Mathematical Statistics*, **29**, 614.

TYC (2011). http://www.tyndall.ac.uk/ (accessed December, 2011).

UCAR (2011). https://www.meted.ucar.edu/ (accessed December, 2011).

Ugarte, M. D., Militino, A. F., and Arnholt, A. T. (2008). *Probability and Statistics with R.* Boca Raton: Chapman and Hall/CRC.

Ulbrich, C. W. and Atlas, D. (1977). A method for measuring precipitation parameters using radar reflectivity and optical extinction. *Annals of Telecommunications*, **32**(11–12), 415–421.

UNEP/GRID-Arendal (2005). *Precipitation changes: trends over land from 1900 to 2000.* Maps and Graphics Library, http://maps.grida.no/go/graphic/precipitation_changes_trends_over_land_from_1900_to_2000 (accessed September, 2011).

USAID (2007). *Adapting to Climate Variability and Change: A Guidance Manual for Development Planning.* USAID.

USEPA (2008). *National Water Program Strategy Response to Climate Change.* Office of Water (4101M), EPA 800-R-08–001, USA.

USGS (2009). *Climate Change and Water Resources Management: A Federal Perspective.* USGS Circular 1331, US Department of the Interior.

US Weather Bureau (1976). *Hydrometeorology.* Reports 55A, 56, and 57.

Van Rheenen, N. T., Wood, A. W., Palmer, R. N., and Lettenmaier, D. P. (2004). Potential implications of PCM climate change scenarios for Sacramento-San Joaquin River Basin hydrology and water resources. *Climatic Change*, **62**, 257–281.

Vasiliev, I. R. (1996). Visualization of spatial dependence: an elementary view of spatial autocorrelation. In *Practical Handbook of Spatial Statistics*, Boca Raton: CRC Press.

Venables, W. N. and Ripley, B. D. (2002). *Modern Applied Statistics with S-plus.* New York: Springer-Verlag, 495.

Vergni, L. and Todisco, F. (2011). Spatio-temporal variability of precipitation, temperature and agricultural drought indices in Central Italy. *Agricultural and Forest Meteorology*, **151**(3), 301–313.

Vicente-Serrano, S. M., Saz-Sánchez, M. A., and Cuadrat, J. M. (2003). Comparative analysis of interpolation methods in the middle Ebro Valley (Spain): application to annual precipitation and temperature. *Climate Research*, **24**, 161–180.

Vieux, B. E. (2004). *Distributed Hydrologic Modeling Using GIS*, second edition, Water Science and Technology Library, Vol. 48, Norwell: Kluwer Academic Publishers.

Vieux, B. E. and Bedient, P. B. (2004). Assessing urban hydrologic prediction accuracy through event reconstruction. *Journal of Hydrology*, **299**(3–4), 217–236.

Villarini, G., Ciach, G. J., Krajewski, W. F., Nordstrom, K. M., and Gupta, V. K. (2007). Effects of systematic and random errors on the spatial scaling properties in radar-estimated rainfall. *Nonlinear Dynamics in Geosciences*, 37–51.

Villarini, G., Smith, J. A., Baeck, M. L., and Krajewski, W. F. (2011). Examining flood frequency distributions in the midwest U.S. *Journal of the American Water Resources Association*, **47**(3), 447–463.

Vogel, R. M., Yaindl, C., and Walter, M. (2011). Nonstationarity: flood magnification and recurrence reduction factors in the United States. *Journal of the American Water Resources Association*, **47**(3), 464–474.

Voltz, M. and Webster, R. (1990). A comparison of kriging, cubic splines and classification for predicting soil properties from sample information. *Journal of Soil Science*, **41**, 473–490.

Von Neumann, J. (1941). Distribution of the ratio of the mean square successive difference to the variance. *Annals of Mathematical Statistics*, **12**, 367–395.

Von Storch, H., Zorita, E., and Cubasch, U. (1993). Downscaling of global climate change estimates to regional scales. An application to Iberian rainfall in wintertime. *Journal of Climatology*, **6**, 1161–1171.

Wahl, K. L. and Wahl, T. L. (1995). Determining the flow of Comal Springs at New Braunfels, Texas. *Texas Water '95*, American Society of Civil Engineers, San Antonio, Texas, 77–86.

Walker, G. T. (1908). Correlation in seasonal variation of climate (Introduction). *Memoirs Indian Meteorological Department*, **20**, 117–124.

Walker, G. T. (1918). Correlation in seasonal variation of weather. *Quarterly Journal of Royal Meteorological Society*, **44**, 223–224.

Wallis, J. R., Lettenmaier, D. P., and Wood, E. F. (1991). A daily hydroclimatological data set for the continental United States. *Water Resources Research*, **27**, 1657–1663.

Wang, F. (2006). *Quantitative Methods and Applications in GIS.* Boca Raton: CRC Press.

Washington, W. M., Weatherly, J. W., Meehl, G. A., *et al.* (2000). Parallel climate model (PCM) control and transient simulations. *Climate Dynamics*, **16**, 755–774.

Watson, D. and Adams, M. (2011). *Design for Flooding: Architecture, Landscape and Urban Design for Resilience to Climate Change.* Hoboken: John Wiley.

WCT (2011). http://www.ncdc.noaa.gov/oa/wct/ (accessed December, 2011).

WDC (2011). http://www.ncdc.noaa.gov/oa/wdc/ (accessed December, 2011).

Webb, R. H. and Betancourt, J. L. (1992). *Climatic Variability and Flood Frequency of the Santa Cruz River, Pima County, Arizona.* US Geological Survey Water-Supply Paper 2379.

Webster, R. A. and Oliver, M. A. (2007). *Geostatistics for Environmental Scientists.* Chichester: John Wiley and Sons.

Wei, F., Xie, Y., and Mann, M. E. (2008). Probabilistic trend of anomalous summer rainfall in Beijing: role of interdecadal variability. *Journal Of Geophysical Research*, **113**, D20106, doi:10.1029/2008jd010111.

Weisner, C. J. (1970). *Hydrometeorology.* London: Chapman & Hall.

Weiss, L. L. (1964). Ratio of true to fixed-interval maximum rainfall. *Journal of Hydraulics Division, ASCE*, **90**(1), 77–82.

WEKA (2011). *Weka 3.6.0 Manual*, Waikato Environment for Knowledge Analysis, http://www.cs.waikato.ac.nz/ml/weka/ (accessed December, 2011).

Wentz, F. J. I., Ricciardulli, K., Hilburn, K., and Mears, C. (2007). How much more rain will global warming bring? *Science*, **317**, 233–235.

Westerberg, I., Walther, A., Guerrero, J. L., *et al.* (2009). Precipitation data in a mountainous catchment in Honduras: quality assessment and spatiotemporal characteristics. *Theoretical Applied Climatology*, doi: 10.1007/s00704-009-0222-x.

Widmann, M. and Bretherton, C. S. (1999). Validation of mesoscale precipitation in the NCEP reanalysis using a new grid-cell data set for the northwestern United States. *Journal of Climate*, **13**, 936–950.

Wigley, T. M. L. and Jones, P. D. (1985). Influence of precipitation changes and direct CO2 effects on streamflow. *Nature*, **314**, 149–151.

Wilby, R. L. and Fowler, H. J. (2011). Regional climate downscaling. In *Modeling the Impact of Climate Change on Water Resources*, F. Fung, A. Lopez, and M. New (editors), Chichester: Wiley-Blackwell.

Wilby, R. L. and Wigley, T. M. L. (1997). Downscaling general circulation model output: a review of methods and limitations. *Progress in Physical Geography*, **21**, 530–548.

Wilby, R. L., Dawson, C. W., and Barrow, E. M. (2002). SDSM: a decision support tool for the assessment of regional climate change impacts. *Environmental Modelling & Software*, **17**, 147–159.

Wilby, R. L., Charles, S. P., Zorita, E., *et al.* (2004). *Guidelines for Use of Climate Scenarios Developed from Statistical Downscaling Methods.* Prepared for consideration by the IPCC at the request of its Task Group on Data and Scenario Support for Impacts and Climate Analysis (TGICA).

Wilks, D. S. (1999). Multisite downscaling of daily precipitation with a stochastic weather generator. *Climate Research*, **11**, 125–136.

Wilks, D. S. (2006). *Statistical Methods in the Atmospheric Sciences*. Burlington: Academic Press.

Wilks, D. S. and Wilby, R. L. (1999). The weather generation game: a review of stochastic weather models. *Progress in Physical Geography*, **23**, 329–357.

Willmott, C. J. (1982). Some comments on the evaluation of model performance. *Bulletin American Meteorological Society*, **63**(11), 1309–1313.

WMO (1982). *Methods of Correction for Systematic Error in Point Precipitation Measurement for Operational Use (B. Sevruk)*. Operational Hydrology Report No. 21, WMO No. 589, Geneva.

WMO (1983). *Guide to Climatological Practices*, second edition. WMO No. 100, Geneva.

WMO (1984). *International Comparison of National Precipitation Gages with a Reference Pit Gage (B. Sevruk and W. R. Hamon)*. Instruments and Observing Methods Report No. 17, WMO/TD No. 38, Geneva.

WMO (1985). *International Organizing Committee for the WMO Solid Precipitation Measurement Intercomparison: Final report of the first session*. Geneva.

WMO (1986a). *Manual for Estimation of Probable Maximum Precipitation*. Operational Hydrology Report No. 1, WMO No. 332, Geneva.

WMO (1986b). Papers presented at the *Workshop on the Correction of Precipitation Measurements*, Zurich, Switzerland, 1–3 April 1985, B. Sevruk (editor). Instruments and Observing Methods Report No. 25, WMO/TD No. 104, Geneva.

WMO (1988). *Technical Regulations, Vol. I – General Meteorological Standards and Recommended Practices; Vol. II – Meteorological Service for International Air Navigation; Vol. III – Hydrology*. WMO No. 49, Geneva.

WMO (1989). *Calculation of Monthly and Annual 30-Year Standard Normals*. WMO/TD No. 341, WCDP No. 10, Geneva.

WMO (1989a). *Catalogue of National Standard Precipitation Gages (B. Sevruk and S. Klemm)*. Instruments and Observing Methods Report No. 39, WMO/TD No. 313, Geneva.

WMO (1989b). *International Workshop on Precipitation Measurements, St Moritz, Switzerland, 3–7 December 1989*, B. Sevruk (editor). Instruments and Observing Methods Report No. 48, WMO/TD No. 328, Geneva.

WMO (1990). *On the Statistical Analysis of Series of Observations*. WMO/TN No. 143, WMO No. 415, Geneva.

WMO (1994). *Guide to Hydrological Practices: Data Acquisition and Processing, Analysis, Forecasting and Other Applications*. Geneva.

WMO (1996). *Climatological Normals (CLINO) for the Period 1961–90*. WMO No. 847, Geneva.

WMO (2000). *Detecting Trend and Other Changes in Hydrological Data*, Z. W. Kundzewicz and A. Robson. (editors), WMO report-TD-No. 1013, Geneva.

WMO (2004). *Handbook on CLIMAT and CLIMAT TEMP Reporting*. WMO/TD No. 1188, Geneva.

WMO (2006). *Summary Statement on Tropical Cyclones and Climate Change*. http://www.gfdl.gov/~tk/manuscripts/IWTC_Summary.pdf.

WMO (2007). *The Role of Climatological Normals in a Changing Climate*. WMO/TD No. 1377, WCDMP No. 61, Geneva.

WMO (2008). *Guide to Meteorological Instruments and Methods of Observation*. Report No. 48, WMO/TD No. 328, Geneva.

WMO (2009a). *Guidelines on Analysis of Extremes in a Changing Climate in Support of Informed Decisions for Adaptation*. Geneva.

WMO (2009b). *Guide to Hydrological Practices: Volume II Management of Water Resources and Application to Hydrological Practices*. WMO No. 168, Geneva.

Wood, A. W., Maurer, E. P., Kumar, A., and Lettenmaier, D. P. (2002). Long-range experimental hydrologic forecasting for the eastern United States. *Journal of Geophysical Research, Atmospheres*, **107**(D20), 4429.

Wood, A. W., Leung, L. R., Sridhar, V., and Lettenmaier, D. P. (2004). Hydrologic implications of dynamical and statistical approaches to downscaling climate model outputs. *Climatic Change*, **15**(62), 189–216.

Woodley, W. and Herndon, A. (1969). A raingauge evaluation of the Miami reflectivity-rainfall rate relation. *Journal of Applied Meteorology*, **9**(2), 258–264.

Woodley, W. L., Simpson, J., Biondini, R., and Berkeley, J. (1977). Rainfall results, Florida area cumulus experiment 1970–75, *Science*, **195**(4280), 735–742.

Woodridge, J. (2003). *Introductory Econometrics: A Modern Approach*. South-Western College Publication.

Wu, R., Hu, Z. Z., and Kirtman, B. P. (2003). Evolution of ENSO-related rainfall anomalies in East Asia. *Journal of Climate*, **16**, 3742–3758.

Xia, Y., Fabian, P., Stohl, A., and Winterhalter, M. (1999). Forest climatology: estimation of missing values for Bavaria, Germany. *Agricultural and Forest Meteorology*, **96**, 131–144.

Xia, Y., Fabian, P., Stohl, A., Winterhalter, M., and Zhao, M. (2001). Forest climatology: estimation and use of daily climatological data for Bavaria, Germany. *Agricultural and Forest Meteorology*, **106**, 87–103.

Xu, C.-Y. and Singh, V. P. (1998). A review on monthly water balance models for water resources investigations. *Water Resources Management*, **12**, 31–50.

Xue, Y., Smith, T. M., and Reynolds, R. W. (2003). Interdecadal changes of 30-yr SST normals during 1871–2000. *Journal of Climate*, **16**, 1601–1612.

Yang, D., Goodiso, B. E., Ishida, S., and Benson, C. S. (1998). Adjustment of daily precipitation data at 10 climate stations in Alaska: application of World Meteorological Organization intercomparison results. *Water Resources Research*, **34**, 241–256.

Yeh, S.-P., Ou, S.-H., Doong, D.-J., and Kao, C. C. (2006). Joint probability analysis of waves and water level during typhoons. *Third Chinese–German Joint Symposium on Coastal and Ocean Engineering*, National Cheng Kung University, Taiwan, November 8–16, 2006.

Young, B. C. and McEnroe, B. M. (2003). Sampling adjustment factors for rainfall recorded at fixed time intervals. *Journal of Hydrologic Engineering*, **8**(5), 294–296.

Yue, S. (2000). Joint probability distribution of annual maximum storm peaks and amounts as represented by daily rainfalls. *Hydrological Sciences Journal*, **45**(2), 315–326.

Yue, S. and Pilon, P. (2004). A comparison of the power of the t test, Mann-Kendall and bootstrap tests for trend detection. *Hydrological Sciences*, **49**(1), 21–37.

Yue, S., Pilon, P., and Cavandias, G. (2002). Power of the Mann–Kendall and Spearman's rho tests for detecting monotonic trends in hydrological series. *Journal of Hydrology*, **259**(1–4), 254–271.

Yukimoto, S., Noda, A., Kitoh, A., *et al.* (2001). The new Meteorological Research Institute coupled GCM (MRI-CGCM2): model climate and variability. *Papers in Meteorology and Geophysics*, **51**(2), 47–88.

Yule, G. U. (1912). On the methods of measuring association between two attributes. *Journal of the Royal Statistical Society*, **LXXV**, 579–652.

Zadeh, L. (1965). Fuzzy sets. *Information Control*, **8**, 338–353.

Zhang, C. and Zhang, S. (2002). *Association Rule Mining: Models and Algorithms*. Lecture Notes in Artificial Intelligence, New York: Springer.

Zhang, L., Liu, C., Xu, C. Y., Xu, Y. P., and Jiang, T. (2006). Observed trends of water level and streamflow during past 100 years in the Yangtze River Basin, China. *Journal of Hydrology*, **324**(1–4), 255–265.

Zhang, P., Steinbach, M., Kumar, V., *et al.* (2005). Discovery of patterns in earth science data using data mining. In *Next Generation of Data-Mining Applications*, M. M. Kantardzic and J. Zurada (editors), Hoboken: Wiley-IEEE Press, 167–187.

Zhang, X., Harvey, K. D., Hogg, W. D., and Yuzyk, T. R. (2001). Trends in Canadian streamflow. *Water Resources Research*, **37**(4), 987–998.

Zhao, Q. and Bhowmick, S. S. (2003). *Association Rule Mining: A Survey*. Technical report, CAIS, Nanyang Technological University, Singapore, Report No. 2003116.

Zimmermann, H. J. (1987). *Fuzzy Sets, Decision Making, and Expert Systems*. Dordrecht: Kluwer Academic Publishers.

Zimmermann, H. J. (1991). *Fuzzy Set Theory and Its Applications*, second edition. Dordrecht: Kluwer Academic Publishers.

Index

accumulated rainfall, 123, 125, 126
adaptation, 236, 247
Akaike's information criterion (AIC), 59, 82
Anderson–Darling test, 129, 130, 199, 200, 211
annual exceedence probability (AEP), 213
annual maximum duration series, 209
annual maximum series (AMS), 222
anomalies, 39, 169, 171, 190, 193
antecedent moisture condition, 121, 123, 227
antecedent precipitation index, 132
anthropogenic, 1, 117, 247
area-weighting method, 104
artificial neural networks (ANN), 48, 81
association rule mining (ARM), 77, 78
average recurrence interval (ARI), 213

baseflow, 131
baseflow separation, 131
bias correction, 27, 28, 30, 31, 32, 154
bias-correction and spatial disaggregation (BCSD), 152, 154
bilinear interpolation, 104, 105
binarization, 86
binary variables, 73, 74, 75, 77, 86, 90, 91, 108, 109
bivariate normal distribution, 127, 128
block approach, 37
Boolean distance measures, 86
bootstrap method, 93
Box–Cox transformation, 127, 129, 130, 199

calibration, 49, 50, 95, 96, 138, 150, 151, 233, 234, 235, 236
Canberra distance, 85
categorical value evaluation, 98
change factor approach, 148, 152
change-point, 144
Chi-square test, 144, 200
classification methods, 83, 215
climate change, 1, 5, 169
climate change information, 189, 241, 242, 247, 248
climate indices, 219
climate normals, 219
climate variability, 1, 169, 175, *see* climate change
    natural variability 6
climatic oscillations, 169, 175, 177
    Atlantic Multi-decadal Oscillation (AMO), 6, 135, 175, 177, 187, 190, 191
    Atlantic Oscillation (AO), 169, 170
    El Niño Southern Oscillation (ENSO), 6, 7, 117, 169, 170, 171, 185, 187, 189–91, 210
    Inter-decadal Pacific Oscillation (IPO), 170
    Madden–Julian Oscillation (MJO), 169
    North Atlantic Oscillation (NAO), 169, 170, 175, 187, 190, 191
    Pacific Decadal Oscillation (PDO), 6, 7, 169, 175, 187, 190, 191
    Pacific North American (PNA), 170
    Southern Oscillation (SO), 170
    Southern Oscillation Index (SOI), 170, 171, 189
climatological mean estimator, 52, 215

closeness, 88
clustering, 34
clusters, 75, 84
coefficient of correlation weighting method, 58, 72, 77, 89, 91, 94
concordance, 98
confidence, 78, 79
confidence intervals, 98
contingency table, 98
continuous simulation, 98, 227
convective storms, 115, 116
correlation coefficient (CC), 30, 82, 96
correlation distance, 85, 86
correlogram, 82
cosine distance, 85
cross-validation, 95, 97, 108, 178
curve number, 121, 122, 123
cyclonic precipitation, 135
Czekanowski measure, 87

D'Agostino–Pearson test, 130, 199, 200
data mining, 78
data scarcity, 117, 119
deep moist convection (DMC), 116
Delaunay triangulation, 53, 63
dependence analysis, 123, 139
dependence parameter, 136
dependence value, 136
design storms, 226
desk study approach, 135
Dice distance, 87
distance measures, 84
distance metrics, 84, 86, 87, 88
downscaling, 148
drop-size distribution (DSD), 14, 15, 16
dynamic downscaling, 148, 151

El Niño, 170, 171, 175, 185, 187, 189, 190
El Niño Southern Oscillation, 170, 213
emission scenario, 152, 242, 247
episodic events, 135
epochal method, 130
equal weights method, 105
error performance measures, 26
error rate, 98
error statistics, 29, 99
errors
    random, 209
    systematic, 209
errors in reflectivity, 15
Euclidean distance, 84, 85, 91, 106, 108
exact interpolation, 50
exceedence probability, 201
Expert Team on Climate Change Detection and Indices (ETCCDI), 217
exploratory data analysis (EDA), 195
extended reconstructed sea surface temperature (ERSST), 171

fixed function set genetic algorithm method, 70, 72
flash floods, 133
frequency factors, 198, 206
friction distance, 73, 75
fuzzy constraints, 229, 244, 245
fuzzy linear programming (FLP), 243
fuzzy mathematical programming, 229
fuzzy membership function, 230, 244

G statistic, 34
gage mean estimator, 51
General Circulation Model (GCM), 148, 149, 150, 151, 152, 154, 155, 156, 222, 241, 242
generalized extreme value distribution (GEV), 208
genetic algorithms (GA), 70, 108, 129
geographically weighted optimization (GWO), 91
geostatistical interpolation, 36
geostatistical spatial interpolation, 66
global gridded datasets, 109
global interpolation, 50, 101
global polynomial functions, 59, 101
global precipitation data sets, 41
global precipitation measurement (GPM), 32
goodness-of-fit test, 93, 130, 199, 204, 211
Gower distance, 85
Gringorten formula, 201
groundwater moisture, 235
groundwater table, 117

Hamming distance, 88
heteroscedasticity, 59
homogeneity, 106, 109, 135, 196, 209, 212
homoscedasticity, 59
Huff's methodology, 117
hurricanes, 135, 139, 145, 174, 177, 221
HYDRO-35, 208
hydroclimatology, 115
hydrologic design, 225, 226, 227
Hydrologic Research Analysis Project (HRAP), 32, 104, 105
hydrometeorological floods, 2, 115, 120
hydropower, 245, 246

imputation variance, 93
independence, 209
index for observation, 89
index storm, 201
inductive model development, 70
inexact interpolation, 50
infilling, 106
intensity–duration–frequency (IDF), 206, 208, 226, 229, 231, 232
inter-event time definition (IETD), 242, 246
inverse distance weighting method, 27, 54, 63, 64, 72, 76, 91, 101, 105, 108
inverse exponential weighting method, 28, 58
iso-pluvial, 209

Jaccard distance, 87
jack-knife re-sampling, 94
joint cumulative probability, 126
joint probability analysis, 127, 135
joint probability approach, 141
joint probability method, 136, 138

Kappa statistic, 98
Kendall's tau, 139
kernel density estimation (KDE), 177, 183, 186, 211
K-means, 84
K-means clustering method, 89
K-nearest neighbor, 84
Kolmogorov–Smirnov (KS) test, 26, 129, 130, 157, 199, 200, 211
kriging, 66, 68, 69, 83
Kulzinsky distance, 88

L-moments, 198, 201, 204, 206, 222
La Niña, 170, 171, 174, 175, 185, 187, 190
lag time, 123, 125, 126, 133, 139
Latin-hypercube sampling, 123
lead times, 133, 134
Lilliefor test, 130, 199, 200
local filters, 101
local interpolation, 50, 101, 108
local polynomial functions, 59, 101
log-likelihood function, 129
low-impact development (LID), 246

macroscale, 115, 116
Mahalanobis distance, 85
Manhattan, 85
Manning's equation, 231
Mann–Kendall (MK) test, 118, 214
    with pre-whitening, 214
    with trend free pre-whitening, 214, 217
Mann–Kendall tau, 144
Mann–Kendall tau test, 203
Markov-chain probabilities, 26
mathematical programming, 73, 77, 90, 108
maximum area method, 105
maximum distance, 85
maximum likelihood estimation (MLE) method, 144, 198, 201, 202, 204, 206, 222
mean absolute deviation index (MADI), 201
mean absolute error (MAE), 96
mean error (ME), 96
mean imputation, 92, 94
mean sea level pressure (MSLP), 170
mean square deviation index (MSDI), 201
mesoscale convective complexes, 115
meteorologically homogeneous precipitation areas, 124
method of moments, 198, 202, 206
Minkowski distance, 85
missing at random, 49
missing completely at random, 49, 109
missing data, 48, 49, 77, 84, 89, 90, 92, 94, 106, 108
missing not at random, 49
mixed integer non-linear programming (minlp), 77, 89, 109
modifiable areal unit problem (MAUP), 77
Monte Carlo, 123, 137
moving window, 22, 101, 104, 209, 215
multi-model assessment, 99
multiple best estimators, 77
multiple imputation, 92, 94
multiple linear regression, 58, 59, 65

natural neighbor interpolation, 63
nearest neighbor, 84, 104
nearest neighbor weighting method, 63
NEXt generation RADar (NEXRAD), 13, 14, 19, 20, 27, 49, 90, 105
non-linear least squares (NLS), 93
non-negative least squares, 59, 108
non-parametric, 118, 131, 144, 199, 211, 212
non-stationarity, 144, 145, 210, 221, 222, 247
normal ratio method, 63
North Pacific Index (NPI), 170
nugget, 67

Occam's razor, 72
Oceanic Niño Index (ONI), 171
operating policy, 244, 246, 248
optimal coefficients, 22, 72
optimal density, 36
optimal K-nearest neighbour classification method, 88
optimal monitoring network design, 36
optimal weighting method, 88, 91
optimal weights, 89
optimization solvers, 107, 129

optimum network design, 37
ordinary kriging, 81, 82, 83, 108
overestimation, 15, 78, 80, 109

parameter estimation, 197, 201, 202, 234
parametric, 199
partial duration series (PDS), 130, 213, 222
peaks over threshold (POT), 117, 130, 131, 196, 213
Pearson correlation coefficient, 139
Pearson distance, 87
penalty function, 243, 245
percentage of correct state for all time intervals (P_A), 97
percentage of correct state for dry time intervals (P_d), 96
percentage of correct state for wet time intervals (P_w), 96
performance evaluation, 30
performance measures, 99, 156
plotting position, 196, 203
point estimation, 50, 51
point measurements, 11, 49, 53, 110
positive definiteness, 69
positive kriging, 68
potential evapotranspiration, 234
precipitation bands, 115
precipitation measurements
    error, 11
        random error, 11
        systematic error, 11
    history of, 10
    radar, 10, 11, 13, 14, 31, 45, 49, 90, 213, 214
        reflectivity, 16
    rain gage, 10, 11, 12, 13, 37, 90, 109
    satellite, 10, 11, 31, 45, 214
precipitation monitoring network, 34
precipitation monitoring network density, 34
precipitation regimes, 116
precipitation thresholds, 133
precipitation-elevation Regression on Independent Slopes Model (PRISM), 59
principle of parsimony, 72
probabilistic intensity duration and frequency (PIDF), 247
probability of exceedence, 126, 127, 129
probability-plot correlation coefficient (PPCC) approach, 200, 201
probable maximum precipitation (PMP), 118, 214
proximity measures, 88, 90

quadrant method, 76
qualitative evaluation, 98
quantile plot, 196
quantile–quantile plot (Q–Q plot), 26

radar reflectivity, 14
rain radar retrieval, 19
Rainfall Analysis and INterpolaton (RAIN), 110
rainfall distributions, 228
random variates, 94
rational peak discharge, 230
reanalysis data sets, 42
recursive digital filters, 132
reflectivity, 14, 15, 17
regional frequency analysis, 201, 207, 209
regional statistics, 7, 106, 209
regression analysis, 137
regression models, 58
re-sampling, 93, 94, 104
residuals, 59, 98
return period, 129, 136
revised nearest neighbors, 64
river regimes, 116
robust regression, 65
Rogers and Tanimoto distance, 87
root mean square error (RMSE), 96
runs test, 131, 212

Russell and Rao distance, 87
RX1D, 6
RX5D, 6

sampling adjustment factor (SAF), 209
Satterthwaite's modified t-test, 211
sea surface temperature (SST), 170, 171, 175, 189–91
semi-variogram, 66, 68, 69, 81, 82, 83, 106
Sen's slope, 144, 203, 214, 215
sensitivity, 98
shallow groundwater, 120, 121
Shapiro–Wilk test, 130, 199
sill, 67, 69
similarity measures, 85, 86
simple matching metric, 86
single best classifier, 83
single best estimator (SBE), 51, 77, 80, 105, 108, 109
site-specific statistics, 7, 106, 209
soft-computing, 243
soil conservation service, 121, 122
soil moisture, 235
soil moisture accounting, 121, 123
soil storage capacity, 121, 122
Sokal–Michener distance, 88
spatial analogue, 152
spatial autocorrelation, 63
spatial domain filters, 104, 105
spatial downscaling, 148, 149, 151
spatial interpolation, 36, 48, 49, 68, 78, 97, 101, 106, 108, 109
spatial variability, 82
Spearman's rho, 139
Special Report on Emission Scenarios (SRES), 152
specificity, 98
squall lines, 115
squared Euclidean, 85
standard normals, see climate normals
stationarity, 6, 7, 77, 144, 145, 196, 209, 210, 222, 247, 251, 253, 259, 260
    hydrologic stationarity (HS), 210
statistical distributions, 197
    bivariate normal distribution, 128, 129, 130, 137, 141
    conditional distributions, 130
    exponential distribution, 201
    extreme value distribution, 197, 206
    extreme value Type III distribution, 197
    gamma type distribution, 197, 201, 202, 206, 210
    Gaussian distribution, 202, 206
    generalized extreme value distribution (GEV), 144, 197, 204
    generalized logistic distribution (GLO), 197
    generalized normal distribution (GNO), 197
    generalized Pareto (GPA), 197
    joint probability distribution, 130
    log-normal distribution, 197
    normal distribution, 130, 197, 210
    Pearson Type III distribution, 197
    three-parameter log-normal distribution, 197
    Wakeby, 197
statistical downscaling, 148
statistical evaluation of extreme precipitation (SEEP), 222
statistics preserving spatial interpolation, 106, 107
step-wise regression, 65
stochastic interpolation, 48, 106, 109
stochastic regression, 92
storm transposition, 119
stormscale, 115
streamflow partitioning method, 132
summary statistics, 130, 156, 209
support, 78, 79
support vector machine-logistic regression-based copula (SVMLRC), 84
support vector machines, 83, 84
surface generation, 50, 100, 101
synoptic scale, 115, 116

Taylor diagram, 26, 100
Taylor–Russell diagram, 189
teleconnections, 6, 7, 135, 169, 177, 183, 189, 190, 213, 239, 247, 256
temporal analogue, 152
temporal downscaling, 148, 149
temporal interpolation, 65, 94, 108
Thiessen polygons, 53, 57, 63, 104
thin-plate splines, 62
thin-plate splines with tension, 62
Thomas Model, 233, 234, 235
Thornthwaite and Wilm, 235
time of concentration, 230, 232
transfer functions, 149
transition probabilities, 164
Trans-Niño Index (TNI), 171
travel time, 232
trend surface models, 60
Tropical Rainfall Measuring Mission (TRMM), 32, 46

tropical systems, 115
t-test, 211

underestimation, 15, 78, 80, 96, 109
universal approximation-based ordinary kriging (UOK), 81

validation, 50, 95, 235, 236
variogram cloud, 83, 101
Voronoi polygons, *see* Thiessen polygons

Waikato Environment for Knowledge Analysis, 79, 80
water balance model, 5, 233
Weather Climate Toolkit, 19
weather generators, 149, 151, 162
weather typing, 149, 150

Yule distance, 88

Z–R relationship, 14, 15, 16, 17, 19, 20, 23